Old Materials, New Climate

Old Materials, New Climate: Traditional Building Materials in a Changing World is an accessible guidebook to understanding historic materials – how they were traditionally made, how they survived the test of time, and how changes in climate are now impacting materials in new ways.

Protecting historic buildings from a rapidly changing and unpredictable climate requires an understanding of how climate affects weather and how weather affects the durability of the most widely used traditional materials – wood, adobe, brick, lime, concrete, metal, and paint. This resource examines how gradual and dramatic changes in climate threaten to accelerate normal weathering and presents strategies to safeguard historic materials for future generations. Illustrated case studies explore how weather is affecting materials in specific historic buildings in climate zones in the United States and across the globe.

Drawing on the work of experts in conservation, biology, chemistry, and environmental impacts, this book is an invaluable resource for any student, preservationist, architect, or contractor interested in expanding their knowledge of materials and why they perform as they do.

Susan Pranger has focused on preservation and adaptive reuse for most of her career as an Architect and Educator, gaining a fascination for historic buildings, and a deep respect for the owners, advocates, consultants, and craftsmen who ensure their survival. Susan is a licensed Architect and LEED accredited professional with 40 years of experience in professional practice, 12 years as chair of the Boston Landmarks Commission, and over 10 years teaching in the Sustainable Design and Historic Preservation at the Boston Architectural College (BAC), an institution with a long tradition of combining education with practice. In 2015, Susan received a Master of Design Studies in Sustainable Design from the BAC, exactly 40 years after receiving a Bachelor of Science in Architecture from the University of Detroit, Michigan.

"A must-have reference for the heritage community in these critical times. This long overdue review of the impact of climate change on the full range of historic building materials is brilliantly addressed through technical summaries and compelling case studies from across the planet. This is the rare book that is both a delight to read and a go-to source of important information."

Jean Carroon, *FAIA, LEED fellow, former Chair of the AIA Historic Resources Committee*

"Susan Pranger has presented us with just what we need to understand the threats to heritage from a changing climate … real materials and real sites, with clear descriptions, detailed photographs, and thoughtful tabulations … all so useful."

Peter Brimblecombe, *Emeritus Professor School of Environmental Sciences, UEA Norwich*

"Pranger brilliantly delves into the intricate relationship between weather, climate, and the timeless building materials that shape history. Touring these notable case studies illuminates the impact of climate change and empowers professionals to protect the past while designing a sustainable future. A vital addition for libraries of architects and preservationists, it is a roadmap to protecting our heritage in a warming world. Uncover the secrets of material resilience and continuity in this must-read masterpiece."

Eric Corey Freed, *Director of Sustainability, Cannon Design, Architect, Author, Speaker*

"This timely publication gives an overview of traditional materials and how they perform in a variety of circumstances and futures. It offers guidance on how an older material palette can contribute to carbon reduction in the built environment and the changes needed to address future climate scenarios in order to keep historic buildings relevant."

Roger Curtis, *MRICS, FSA, Head of Technical Resources, Historic Environment Scotland*

"*Old Materials, New Climate* should be a standard reference for preservationists considering the effects of climate change on the historic structures they are tasked with saving."

Christopher H. Marston, *U.S. National Park Service*

"Susan Pranger's well-researched book examines the effect climate has on building materials, including concrete! New history emerged about the concrete work for Frank Lloyd Wright's Unity Temple, offering a greater understanding of the material and how we can best preserve it."

Heidi Ruehle, Executive Director, *Unity Temple Restoration Foundation*

"Thoroughly researched and expertly written, Susan Pranger's *Old Material, New Climate* combines an expansive survey of historic building materials, methods, and techniques, and the growing crisis our collective built heritage will face in a future of unreconcilable climatic change."

Benjamin R. Curran, *The Campaign for Historic Trades*

"*Old Materials, New Climate* addresses the intersection of preservation and sustainability in thorough detail from various global perspectives, covering a variety of materials from wood to concrete. Anyone looking for a more nuanced understanding of how traditional materials function in our rapidly changing climate should add this to their library."

Eleni Glekas, *Director, Historic Preservation and BAC/NCA/UOBS University Partnerships*

"Monumental work! Susan Pranger has given us a comprehensive look at historic preservation from the perspective of the construction materials encountered in typical preservation projects. Students and those in the general public with an interest the craft of building will find *Old Materials, New Climate* absorbing reading, especially the descriptions of the technical intricacies and drama of actual preservation projects such as the San Antonio Missions or the lighthouses of the Pacific Northwest. Activists and professionals in the preservation field will find a deeper utility in the encyclopedic chapters on specific materials and case studies that offer concrete examples of how to manage preservation projects to the highest technical and professional standards."

Curt Lamb, PhD, M.Arch. *former Director of the Boston Architectural College's Masters in Historic Preservation and Masters in Sustainable Design degree programs*

"[...] a foundational document with chapters that people should reread in practice as their projects raise problems with existing materials...a strong introduction to the real world of traditional materials."

Henry Moss, *AIA, Principal, Bruner/Cott Architects*

Old Materials, New Climate

Traditional Building Materials in a Changing World

Susan Pranger

Routledge
Taylor & Francis Group
NEW YORK AND LONDON

Designed cover image: Photo by Susan Pranger, Chetro Ketl, Chaco Canyon, New Mexico (2011)

First published 2024
by Routledge
605 Third Avenue, New York, NY 10158

and by Routledge
4 Park Square, Milton Park, Abingdon, Oxon, OX14 4RN

Routledge is an imprint of the Taylor & Francis Group, an informa business

Library of Congress Cataloging-in-Publication Data
Names: Pranger, Susan, author.
Title: Old materials, new climate : traditional building materials in a changing world / Susan Pranger.
Description: New York, NY : Routledge, 2024. | Includes bibliographical references and index. |
Identifiers: LCCN 2023031293 | ISBN 9780367749569 (hardback) | ISBN 9780367749576 (paperback) |
ISBN 9781003160489 (ebook)
Subjects: LCSH: Building materials--Climatic factors.
Classification: LCC TA403.6 .P685 2024 | DDC 691--dc23/eng/20231130
LC record available at https://lccn.loc.gov/2023031293

ISBN: 978-0-367-74956-9 (hbk)
ISBN: 978-0-367-74957-6 (pbk)
ISBN: 978-1-003-16048-9 (ebk)

DOI: 10.4324/9781003160489

Typeset in Minion Pro
by KnowledgeWorks Global Ltd.

To Dan

Contents

Figures

Foreword

In 1803, Lazare Carnot proposed that "in any natural process there exists an inherent tendency towards the dissipation of useful energy".

The concept of preservation is complex and varied. It assigns a value to an object that may be defined in physical terms, or as a reflection of cultural and historical significance. If we look at preservation in an architectural sense, it embodies these values and treats examples as worthy objects, as representative of technique, as icons of a particular point in time.

And this is one of the paradoxes that we face. All objects are part of a continuum, and that includes both synthesis and decay. From the moment an object is finished, it begins to decay. In fact, the process of creation is often a studied erosion of material. In the alchemic tradition, all that we build are products of earth, and they are subject to the forces of fire, air, and water.

This study looks at the nature of materials and tries to place them along a chronology of building techniques that have been developed and refined, sometimes lost. In our observations, we see a wide variety of approaches. Some wish to freeze the moment in time, whether it be in pristine condition or a ruin. Some seek to retain the fabric of the object and continue to use it when its function and role in society have changed dramatically. For some, the object itself is less important than the moment it commemorates. In all these cases, there is the question of use. Relic? Museum piece? Urban fabric? Restoration?

Our history shows that we build upon the bones of former civilizations. Stones of monuments are rubble infill for the next generation. Castles and fortresses prove to be ephemeral promises of security in a world of expanding technology. The world is filled with glories past, and they all come to the realization of Shelley's Ozymandias, "Look on my works, ye mighty, and despair!"

Preservation is a privilege that recognizes something beyond the utility of existence and defies the cannibalistic consumer nature of humankind. It is an opportunity to respect our past and revel in the surviving artistry of craftsmen and the societies that inspired them.

Daniel Thomas, 2023

Bibliography

Carnot, L. N. M. 1803. *Fundamental Principles of Equilibrium and Movement.* Paris: *Paris Bachelier.*

Shelley, P. B. 1818. "Ozymandias." In *The Examiner, London*. January 11, 1818. No. 524, page 24. https://en.wikipedia.org/wiki/Ozymandias#/media/File:Ozymandias_The_Examiner_1818.jpg

Preface

This book began with the goal of summarizing 40 years of experience in architecture, preservation, sustainability, and traditional materials and quickly became a 10-year journey of new exploration. My intent was to explain in plain language how materials are made, how they naturally deteriorate, and how changes in weather and climate affected those processes. While researching these topics, I discovered another theme: the role of caretakers, educators, researchers, craftsmen, and professionals, each with a passion for historic materials and craftsmanship, skills that are rapidly being lost.

The Context chapters are not intended to be comprehensive. The chapter on Climate provides an introduction to the range of issues that affect weather and materials but does not attempt to identify or explain all of the causes and mechanisms of climate change. Similarly, the evolution and history of preservation movements, and a complete inventory of national and global movements are beyond the scope of this book. Instead, the chapter of Preservation Philosophy attempts to introduce a range of approaches and philosophies that affect decisions regarding conservation of materials.

The material chapters attempt to balance the big picture with the vital details. Case studies in the United States and abroad illustrated some, but not all, of the materials, climate zones, and issues. This is not a "how-to" book: it will not provide directions to restore materials. Instead, the goal is to provide a wide audience with a slightly more in-depth understanding of traditional materials, why they behave the way they do, and what we might expect in the future.

The content of *Old Materials, New Climate* is drawn from my own travel and experience, and from the expertise and research of a multitude of papers, books, websites, and many individuals with far more understanding than I will ever have. Despite my and their best efforts, it is likely that this compilation will include omissions and misinterpretations, for which I take full responsibility. I encourage you to accept this book in the spirit that it is offered: as the beginning of your own exploration of materials.

Susan Pranger, 2023

Acknowledgments

Credit for the completion of this book is due first and foremost to Daniel Thomas, my husband, proofreader, critic, photographer, and travel companion. Special thanks also to Margaret Rosewitz for proofreading and editing the entire text. My brothers Ron, Tim and Greg Pranger, and good friends Malcom and Ann Jollie provided valuable logistical support during our research trips to their home states of Arizona, Michigan, Illinois and Washington, respectively.

My sincere thanks to the many individuals who were generous with their time and knowledge, provided invaluable insight and access to historic sites, and caught many of my mistakes. Any remaining errors are entirely my own.

The visit to Japan would not have been possible without the advice and translations provided by Naoko Takayanagi of the Japan Society of Boston, the experience shared by Masahiko Harada, Kouichi Hirota, and Yasuko Murata, of the Association for Preservation of the Nikko World Heritage Site Shrines and Temples, and the assistance of Mikiko Hayashi, Peter Brimblecombe, and Yoshi Ichiro. The European case studies were enhanced by the knowledge shared by Ingmar Kroken and Berit Bakosgjeelten in Norway and Andy Stockdale, Mitchell Fotheringham, Roger Curtis and Roz Artis in Scotland.

Khalid Ibrahim and Zahid Usman contributed the entire chapter on Lahore, Pakistan, including all text, photographs, maps, and diagrams.

Robert Armbruster, Bjorn Miller, Bob Joyce, and Heidi Ruehle spent hours explaining the history and condition of concrete in Chicago and Oregon. Christopher Marston, Jan Lewandoski, Kevin Groom, Nicole Gratton, Rebecca Burrow, Ron Anthony, Suzana Radivojevic, and Timothy Andrews provided insight into the character of wood and the life of covered bridges across the United States. Alex McMurry, Amy McAuley, Christy Sweet, Deb Edwards, Susan Licht, and Thomas Linegar provided access and shared their knowledge of West Coast lighthouses. Angela Wyrembelski, Richard Bardelli and Ron Staley provided access and background on the renovation of Michigan Central Station.

Alayna Jordan, Bob Vint, Eric Means, Ivan Myjer, Matilde Rubio, Starr Herr-Cardillo, and Tim Lewis provided invaluable access and insight into the mission churches of San Antonio, San Xavier, and San José de Tumacácori. Campbell Waterhouse and Judith Selwyn provided insights into the materials of the Old State House in Boston, while Christina Polwatka, Ben Haavik, and Bruce Blanchard clarified the history and condition of the Spencer Peirce Little House. My colleagues at the Boston Architectural College, Michael Fiorillo, Eleni Glekas and Curt Lamb, provided invaluable support.

Photographs were contributed by the following individuals or organizations:

Robert Armbruster
Nicole Gratton
Kevin Groom
Gunny Harboe, Architects
Betsy Holland
Khalid Ibrahim
Oregon Department of Transportation
Susan Pranger
Tim Pranger
Suzana Radivojevic
Margaret Rosewitz
Daniel Thomas
Unity Temple Restoration Fund
Zahid Usman
Washington State Parks and Recreation
Angela Wyrembelski

PART 1

Context

CHAPTER 1.1

The Shifting Climate

A Climate Overview

The climate has always been changing. The earth's orbit changes over the course of a predictable 30,000-year cycle, changing the amount of sunlight available, and leading to a series of ice ages. Volcanic activity and tectonic shifts slowly cycle carbon through the planet's atmosphere, oceans, rocks, and soils. Thermoclines, the currents in the ocean, atmospheric currents in the lower atmosphere, and the oscillations of El Niño and La Niña, all naturally fluctuate from year to year, influencing the timing and location of rain and tropical storms. In the past, these changes have been gradual. Today, these are happening much more rapidly and are becoming harder to predict.

The Intergovernmental Panel on Climate Change (IPCC) has recently reported that global surface temperatures have increased 1.1°C above levels recorded between 1850 and 1900, and that the rate of increase was higher in the last 50 years than in any period over the past 2,000 years. Carbon dioxide levels are now higher than at any period in the last two million years. Globally, there has been an increase in drought, extreme temperatures, fire weather, compound flooding, heavy precipitation, glacier retreat, sea level rise, and upper ocean acidification. An increase in greenhouse gases in the atmosphere is reducing the ozone in the atmosphere that blocks damaging ultraviolet light. Changes are occurring more rapidly than expected because of feedback loops and other unanticipated reactions to decreased ozone, increased carbon dioxide, and higher temperatures.

If the rate of global climate change is hard to predict, changes to the weather are even more so. Climate is the average weather over a period of time, usually three decades, while weather refers to the actual conditions of temperature, wind, and

DOI: 10.4324/9781003160489-2

precipitation. There is a significant difference between the statistical averages of climate and the specific changes in temperature and moisture that create daily and seasonal weather. Each incremental increase in global temperature is expected to cause much more extreme regional weather conditions and increase regional differences, but the changes do not follow a direct line.

Patterns of rainfall are also affected by global water cycles in the ocean and in the atmosphere, which are affected by temperature. The ocean is becoming more acidic with higher salinity. Heat waves and droughts are expected to be more frequent, as are regional increases in tropical cyclones and extratropical storms.

The boundaries of established climate zones and the expected patterns of weather within each zone are gradually changing, but extreme events outside the norm are becoming more common. Regional weather patterns are much more varied, extreme, and unpredictable than the global or regional averages indicate. Consequences for biological ecosystems will be more abrupt as conditions reach tipping points.

Emissions and the accompanying increases in temperature have an exponential impact on major climate systems such as the global water cycle, global monsoons, and extremes of wet and dry weather.

All these factors are interdependent. This causes tremendous uncertainty about what weather conditions will be and how historic materials and assemblies will respond. In many locations, changes in the climate will simply increase the severity or frequency of stresses that the materials and buildings are already experiencing. However, small changes can tip weather impacts from manageable to catastrophic and can introduce new and unanticipated challenges. The impact of climate change on materials will depend on the local weather conditions and the character of the materials themselves.

The Carbon Cycle

There is a direct correlation between temperature and carbon dioxide in the atmosphere and ocean, and the creation and destruction of carbon-based materials. The carbon cycle describes how carbon moves between the atmosphere, the lithosphere (rocks), soil, and water (rain, rivers, and the ocean) as it seeks a chemical equilibrium. The slow cycle moves 10 to 100 million metric tons of carbon from the earth to atmosphere to ocean and back again over hundreds of millions of years. The fast cycle is much more rapid, tied to the life span of organisms in the biosphere that absorb carbon as they grow, and release it as they die and decay. The fast cycle currently produces 1,000 to 100,000 million metric tons of carbon per year and has greater potential to create imbalances in carbon dioxide. Short-term imbalances can accumulate very quickly, while the natural processes to restore the balance take much longer. The earth will eventually rebalance itself, but in the interim, the impact on organic and inorganic material could be devastating.

The Slow Cycle

The slow carbon cycle begins when the components of rock are broken down by extreme heat and the pressure of moving tectonic plates into silicates and carbon dioxide. Volcanic activity releases carbon dioxide into the atmosphere and silicate particles fall to the earth. The carbon dioxide is absorbed by the ocean or combines with water in the atmosphere to create a weak carbonic acid that falls to the earth as slightly acidic rain. The acid rain chemically erodes rock and stone, releasing ions of calcium, magnesium, potassium, or sodium that are washed away and carried by rivers to the ocean.

Plankton and shell-building organisms in the ocean create calcium carbonate by combining ions of carbon dioxide and bicarbonates. As these organisms die, calcium carbonate shells and sediment fall to the ocean floor and are compressed into sedimentary limestone or metamorphic marble that store 80% of the carbon stored in rock. This process of capturing carbon dioxide takes place over millions of years and can store carbon for hundreds of millions of years.

The Fast Cycle

The fast cycle is tied to the cycle of organic life. Plants and microscopic ocean phytoplankton combine carbon dioxide with water to create sugar and oxygen. The carbon dioxide is released back into the atmosphere when plants or other organisms break down the sugars, die and decay, or are consumed by fire. Carbon dioxide levels in the atmosphere are so impacted by this process that they fluctuate seasonally, with lower levels when plants are growing and absorbing carbon dioxide and higher levels as plants are dying and decaying.

Organic compounds that are compressed to create oil, natural gas, or coal release carbon back into the atmosphere when burned – by volcanic activity or by human activity. However, humans add 100 to 300 times more carbon to the air than volcanoes each year, primarily by burning fossil fuels.

The destruction of plant life and the burning of fossil fuels reduce the amount of carbon dioxide stored in plants and increase the amount stored in the atmosphere and ocean. The slow cycles that store carbon dioxide in the ocean, rock, and soil cannot respond quickly enough to restore this imbalance.

The Ocean

Carbon dioxide moves back and forth between the atmosphere and the ocean water. As carbon dioxide is dissolved in the ocean, it creates carbonic acid, making the ocean more acidic (or less alkaline). This reaction releases hydrogen, which combines with carbonate from rock weathering to create bicarbonates, reducing the supply of carbonates available for shell-building organisms.

In the past, the ocean released carbon dioxide into the atmosphere in balance with the carbon it received from rock erosion. Since the Industrial Age, the ocean has been absorbing much more carbon from the atmosphere than it can release. The ocean can absorb 85% of the carbon that human activity is adding to the atmosphere, but this is a slow process that could take a million years. In the short term, a more acidic ocean will dissolve shells more rapidly – good for creating limestone but bad for marine life.

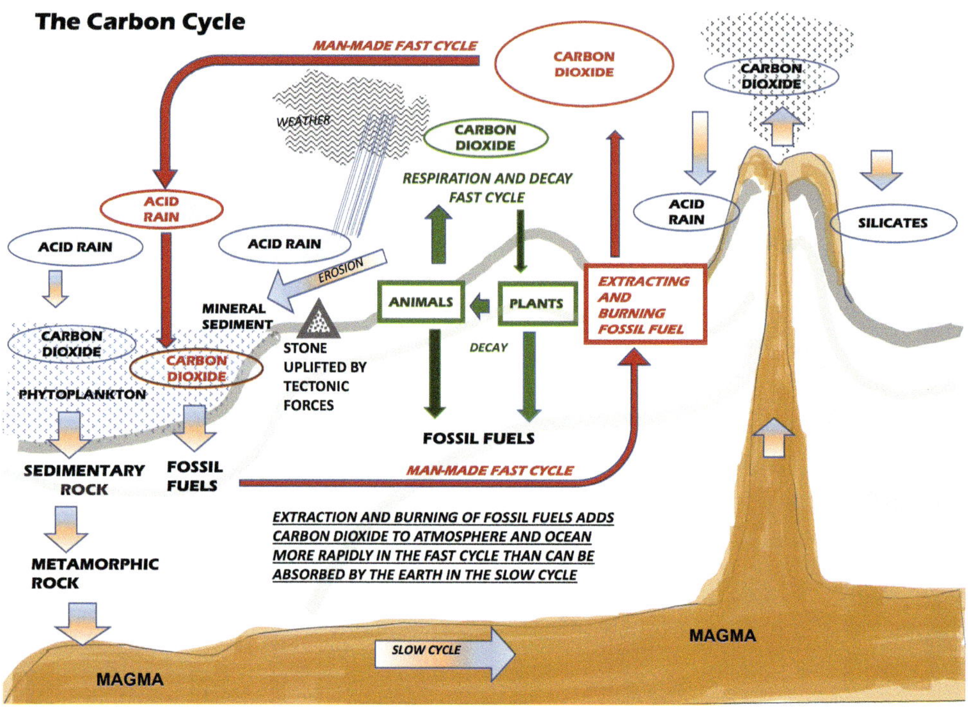

Figure 1.1.1 The carbon cycle. (Diagram by Susan Pranger, 2023.)

Regional Weather

Although climate change is a global phenomenon, the effects will be experienced differently in each region of the world. Latitude, altitude, topography, elevation, surface vegetation, water, and hardscape, and level of urbanization all impact local weather patterns and how that weather will impact the natural and human-made environment.

Temperatures and UV exposure will continue to rise across the globe, with an increase in the number and duration of heat waves, although the impacts and rate of increase will not be consistent across the globe. The impact of temperature increases on biological growth will be most dramatic where there is a reduction in freezing temperatures that normally limit many pest populations.

Globally, wet areas are expected to get wetter and dry areas to get dryer. Both are expected to experience more days of intense rainfall. An increase in temperature of a single degree can increase the water-carrying capacity of air by 7%. When it does rain or snow, there is more moisture in the air available to precipitate.

Relative humidity is about 5% lower in urban areas because the temperatures are higher and there is less transpiration from plants and open water. However, rainfall is about 10% higher because water vapor is condensed by the higher concentrations of particles in the air. Fog is also more prevalent in humid cities, especially in the winter. Fog has a larger surface area and moves slowly, which allows it to accumulate higher concentrations of pollutants than dry air or rain.

The interaction between global warming and wind speed is complex and difficult to predict because of the lack of data and other factors that vary locally, such as topography, vegetation, air pollution, thermal and pressure gradients, and urbanization.

An increase in the frequency and distribution of forest fires has been tied to changes in the climate, driven by higher temperatures, reduced rainfall, and high winds. These fires have thrown huge amounts of particulates into the air. These particulates travel across the globe to more humid and more industrial areas where they contribute to local air pollution.

Changes to the Ecosystem

Climate change, urbanization, and globalization have all combined to increase the habitat and lifespan of insects, bacteria, and fungi. Increasing temperatures and humidity, and a reduction in freezing weather, have weakened the weather constraints that historically limited the growth and spread of many forms of biological growth while globalization is expanding access to new, untouched environments.

Temperature and humidity affect the life cycle, population size, range, and geographic distribution of insects, and can also create a mismatch between insects and their place in the overall ecology. Insects might move or change the timing of their lifecycle in response to the weather, while other plants and insects may adapt on a different timeline, take longer to adapt or fail to adapt. Favorable changes in habitat may increase populations or introduce new populations. Unfavorable changes may decrease pest populations but may also decrease beneficial insects that keep others in balance.

The Impact of Climate on Traditional Materials

The impact of climate on any material, traditional or not, depends on the interaction between the individual material and the weather – moisture, temperature, solar radiation, and wind. The type and extent of deterioration will depend on many factors, including local conditions, the length and frequency of weather impacts, and how the material is used and maintained. All the possible variations cannot be predicted, and the interactions between materials and climate will change as the climate changes.

However, an understanding of how different materials respond to the various impacts of heat, water, and biological infestation is the first step to predicting how a specific material might respond to future changes in the environment. Understanding of how materials are created and how they naturally deteriorate provides insight into how that deterioration might increase in response to more intense weather conditions and biological attack. Understanding how materials have resisted those risks in the past in various climates will suggest strategies for the future.

Bibliography

Adair, Jordan, Marisa Cicale, Mark Hofberg, John Junghans, H. Kerrick, George Luo, Michael Mercado, S. Oliver, S. O'Neill, Dan Remer, Matthew Schulzinger, Ryan Shofnos, H. Tolley, and Wesley Tse. 2010. "The Effect of Wood Burial and Submersion on Decomposition: Implications for Reducing Carbon Emissions." http://hdl.handle.net/1903/10081 or https://www.semanticscholar.org/paper/The-Effect-of-Wood-Burial-and-Submersion-on-for-Adair-Cicale/e82bf3f04e08c085825119bc5dcb7a7115650311.

Andrady, Anthony L., Hamid S. Hamid, and Ayako Torika. 2003. "Effects of Climate Change and UV-B on Materials." *Photochemical and Photobiological Sciences* 2 (1): 68–72.

Anthony, Ronald W. 2007. "Practice Points Number 3: Basics of Wood Inspection." *APT Bulletin, The Journal of Preservation Technology* (The Association for Preservation Technology International) XXXVIII (2–3). https://www.apti.org/assets/docs/03-Anthony.pdf.

Atisd, P. A., N. Bellouin, E. Coppola, and L R. G. Jones. 2021. *Climate Change 2021: The Physical Science Basis.* Technical Summary, Cambridge, United Kingdom: Cambridge University Press. https://doi.org/10.1017/9781009157896.002.

Atkins. 2013. "Assessment of Heritage at Risk from Environmental Threat." *English Heritage.* November 5. http://www.english-heritage.org.uk/publications/assessment-heritage-at-risk-from-environmental-threat/.

Blunden, Jessica and Derek S. Arndt. 2014. "BAMS The State of the Climate in 2013." *Bulletin of the American Meteorological Society* 95: S1–S279. https://doi.org/10.1175/2014BAMSStateoftheClimate.1.

C2Es. n.d. "Heavy Precipitation." http://www.c2es.org/science-impacts/extreme-weather/extreme-heat.

IPCC. 2023. "Summary for Policymakers." In *Climate Change 2023: Synthesis Report.* Geneva, Switzerland: IPCC.

McSweeney, Robert. 2014. "The Carbon Brief." *Pacific Winds Change the Speed of Global Warming, Says New Study.* http://www.carbonbrief.org/blog/2014/12/pacific-winds-change-the-speed-of-global-warming-says-new-study/.

NASA. 2011. *Carbon Cycle.* June 16. https://earthobservatory.nasa.gov/features/CarbonCycle/page3.php.

———. 2023. *Global Climate Change. The Effects of Climate Change.* https://climate.nasa.gov/effects/.

Ren, Diandong. 2010. "Effects of Global Warming on Wind Energy Availability." AIP *Journal of Renewable and Sustainable Energy.* http://scitation.aip.org/content/aip/journal/jrse/2/5/10.1063/1.3486072.

CHAPTER 1.2

The Preservation Perspective

Introduction

Preservation of buildings and conservation of materials are an effort to retain our heritage but cannot exist in isolation, separate from social and ecological movements. The interpretation of heritage, integrity, and authenticity has expanded to include criteria of equity, history, and place that reflect differences in global cultures, economy, and ecology. The call for social, racial, and economic equity requires recognition of tangible and intangible heritage of the commonplace, poor, and disadvantaged.

"Significance" and "character-defining feature" may include less tangible, more abstract themes such as craftmanship, religious, ethnic, or cultural attachment, and social narrative. The significance of a place may not be tied to the physical fabric of materials so much as it is tied to the memories and stories of the people who cherish the place. Preservationists who are committed to identifying and implementing social justice for all races, genders, and cultures have an opportunity and a responsibility to consider how social history and culture, and connection to place are expressed and preserved, and how these goals affect decisions about material conservation.

Evolving climate patterns are already pushing preservation stakeholders into unfamiliar and uncomfortable territory as they face the practical and more extreme threats of a dramatically changing environment. Rigid adherence to principles of authenticity of construction and integrity of place may leave increasing numbers of properties at risk of total destruction. Buildings and sites that tell the history of the racially and economically disadvantaged, immigrants, African Americans, and Native Americans are usually constructed of less durable materials and have seen less maintenance. They are less likely to have survived the ravages of time and

DOI: 10.4324/9781003160489-3

weather, displacement, and urban renewal and face new risks as the climate becomes more severe.

Conflicting strategies have always been promoted as the most appropriate to preserve history. Some stress the importance of conservation – retaining the original fabric or structure as it currently exists – while others advocate for varying degrees of restoration, replacement, or reconstruction to a specific point in time. Still others advocate for evolution or adaptive reuse that recognizes the passage of time.

The "best" approach depends upon the aspirational goals and practical realities of each project. The best approach is one that identifies and meets the cultural and social priorities of the impacted stakeholders, preserves the elements that tell the important story, and recognizes the reality of economic and climatic constraints. As climate change challenges the ability to preserve the physical fabric of historic structures – the materials – it is useful to consider how the preservation movement has evolved, and how it can continue to evolve so that it is relevant in the future.

National Preservation Movements

Prior to the late 19th century, preservation and conservation were private affairs, managed by property owners or dictated by a hierarchy to acknowledge the legacy of individuals or nations. In the late 19th century and throughout the 20th century, organizations began to form to promote the protection of natural and cultural heritage and history for the "public good". The desire to preserve sites of national importance developed in conjunction with a growing sense of national identity.

The original focus of these organizations was one of advocacy, creating criteria to identify and protect significant properties. Later criteria were developed to identify and protect historic buildings and monuments. In the 21st century, these criteria have been adapted to include a broader definition of heritage.

In some cases, organizations acquired ownership and responsibility for properties. Great Britain and the United States both began acquiring properties under government departments but found this approach too ambitious and cumbersome, and quickly shifted to a role of oversight of properties owned by individuals or non-profits. More common was the establishment of governmental agencies that provided advocacy, criteria, and oversight of the conservation or restoration of properties that were owned privately or by non-profits. Standards addressed preservation of the physical fabric and character-defining features of historic structures.

Some national movements have responded to unique challenges with more flexible approaches. Japan has always valued both physical fabric and the underlying craftsmanship, recognizing the historic value of complete recreations. Germany developed guidelines to balance the desire for preservation against the economic and physical realities of infrastructure damaged by war. Australia has developed broad guidelines to respond to the intangible culture of their indigenous population.

The United States

The National Park Service (NPS) was originally established in 1916 to manage western land. The creation of the Historic Sites Act of 1935 shifted their focus to the acquisition of properties that commemorated or illustrated the history of the United States. This mission quickly became unwieldy and shifted to a policy of listing properties owned privately or by non-profits, and supporting them with technical assistance, grants, and tax credit programs. The National Trust for Historic Preservation (NTHP) – a private, non-profit organization – was established in 1949 to provide ownership, education, and advocacy for historic buildings, neighborhoods, and heritage. Today, the NPS, a federal agency, owns National Parks and a few National Historic Landmarks, but most historic properties and districts are owned privately or by non-profits.

The National Historic Preservation Act (NHPA) of 1966 created the National Register of Historic Places (NRHP) and a system of State Historical Preservation Officers (SHPO) that work with the NPS to provide oversight of changes to any historic property that is receiving federal funding, regardless of ownership.

The NPS focused on defining "significance" until 1978 when "the Secretary of the Interior's (SOI's) Standards for the Treatment of Historic Properties" were established. The SOI's standards apply to federally assisted projects and provide a template for the guidelines used by most local and state landmarks and districts in the United States. Legislation was passed to allow municipalities and states to designate and protect privately owned properties as individual landmarks and historic districts. Landmark designations carry significant protections, either specific to the site or referencing the SOI's standards. The SOI's standards and other historic designations provide specific guidelines and restrictions on the repair or replacement of existing historic materials.

Great Britain

The Society for Preservation of Ancient Buildings (SPAB) was established in 1877 by William Morris and inspired by John Ruskin in reaction to restoration work that they felt was detrimental to the existing historic fabric. Their role was primarily one of education and advocacy. England began acquiring sites of historic importance under the Office of Works in 1882 and established a National Trust in 1894. In 1913, Parliament charged the Office of Works with acquiring a collection of the greatest sites and buildings that told the story of Britain, originally focusing on prehistoric and medieval sites. By 1933, there were 273 sites including Stonehenge and a number of castles, abbeys, and forts with a focus on both conservation and education. In 1968, a new Town and Country Act required that SPAB be notified of alterations or demolitions to any buildings dated prior to 1720. In 1983, England's national collection was transferred to English Heritage, which split into two parts in 2015. English Heritage continues to manage over 400 historic buildings with a focus on inspiration, conservation, involvement, and financial sustainability, while Historic England acquired responsibility for policy, planning, and grants. Cadw was established as the environment service for Wales in 1984 and Historic Environment Scotland (HES) was established in 2015 to manage properties, policies, and grants in Scotland.

Canada

Canada provides oversight of its historic heritage at the federal, provincial, and municipal levels. At the federal level, the "National Historic Sites of Canada" designates sites with national historic significance, while the "Federal Heritage Buildings" include buildings owned by the federal government. Provinces and municipalities each have distinct systems for designating and overseeing sites owned by the provincial or municipal governments, individuals, or non-profits. The National Trust for Canada, a not-for-profit charity, established in 1973 advocates for heritage policies and laws. The federal, provincial, and territorial (F/P/T) governments have collaborated to develop the Canadian Register of Historic Places (CRHP) and the "Standards and Guidelines for the Conservation of Historic Places in Canada".

Germany

The urban planning movements in West Berlin after World War II responded to the need to rebuild both housing and infrastructure initially by abandoning deteriorated city centers and constructing new housing estates in the undeveloped city perimeter. However, the high cost of new infrastructure quickly resulted in a return to the city center, with a focus of demolition and new construction destroying the physical, historic, and social fabric of existing communities.

At the end of the 1960s, the Behutsamen Stadterneuerung or "Cautious Removal" movement illustrated the economic and social benefits of "urban renewal areas with the aim of intensifying the modernization of old buildings…" The combined forces of public and professional pressure eventually led to the adoption by the Berlin House of Representatives in 1983 of legally binding redevelopment concepts based on the 12 principles of "cautious urban renewal". These principles focused not only on "preserving the substance", and reducing demolition, but also on the "participatory rights and material rights of those affected by social planning" (IBA 2023).

Australia

The Australia National Committee of International Council on Monuments and Sites (Australia ICOMOS) adopted the Burra Charter in 1979, adapting ICOMOS principles to reflect the tangible and intangible heritage in Australia. The Burra

Charter has become an international model for acknowledging the less tangible aspects of heritage of the indigenous people of Australia.

The Burra Charter acknowledges that conservation will vary depending on circumstances, and may include a combination of maintenance, preservation, restoration, reconstruction, adaptation, and interpretation, and may include retention of a later addition that contributes to the significance of the place. Furthermore, it states that conservation may include the "retention or reintroduction of a use; retention of associations and meanings...".

Japan

The Meiji government's Decree of 1871 established a shrine system that divided all shrines into official Imperial or National government shrines, or "other" shrines, of minor, medium, or major importance. In 1897, the Law on the Preservation of Shrines and Temples established additional legal protections.

This system was set aside in 1946 by administrators of the World War II Occupation, who considered it as a symbol of State Shintoism. In 1950, the Law for the Protection of Cultural Properties designated sites as either National Treasures or Important Cultural Properties and established restrictions for alterations and oversight by the national government. Local governing bodies determine the type of materials, the conservation methods, and the extent of replacement that are appropriate for each shrine or temple location.

World Cultural and Natural Heritage

The widespread destruction of World War II and the following growth of global cooperation led to an increasing acknowledgment that heritage transcends political borders. Led by the United Nations, countries collaborated to identify universally important natural and cultural sites and criteria that recognized differences in cultural attitudes and priorities.

The International Centre for the Study of the Preservation and Restoration of Cultural Property (ICCROM) was established in 1956 to research and improve restoration methods through an international network of experts and specialized institutions. In 1959, a center was established in Rome to provide training in conservation and an international library for research and documentation. ICCROM collaborates with partner organizations on sustainable management and long-term conservation that minimize damage or risk to the original fabric of cultural resources.

In 1972, United Nations Educational, Scientific and Cultural Organization (UNESCO) established an international treaty – the World Heritage Convention – to identify, protect, and preserve cultural and national heritage. World Heritage List properties must meet criteria for "outstanding universal value", authenticity and integrity, and must have an adequate system for protection and management. These criteria cover both cultural and national heritage, and include both architectural and technological creations and settlements, and sites that are "directly or tangibly associated with events or living traditions, with ideas, or with beliefs, with artistic and literary works or outstanding universal significance".

International Council on Monuments and Sites (ICOMOS) was established in 1965 as the principal advisor to UNESCO on cultural and World Heritage matters. ICOMOS explicitly acknowledges in its "Ethical Principles":

- The "diverse tangible and intangible values" of cultural heritage that "hold different meanings for different groups and communities".
- The role of heritage as an economic, social, and cultural driver of sustainable local and global development".
- The value of community involvement and collaboration in conservation decision making.
- The importance of exploring "all viable options…and that chosen options are justified".

Shared Principles

Defining Significance

The term "historic" could be applied to the entire stock of older existing buildings that contribute to a sense of identity and place. Buildings with an official designation of "Landmark" or "World Heritage" must have a higher level of historical, architectural, engineering, archeological, and/or cultural significance. Although the distinction is somewhat artificial, understanding how and why a building is significant will help prioritize our use of resources and how we approach preservation and conservation.

Buildings of significance are usually aesthetically or architecturally significant in their own right or have physical attributes of historic or cultural significance. However, relatively modest buildings are increasingly being recognized because they are the best remaining representatives of a historic person or event. The Malcolm X – Ella Collins-Little House, a modest home, was designated a Boston Landmark for its significance "as the primary [and only extant] residence of preeminent black nationalist leader Malcolm X during a three-year period associated with his turbulent coming of age…" (BLC 1998).

It has recently been suggested that historic or cultural significance can exist independent of a physical structure or location. The term "heritage" is increasingly applied to both tangible and intangible aspects of history. In England, The National Planning Policy Framework (NPPF) defines significance as "the value of a heritage asset to this and future generations because of its heritage interest". That interest may be archaeological, architectural, artistic, or historic.

Australia's Burra charter even more broadly defines cultural significance as "aesthetic, historic, scientific, social, or spiritual value for past, present, or future generations" as embodied in its "fabric, setting, use, associations, meanings, records, related places, and related objects". Burra Charter is significant for its expanded definition of setting as the "extended environment of a place that contributes to its cultural significance and distinctive character…Setting may include historical and contemporary relationships, such as use and activities, social and spiritual practices, and relationships with other places, both tangible and intangible" (Australia ICOMOS 2013).

Character-Defining Features

An understanding of "character-defining features" should guide decisions regarding repair, replacement or modifications of historic materials, components, or the structure as a whole. The significance of specific materials should guide the decision to retain or replace material and to use identical or substitute materials. Significance will determine whether details can be modified to better resist weather conditions or whether the entire structure can be relocated to avoid more severe hazards of sea level rise. The importance of preserving the original fabric will determine the acceptable level of risk. If a material is irreplaceable, then a higher priority must be placed on protecting that element from damage.

Integrity and Authenticity

Another aspect of historic significance that is constantly being redefined is "authenticity" and "integrity". These terms are often used interchangeably but are subtly different. Both are used to describe original physical fabric and/or original location. However, UNESCO offers a more accurate definition of integrity as "measure of wholeness and intactness of the natural or cultural heritage and its attributes" and a more subjective measure of authenticity as whether the property "credibly and truthfully reflects the value attributed to the property". Inclusion in the World Heritage list requires that sites be judged authentic but also notes that "…judgments about value attributed to cultural heritage, as well as the credibility of related information sources, may differ from culture to culture, and…must be considered and judged primarily within the cultural contexts to which it belongs" (UNESCO 2012).

Attributes of authenticity can include tangible aspects of form, design, location and/or setting, and intangible features such as use and function, traditions, language, spirit and feeling, and other forms of intangible heritage. In other words, physical fabric and location are not the only measures of authenticity. When the significance of a place is related to a historical event or person, the location might retain its importance of association even if it no longer looks the way it did during the period of significance. Similarly, a building may still represent a historical event or person even if it has been relocated. As climate change compromises specific materials and locations, the question of authenticity and integrity may require a more flexible interpretation.

Interpretations of significance, authenticity, and integrity are especially important when considering how to tell the story of disenfranchised populations because many buildings and sites that represent the poor and transient have been modified or destroyed over time. The memory of an event or person may be all that remains at a historic location. This is not to say that retention of existing material is not important, only that it should be a conscious decision that is supported by preservation goals and the importance of the original fabric.

Conservation

Just as there are many opinions on what is historically significant, there are different philosophies about what constitutes conservation. The NPS defines *Preservation*, *Rehabilitation*, *Restoration*, and *Reconstruction* as different approaches to conservation, each appropriate under different circumstances. Different strategies can be applied at different scales. For example, a specific element might be preserved while reconstructing a building or specific elements might be replaced to preserve the building. An entire building might be replaced with new materials while preserving the history, architecture, and craftsmanship, as has been the tradition in Japan.

There are additional categories that could be called types of conservation: *adaptive reuse*, or what English Heritage calls "*constructive conservation*", *relocation of original fabric*, and *digital reconstruction*. Each of these categories acknowledges the need to make changes to allow the "continued use and enjoyment" of a historic building that might otherwise be lost.

Selecting the right approach to conservation is dependent on answering the following questions: What are you trying to preserve…an artifact or a story? The approach to conservation will be different if the goal is to preserve the original fabric, tool marks, or craftsmanship than the approach when the priority is to tell the story about a particular life or history. Are you preserving the building as it exists today or as it did at some point in the past?

Approaches to Conservation

The US Secretary of the Interior's Standards (preservation, rehabilitation, restoration, and reconstruction) are representative of similar approaches used across the globe and can be applied separately to all scales of a project, from the overall structure to the individual materials.

Preservation

The NPS defines preservation as placing "a high premium on the retention of all historic fabric through conservation, maintenance, and repair. It reflects a building's continuum over time, through successive occupancies, and the respectful changes and alterations that are made" (NPS 2023).

Canada's standards for preservation have a similar focus on "short-term and interim measures to protect and stabilize the place, as well as long-term actions to stave off deterioration or prevent damage...though routine maintenance and small repairs...". Canada's standard #8 does allow for replacement of a part of an element where sufficient evidence exists to accurately replicate the missing or deteriorated part (Canada's Historic Places 2010).

The Society for the Protection of Ancient Buildings (SPAB) encourages a preservation approach to conservation, emphasizes retention of existing fabric and location to the greatest extent possible, and avoids hiding repairs or changes that are part of the history of a building. Existing fabric should not be removed to restore a building to a previous idealized state and restoration of the existing fabric should be minimal and non-invasive, only as required to prevent future deterioration. In other words, all new work should be in the service of preserving what exists of the building today and celebrating the building as it has evolved over time.

Historic England and HES both place a high value on preservation of original fabric and have invested substantial resources in education and training to maintain or replicate historic materials and techniques. In many cases, they have determined that original materials, such as lime mortars and washes, perform better than new synthetic materials.

The decision to repair the adobe without restoring missing stucco at San José de Tumacácori is an example of preservation of a moment in time.

Rehabilitation

Rehabilitation "emphasizes the retention and repair of historic materials but allows more latitude or replacement". The Canadian standard is similar but acknowledges that the "replacement may be an accurate replica of the missing feature" or "a new design compatible with the style, era and character of the historic place" (Canada's Historic Places 2010).

New material should be identifiable as such and alternative materials can be used if the original material is not available or economically feasible. However, substitute materials should not be used to improve performance if the original materials will perform adequately. Character-defining features should not be removed or relocated to accommodate new functions, especially if doing so gives a false impression of the original work.

The replacement of terra cotta with fiberglass at Michigan Central Station is an example of a rehabilitation.

Restoration

Restoration "focuses on the retention of materials from the most significant time in a property's history, while permitting the removal of materials from other periods" (NPS 2023).

Both the NPS and Canada require that restoration of "missing features…must be based on clear evidence and detailed knowledge of the earlier forms and materials" (Canada's Historic Places 2010). The existing configuration and location of elements from other periods should be thoroughly documented before removal. ICOMOS Australia makes a further distinction between *restoration*, which removes additions and reincorporates existing elements, and *reconstruction*, which introduces new material.

Typically, restoration requires a determination that the interpretation and presentation of a particular moment or period in time is more important than subsequent modifications to a building; it also requires interpretation of physical and historical evidence to recreate components of a building that may no longer exist. It was this requirement for interpretation that led to objections by William Morris and SPAB. They objected first to the underlying premise that a point in the past was more important than the subsequent accumulation of history. They also maintained that there was potential for misinterpretation of the physical and historical evidence.

The removal of cement plaster from historic lighthouses is an example of restoration.

Reconstruction

There may be circumstances where the desire to protect a building and/or its cultural heritage is more important than the desire to retain the integrity of the original fabric or location. "Reconstruction … establishes limited opportunities to re-create a non-surviving site, landscape, building, structure, or object in all new materials" (NPS 2023). Reconstruction is not considered conservation by the Canadian standard, but it is a time-honored method of conserving the traditional structures of Japan.

The complete reconstruction of temples in Japan and stave churches in Norway are examples of reconstruction.

Relocation

Relocation of the original fabric is not addressed by either the NPS or Canadian standards because relocation was considered a loss of integrity. However, relocation should be considered a valid option when the only alternative is complete loss to either economic or climatic concerns. Structures are often relocated to avoid demolition due to development. As climate changes become more extreme, leading to changes in sea level, soil conditions, and weather extremes, relocation may be the only alternative to demolition.

The relocation of the Sanborn Covered Bridge in Vermont is an example of relocation.

Adaptive Reuse (Constructive Conservation)

Adaptive reuse may require changes but can support a preservation goal. Adaptive reuse must respect the original construction, architecture, and history of the building, retain distinctive features, and minimize loss of historic fabric. Modifications that allow continued or new uses for a building support conservation and preservation by making the maintenance and use of a building economically viable.

Digital Reconstruction

Digital reconstruction is often considered a strategy of last resort, where a building will be lost. However, photographs and 3D documentation can also provide a template for future repairs or reconstruction, both where features are eroded gradually and where destruction is sudden and unexpected. This strategy allows existing fabric to be maintained as long as possible, without losing the opportunity to accurately reconstruct details in the future.

Conclusions

Choosing an approach to preservation from the array of options is challenging and depends on many factors, starting with those outlined by the SOI's standards:

- Relative importance in history. Is the building a nationally significant resource – a rare survivor or the work of a master architect or craftsman? Did an important event take place in it?
- Physical condition. What is the existing condition – or degree of material integrity – of the building prior to work? Has the original form survived largely intact, or has it been altered over time? Are the alterations an important part of the building's history?
- Integrity of craftsmanship and technique. Is the original fabric intact? Were replacements and repairs made with authentic materials, workmanship, and craftsmanship? Is the original craft and technique an important part of the history of the building or place?
- Proposed use. An essential, practical question to ask is: Will the building be used as it was historically or will it be given a new use?
- Mandated code requirements. Regardless of the treatment, code requirements will need to be taken into consideration.

There will often be a trade-off between the feasibility of retaining existing fabric and preserving the building as a whole. Decision making is further complicated by the lack of certainty about climate related risks and the reality of limited resources. When original fabric is significant and irreplaceable, the tolerance for risk is low and protection of the fabric becomes a high priority. Insurance is not a viable option if replacement is not a satisfactory option. If the material has already been replaced, or if the design or story are more important than the actual materials, then the tolerance for risk may be high and limited resources might be used elsewhere. In the end, decisions about how and when to treat or replace existing historic materials must be made in the context of priorities for the overall building.

Bibliography

ACNT. 2021. *National Trust, About Us*. https://www.nationaltrust.org.au/about-us/.

Armbruster, Robert, interview by Susan Pranger. 2022. (November 21 December 15).

Australia ICOMOS. 2013. "Australia ICOMOS Burra Charter, 2013." https://australia.icomos.org/wp-content/uploads/The-Burra-Charter-2013-Adopted-31.10.2013.pdf.

BLC. 1998. "The Malcom X – Ella Little-Collins House Study Report." *Boston Landmarks Commission*. http://www.cityofboston.gov/Images_Documents/Malcolm%20X%20-%20Ella%20Little-Collins%20House%20Study%20Report_tcm3-23718.pdf.

Canada's Historic Places. 2010. *Standards and Guidelines for the Conservation of Historic Places in Canada*. Her Majesty the Queen in Right of Canada.

Edahiro, Junko. 2013. *Rebuilding Every 20 Years Renders Sanctuaries Eternal – The Sengu Ceremony at Jingu Shrine in Ise*. September 10. http://www.japanfs.org/en/news/archives/news_id034293.html.

English Heritage. 2008. "Climate Change and the Historic Environment." *English Heritage*. January 18. http://www.english-heritage.org.uk/publications/climate-change-and-the-historic-environment/.

English Heritage. n.d. *Our History*. https://www.english-heritage.org.uk/about-us/our-history/.

Historic England. 2021. *Planning for the Historic Environment*. https://historicengland.org.uk/advice/hpg/historic-environment/.

IBA: Internationale Bauausstellungen Altbau. 2023. *12 Principles of Cautious Urban Renewal, A Paradigmatic Change in Urban Development*. https://www.internationale-bauausstellungen.de/geschichte/1979-1984-87-iba-berlin-die-innenstadt-als-wohnort/12-grundsatze-der-behutsamen-stadterneuerung-ein-paradigmatischer-wandel-in-der-stadtentwicklung/.

ICCROM. 2021. *History*. https://www.iccrom.org/about/overview/history.

ICOMOS. 1999. "Nikko (Japan) No 913." Nomination by UNESCO Advisory Board (UNESCO).

ICOMOS. 2014. "ICOMOS Ethical Principles." https://www.icomos.org/images/DOCUMENTS/Secretariat/2015/GA_2014_results/20150114-ethics-asadopted-languagecheck-finalcirc.pdf.

Marinko, Whitney. 2020. *Historic Real Estate: Market Morality and the Politics of Preservation in the Early United States*. Philadelphia: University of Pennsylvania Press.

Merriam Webster. 2021. *Preservation*. https://www.merriam-webster.com/dictionary/preservation#:~:text=%3A%20the%20act%2C%20process%2C%20or,on%20preservation%20continued%20in%201988.

National Park Service (NPS). 2023. *The Secretary of the Interior's Standards for the Treatment of Hisotirc Properties*. Retrieved from NPS Technical Preservation Services: https://www.nps.gov/orgs/1739/secretary-standards-treatment-historic-properties.htm.

———. n.d. "National Park Service." *Choosing an Appropriate Treatment for Historic Buildings.* http://www.nps.gov/tps/standards/four-treatments/standguide/overview/choose_treat.htm.

National Trust for Canada. 2023. *National Trust for Canada*. https://nationaltrustcanada.ca/.

New World Encyclopedia Contributors. 2019. *Shinto.* November 3. https://www.newworldencyclopedia.org/p/index.php?title=Shinto&oldid=1026695.

Pickles, David, and Jeremy Lake. 2017. *Adapting Traditional Farm Buildings, Best Practice Guidelines for Adaptive Reuse*. Historic England. https://historicengland.org.uk/images-books/publications/adapting-traditional-farm-buildings/heag158-adapting-traditional-farm-buildings/.

Slocombe, Matthew. n.d. *The SPAB Approach to the Conservation and Care of Old Buildings.* SPAB.

SPAB. 2009. "SPAB's Purpose." *SPAB*. https://www.spab.org.uk/what-is-spab-/spab-s-purpose/.

Sprinkle Jr., and R. John. 2014. *Crafting Preservation Criteria*. New York: Routledge.

The Australian Collaboration. 2013. *The Natural and Cultural Heritage of Australia.* https://www.australiancollaboration.com.au/pdf/FactSheets/Heritage-FactSheet.pdf.

UK Parliament. 1907. "The National Trust Act." https://nt.global.ssl.fastly.net/documents/national-trust-acts-1907-1971-post-order-2005-and-the-commons-act-2006.pdf.

UNESCO. 2012. "Operational Guidelines for the Implementation of the World Heritage Convention." World Heritage Centre. http://whc.unesco.org/en/guidelines (English).

UNESCO. 2011. "Preparing World Heritage Nominations." UNESCO/ICCROM/ICOMOS/IUCN. https://whc.unesco.org/en/about/.

UNESCO. 2021. *Shrines and Temples of Nikko*. https://whc.unesco.org/en/list/913/.

University Libraries. n.d. *The Protection of Ancient Buildings: Morris as Preservationist.* https://www.lib.umd.edu/williammorris/morris-as-preservationist.

CHAPTER 1.3

Stakeholders and Resources

Introduction

Even when the technical solutions are clear and consistent with the stakeholder's preservation priorities, the ability to implement a conservation plan for materials as part of a preservation plan for a structure may be limited by a lack of influence or control, limited economic resources, social capital, political policies, and governmental restrictions.

Influence and Control

Historic assets may be owned by individuals or organizations, non-profits, or by municipal, state, or federal government agencies. Although owners may appear to control decision making, many other stakeholders often have informal influence or official jurisdiction over zoning, construction, accessibility, energy use, and appearance. Decisions are also influenced by consultants who design and evaluate changes; contractors, conservators, and craftsmen who implement physical changes and by manufacturers who design and supply materials. Governmental agencies control enforcement of building code and zoning requirements, grants, and special funding for preservation projects, and implementation of local and regional projects to manage stormwater, storm surge, sea level, ground water recharge, and heat island.

Owners and Occupants

Owners of historic properties, whether private or non-profit, are also caretakers, developers, and advocates. They have the greatest opportunity and responsibility

DOI: 10.4324/9781003160489-4

for long-term monitoring of historic assets and for identifying future needs and opportunities. A small private owner may have more direct control over decisions but less access to in-house technical expertise. Large-scale owners may have more technical and financial resources, but also more stakeholders, responsibilities, and restrictions. Non-profits may have access to funding that private owners do not, which may result in mutually beneficial partnerships.

Occupants, visitors, and caretakers may not have official control over decisions but can either support or undermine efforts to fund, maintain, or operate historic buildings. This is especially true when there is a lack of communication about systems used to control environmental conditions and conserve energy. Together with the facility managers and custodians, the people who live, work, and regularly visit are most familiar with the day-to-day functioning of a building, and can be a valuable early warning system if they are familiar with the preservation and environmental goals and intended strategies of the building.

Experts and Consultants

Depending on the scale and importance of a project, owners will work with a variety of consultants and contractors to finance, define, and execute renovations and ongoing maintenance. A small project might involve a contractor and architect. A large or complex project can involve construction managers, developers, architects, engineers, historians, conservators to develop repair or restoration strategies, and identify changes to enhance the future durability of the building. The contractors, subcontractors, specialty craftsmen, and suppliers who will execute a project can provide invaluable insight during the decision-making process.

Developers usually finance and oversee the programming, design, and construction of existing buildings. As such, they have a great deal of responsibility for identifying resources, balancing risk, and making decisions. They may have their own preservation goals, be advised by preservation consultants, or work to meet the goals of owners or advocacy groups. Developers will evaluate the financial risks and benefits of responding to the future impacts of climate changes.

The design team, which whether a single individual or a complex team of consulting firms, is responsible for identifying the condition of the building and evaluating future risks and responding with appropriate solutions. The design team will usually help define goals and priorities for conservation, preservation, sustainability, and resiliency. Each consultant, contractor, or supplier should consider the ways that a building can be durable and resilient within the framework of their individual area of expertise. However, the best projects are those where the entire team, including the owner and the construction team, collaborate to develop strategies to make and implement decisions that reinforce the natural adaptability and flexibility of the existing historic building materials and construction.

Investors and Insurers

Investors and insurers have a specific economic interest in the survival of a project, including the relative costs of initial investments and future repairs. Savvy investors acknowledge the value of both historic and sustainable buildings. A resilient building will have less risk of damage due to climate change, and therefore will reduce the risk to investors and insurers.

Insurers are often the first to identify and quantify the financial risk of fire, flooding, hurricanes, and other environmental hazards based on historical statistics. They have been slow to acknowledge the risks of climate change, including sea level rise, in the past but are now responding to climatic events with increases in insurance rates, exclusions and even withdrawal from some markets. Insurers now expect property owners to identify the unique risks to historic assets and to respond accordingly, and often hold developers to higher standards than those required by building or zoning codes. In the wake of Hurricane Sandy and other extreme weather events, utility companies and insurance companies are stepping up to press FEMA for more cautious standards and they are insisting on designs and materials that anticipate future threats.

Advocacy

It is not surprising that the public feels strongly about its historic buildings. Historic buildings represent the physical embodiment of our collective history, clues to who we are and where we came from. The public can champion preservation by advocating for appropriate care of individual buildings or by lobbying for local, state, or federal policies that fund or control changes to historic assets.

Non-profit advocacy groups exist at all scales from municipal to global. Each organization has its own mission, but they all share the goal of preserving historical assets for the future. Many, including English Heritage, the National Trust for Historic Preservation, and the National Park Service have long acknowledged the need for historic buildings to be sustainable in terms of energy and resource conservation but are now also advocating for resiliency in face of the serious risks that global warming and climate will present for historic assets. These organizations can provide technical advice and can be powerful advocates for policies to address resilience at a regional and state level. Smaller organizations may have less capacity to influence policy but are often in a better position to identify and influence planning for individual historic assets.

Resources and Social Capital

Efficient use of resources can be challenging in an environment that is constantly changing and increasingly unpredictable. Necessary funds may come from earned income or an endowment, from individual or institutional donors, from investors, or from public funds or programs. A non-profit's funds will likely come from a combination of sources, with some restricted or limited to a specific period of time or to a specific type of use or project. Private owners are often limited by their own ability to pay or to borrow based on equity in the property.

Equally important is social capital, the ability of a network of stakeholders to influence or implement policy changes or infrastructure projects at a local, regional, or global level. An example would be individuals or organizations that advocate for a regional approach to responding to sea-level change rather than risk damage or inappropriate changes to a historic asset. Another example is the advocacy effort by the National Trust for Historic Preservation and other groups to protect and strengthen the Federal Historic Preservation Tax Incentive Program, a fund that provides funds for major rehabilitations.

Restrictions

Changes to historic buildings are restricted by local and state building and zoning codes, as well as preservation easements and preservation restrictions.

Municipal, state, and federal "authorities having jurisdiction" (AHJ) enforce building, fire, and zoning codes that may reinforce or discourage actions that protect the historic character or improve resiliency. New energy codes require changes to infiltration or insulation that may compromise historic integrity or physically compromise the performance of traditional materials. The new International Building Code, which has been adopted by most states in the United States, includes recommendations through FEMA's Technical Bulletins 2 and 3 for flood resilience. The National Institute of Standards and Technology is beginning research to set standards and metrics for structural resiliency to allow buildings to recover after an extreme weather event.

Local and state historic commissions, supported by the National Park Service, have the most power and visibility as organizations with official jurisdiction and various degrees of control over demolition or change to historic buildings, and as such act as advocates for historic buildings. Preservation easements or covenants require maintenance and restrict changes on a specific building or property, usually in exchange for an investment from the easement holder. In some cases, an owner will pay the holder of the covenant to enforce an easement. These covenant holders will now need to consider the need to protect buildings from new environmental risks.

Conclusions

A comprehensive preservation plan for a monument or building should include a plan for conservation of materials, strategies for the sustainability and resiliency of the building as a whole, and a disaster plan for responding to extreme events. Each of these components must consider and respond to changing weather conditions. Comprehensive and forward thinking planning will require the input and resources of all stakeholders. Individual stakeholders may need to expand their financial and administrative resources and their own capacity for resilience. The unique challenge of anticipating and responding to unpredictable threats will require that stakeholders think outside the box and respond with flexibility and creativity.

Bibliography

CERES. 2014. "First-of-its-Kind Report Ranks U.S. Insurance Companies on Climate Change Responses." *CERES*. http://www.ceres.org/resources/reports/insurer-climate-risk-disclosure-survey-report-scorecard-2014-findings-recommendations/view.

Co2olbricks. 2013. "Integration of Climate Protection and Cultural Heritage Aspects in Policy and Development Plans." Project Report. Co2olBricks – Climate Change, Cultural Heritage and Energy Efficient Monuments.

English Heritage. 2008. "Climate Change and the Historic Environment." *English Heritage*. January 18. http://www.english-heritage.org.uk/publications/climate-change-and-the-historic-environment/.

FEMA. National Flood Insurance Technical Bulletins. https://www.fema.gov/emergency-managers/risk-management/building-science/national-flood-insurance-technical-bulletins.

Kesner, Kieran. 2015. "Boston Globe." *Landmarks Commission rejects landmark status for Prouty Garden*. http://www.bostonglobe.com/metro/2015/04/28/prouty-garden-goes-before-landmarks-commission/q1JdJgMWpQDjhOHGkVtbRK/story.html.

Linden, Eugene. 2014. "How the Insurance Industry Sees Climate Change." *LA Times*. June 16. http://www.latimes.com/opinion/op-ed/la-oe-linden-insurance-climate-change-20140617-story.html.

MW. n.d. "Developer." *Merriam Webster Dictionary*. http://www.merriam-webster.com/dictionary/developer.

NIST. n.d. "Measures of Building Resilience and Structural Robustness Project." *National Institute of Standards and Technology*. http://www.nist.gov/el/building_materials/structures/mbrsr.cfm.

NPS. n.d. "Adaptation for Cultural Resources." *National Park Service*. http://www.nps.gov/subjects/climatechange/adaptationforculturalresources.htm.

NTHP. n.d. "Climate Change." *National Trust for Historic Preservation*. http://www.preservationnation.org/magazine/2009/march-april/climate-change-green-initiatives.html.

———. n.d. "Historic Tax Credits." *National Trust for Historic Preservation*. http://www.preservationnation.org/take-action/advocacy-center/policy-resources/historic-tax-credits.html.

TPS. n.d. "Tax Incentives for Preserving Historic Properties." *Technical Preservation Services, National Park Service*. http://www.nps.gov/tps/tax-incentives.htm.

Wikipedia. 2015. "Real Estate Development." *Wikipedia*. http://en.wikipedia.org/wiki/Real_estate_development.

Solecki, William, Lee Tryhorn, Art DeGaetano, and David Major. n.d. "Chapter 2 Vulnerability and Adaptation." http://www.nyserda.ny.gov/-/media/Files/Publications/Research/Environmental/EMEP/climaid/ClimAID-Vulnerability-Adaptation.pdf.

PART 2

Traditional Materials

CHAPTER 2.1

Common Issues and Strategies

Introduction

While each material responds differently to environmental conditions, there are many common characteristics of building materials, and many common strategies for responding to weather. Porous materials react in similar ways to moisture, heat, wind, and solar exposure. Many strategies for conservation of traditional materials are very specific to the material and climate and are best discussed in the chapters on specific materials and case studies. However, there are a number of common strategies that apply to all materials and most climates.

An emerging issue in the field of preservation is that of research and training. All indications are that a lack of both knowledge and hands-on experience with traditional building materials is compromising many good intentions. It does no good to require authentic materials and practices if appropriate materials are not correctly specified, and qualified craftsmen are not available to execute those directions.

Resiliency, Durability, and Recovery

Durability is a function of interacting factors: the inherent characteristics of a building material, how it is harvested, fabricated, and incorporated into a building, how it is protected and maintained, and the daily weather and extreme events that it must withstand.

Resilience of materials is defined as the ability to *withstand* or to *recover* from the effects of weather. The ability to *withstand*, repel, or endure impacts requires strength or durability, for example the ability of a slate roof to resist wind and rain.

DOI: 10.4324/9781003160489-6

Recovery can refer to "natural recovery," the ability to absorb a weather impact and recover without damage, as when a masonry wall absorbs and releases water without permanent damage. *Recovery* can also refer to *ease of repair*, when damage occurs but is easily repaired, such as, for example, by repairing a coating or replacing a single missing roof shingle.

Traditional construction methods and materials are often resilient because they were designed to recover by absorbing and releasing moisture, and were dependent on easy, but frequent, maintenance and repair. However, historic buildings may very well be the canary in the coal mine of global warming. They are vulnerable to both the changing climate and to man-made interventions because they were built under very different conditions and with less standardization and control. If they still survive today, it is likely because they were adapted to the weather rather than resisting it. To survive in the future, they must adapt to new conditions.

Design and detailing can improve or compromise durability and resiliency of both materials and assemblies. Understanding how and why materials have responded to specific environmental conditions in the past can provide some insight as to how they will perform as conditions change in the future, and how strategies from one location can be shifted to another as the weather patterns change.

The Nature of Traditional Materials

Natural materials and historic man-made materials are less homogenous than most modern manufactured products and can respond to weather and other treatments in less predictable ways. Each material will respond based on their inherent chemical and mechanical properties as well as many other interacting variables, including the context in which it is used, the quality of workmanship of the material and its installation, and the level of maintenance.

However, materials also have common responses to temperature and heat: most expand and contract, swell and dry, or undergo chemical changes. The relative difference in expansion between different materials and between those materials and water, ice, and salt is a major driver of deterioration. The chemical changes are specific to each material but are influenced by temperature, moisture, and other available chemicals.

Reflectivity (the ability to reflect light or radiation) and emissivity (the ability to emit thermal or electromagnetic radiation) vary with the type of material, as well as the surface texture or coating. These qualities in turn affect how materials react to heat and sunlight and can affect the surrounding ambient temperature (heat island effect).

Understanding materials requires a basic understanding of how fluids move within porous materials: how water changes from liquid to vapor as it moves, evaporating and condensing in response to temperature, relative humidity, solar radiation, and wind pressure. This movement is controlled by the porosity and permeability of the material, capillary action, and pressure differentials. The ability of those fluids to carry salts and other chemicals depends on solubility.

Pores are the voids within a material, and can contain air, water vapor, or water. *Porosity* is a measure of the total volume of the pores, as determined by the size and number of pores and is a measure of the total amount of water that a material can absorb. *Permeability* is a measure of how quickly water moves through the material; it is determined by both the size of the pores and how they are connected.

Materials that are less permeable will absorb water and dry more slowly. Many porous materials, including stone, adobe, brick, and wood, will rapidly absorb water saturating the surface, and blocking additional infiltration of water so that excess rainwater sheets off the surface. However, the water running down the wall can still enter cracks and cause concentrated damage. Wind events can cause pressure differentials that will suck or drive water deeper into the stone.

Solubility is a measure of how much of a substance can be dissolved in a solvent without precipitating and will vary with the acidity (pH) and temperature of the solvent, and the type of material being dissolved. Solubility will determine whether salts or other chemicals are dissolved, where they are carried, and when they are precipitated and redeposited. As the water evaporates, or the temperature drops, there is less carrying capacity in the remaining water, and salts or minerals can precipitate within the pores (crypto florescence) or on the surface (efflorescence). A structure with high porosity may dry more quickly, leaving behind salts that accumulate in the pores and exert pressure from within. A material with small pores may dry out more slowly, allowing salts to either be carried to the surface or precipitate and accumulate within the pores.

While rain is the most obvious source of water, moisture can enter materials from all directions. Humidity and vapor can penetrate from the interior or exterior, surface water can rise from below, rainwater can fall on surfaces or enter the top of walls through roofs with inadequate rain collection. Surface flooding can cause erosion, settlement, or deterioration from rising damp. Infiltration by water or water vapor can be driven by gravity, surface tension, kinetic energy (wind driven rain), hydrostatic pressure, and pressure differential.

Penetrating dampness is water vapor that is gradually drawn horizontally into a material through capillary action or hydrostatic pressure in damp, humid conditions. Rising damp is drawn upward from the ground or from surface water through capillary action. Both types of dampness are more likely to occur in porous materials.

Concentrated water from flooding, design flaws, or roof failures can create hydrostatic pressure that "pushes" water through the pores of adjacent material. This pressure can cause moisture to move uphill, under flashing, and through overlapping joints. Heavier-than-usual rainfall can overwhelm designs that worked perfectly well in the past, such as a gutter that cannot handle an increase in rainfall.

The Impact of Weather on Materials

When the combination of weather, design, and actual construction conspire to trap moisture on the surface or within the walls, they can cause moisture or frost damage, salt damage, rot, insect damage, mold, expansion, and other types of deterioration. Of particular concern is the impact of impermeable coatings, vapor barriers, and insulation and central heating and air conditioning that cause materials to retain moisture. The exterior materials of most traditional buildings were designed to breathe, expand, and contract with the weather. Although the buildings were leaky and their materials absorbed water, they also dried out quickly.

Environmental impacts include the regional weather patterns, related qualities of intensity, frequency, and cycles, and cumulative effects. Durability of material can be influenced by the following "deteriorating factors":

- Atmospheric and climatic action (macro and micro)
- Wetting and drying effects
- Soil and ground water action
- Rodent, insect, bacterial, fungal, and plant action
- Water supply
- Electrolytic action
- Contact or association of incompatible materials
- Specific chemical action or chemical changes in materials
- Wear; impact of vibration
- Action of cleaning and cleansing
- Action of domestic and industrial wastes
- Accidental causes, including fires, lightning, earthquakes, and floods

Differential settlement of foundations and walls can be caused by extreme drying or saturation of the soil, a change in the water table, or an increase in surface water and will damage most rigid wall systems.

There is little evidence that current climate changes are increasing the severity or frequency of earthquakes, however stone that is weakened by other climate conditions will be more vulnerable to damage from earthquakes.

Moisture

Moisture can be introduced as humidity, fog, rain, ground water, and surface water. The movement of moisture through materials is affected by temperature, wind, UVB light, salt and other chemicals in the air and water. Daily and seasonal changes in relative humidity, temperature, and air pressure, and changes in pressure from the interior to exterior, can drive the movement of moisture and chemicals into and through materials. Elements of weather can deteriorate materials directly by causing expansion and contraction, freeze-thaw, and abrasion, or indirectly by facilitating chemical changes or increased biological damage from insects, fungi, mold, algae, and bacteria.

Carbon Dioxide and Other Pollutants

Carbon dioxide and other pollutants, including sulfates, chlorides, nitrates, and phosphates have always been released into the air by natural processes such as volcanoes and forest fires. Sulfates and chlorides occur naturally in desert dust and ocean spray. Ground water naturally contains salts, especially chlorides, sulfates and nitrates, and occasionally other rare earth metals. The levels of sulfates and other pollutants have been reduced in cities but carbon dioxide levels have increased.

Corrosive pollutants have increased exponentially from agricultural and industrial processes and the burning of fossil fuels. Chemicals are also introduced to materials during construction or maintenance intentionally or unintentionally, migrating from cleaning materials, adjacent mortar, concrete or other dissimilar materials. The amount of chemicals absorbed by rain from the atmosphere depends on many factors, including the height, size of drops, and temperature. Small, slowly falling raindrops will absorb more particles than large drops that are falling quickly. Water vapor, or fog, has the potential to absorb larger concentrations of pollutants, and to carry them deeper into the stone. Water absorbs carbon dioxide twice as fast at 32°F as it does at 77°F.

Sand, silt, and clay particles suspended in the water can make the water less acidic by absorbing aerosols as they slowly move through the air. Industrial processes that remove dust can reduce this natural cleaning process; construction sites or plowing fields that add dust might reduce the acidity of local rain. Where limestone is common, calcite dust can convert sulfites in the air to gypsum and reduce acidity in the rain.

Material Research and Investigation

The goal of most historic renovations or restorations is to retain original material or use historic materials and techniques. This approach is not just a matter of maintaining historic authenticity. Differences in physical characteristics and configurations can cause differential movements and stresses that damage the original material, especially if the original is softer or more porous, or has a different chemical composition than adjacent patches or coatings. Research and testing may be required to identify new materials that appear and perform like the original, and to develop appropriate repair techniques.

Prior to making decisions, an investigation of the history and existing condition of the materials, building context, and environment should be undertaken. Research should identify the history of the original construction, the history of prior damage, repairs, and changes, and how the environment has impacted the materials over time. Future environmental conditions, such as projected changes in temperature, rainfall, and pest populations, should also be considered. The source of existing moisture and other decay mechanisms should be identified, and the operation and effectiveness of existing control and management systems – pest management, water management, overhangs, details, and coatings – should be evaluated.

Visual Inspections

A visual inspection of exposed surfaces can identify missing components, surface deterioration, or severe cracking, water damage or staining. Peeling paint, rusty nails, mold, and condensation are all clues of an underlying moisture problem. Subsurface damage from insects, fungi, and bacteria can be identified by the presence of insect bore holes, mud tubes, frass, or fruiting bodies of fungus on the surface of vulnerable materials. Probing with a sharp pick or awl can identify voids below the surface. Tapping with a knuckle or hammer can identify voids or deterioration by changes in sound or vibration.

The natural aging or patina of material should be distinguished from decay or deterioration and protected as part of the character and history of the material. Metal patinas often protect the underlying metal. Another critical component of any investigation is to identify the type and condition of existing coatings, whether the coatings are concealing damage or trapping moisture, and whether they are compatible with proposed new coatings. The presence of existing clear coatings might not be immediately apparent on masonry, stone, or concrete substrates.

On-site observations and testing might need to be supplemented by laboratory tests to confirm the composition and condition of existing materials, and identify past and current sources of water, biological damage, and mechanical stress. Inspections and testing may be required of both the existing building materials and of the proposed replacement materials, to confirm that they are sound and compatible. Consider how results may vary under future climate conditions.

Testing

While some tests can be performed with inexpensive equipment and basic training, more sophisticated testing methods should be performed or supervised by consultants who know what tests will provide useful information, how to properly conduct the test, and equally important, how to interpret the results.

Non-destructive tests can be performed in the field or in the laboratory if a component can be removed and replaced. Destructive tests are performed in the field or laboratory depending on the type of equipment and control environment needed. Tests that are technically "destructive" may only require a small sample of original material or might be conducted on a sample of identical material from a non-historic source.

Testing can also provide valuable information about the composition or physical characteristics of a material:

- Type and proportion of component materials, including minerals, aggregates, or chemicals
- Type and condition of binders
- Density and structure of cells or crystals

- Compressive or tensile strength, ductility
- Moisture content

Decay or damage that is not visible on the surface must be identified through a range of non-destructive and destructive testing. Tests that detect differences in a material can be used to identify voids, buried materials or voids, or differences in the density or moisture content. There is a vast array of available tests. An expert should be consulted to determine which tests would be appropriate and useful, and to correctly interpret the results.

Predicting Moisture Movement

Our ability to model the performance of materials and systems is more sophisticated than ever, but the results can be flawed because the materials and systems themselves are flawed. Hand-crafted masonry, for example, is less homogenous than later machine-made masonry, and few vapor barriers are installed without gaps and cracks. Traditionally, architects relied on "dew-point" analysis to predict where condensation would occur in a wall assembly based on the exterior and interior temperatures, and the R value of each layer. However, this system only considers the thermal resistance of the materials and wall assembly at a point in time under a specific set of temperature and humidity conditions.

In recent decades, Wärme und Feuchte instationär (WUFI) analysis has become a tremendous tool for predicting the migration of moisture and temperature through exterior walls under specific conditions of exterior weather, internal temperature and humidity, and wall construction. Unlike earlier models, WUFI considers the impact of material porosity and wall cavities on vapor migration, absorption, capillary action, condensation, and drying time. This information is invaluable in evaluating the impact of adding insulation to historic walls, but there are limitations. WUFI is a model, not a field test of actual conditions; it can compare assemblies but not predict absolute performance. More accurate input data will provide more accurate results, but they must be interpreted by an expert who has expert knowledge of the software and its underlying assumptions and a strong understanding of the physical properties of materials.

Maintenance, Cleaning, and Coatings

A proactive approach of inspection and maintenance will be more effective and more economical over the long run than attempting to restore materials after they have deteriorated or been damaged. Regular inspections and cleaning can identify most potential problems before irreversible damage occurs.

Regular cleaning is not just an issue of aesthetics and should be performed on a regular schedule based on need and access. Left in place, bird droppings, algae, dirt, and debris hold moisture or harmful chemicals, allowing them to react with or penetrate exterior surfaces.

Coatings can protect the underlying substrate from solar, UVB damage, dirt and chemicals in the air, and moisture. Hard, smooth coatings can discourage fungal, insect, or even bird damage. New research has identified additives to increase UV resistance in lacquer and other resins, paint coatings and other synthetic materials. Coatings can reduce, but not prevent, moisture changes from fluctuations in ambient humidity in hygroscopic materials such as wood. However, coatings must also allow the material to dry – to release the moisture that will inevitably breach all defenses.

Today, conservation standards require that both products and procedures more closely mimic the performance of the original materials with regard to strength, porosity, and permeability. Products that allow "retreatability" are preferred to those that are permanent, or reversible. Retreatability allows the possibility to retreat the substrate with either the same or a different consolidant in the future. This philosophy acknowledges that few materials are truly reversible, and that we need to accommodate future discoveries.

Common Strategies for Repair and Modifications

Where possible, materials should be repaired or replaced with identical material, or material that closely matches the original in appearance and composition. This is not simply an issue of historical authenticity. Incompatible materials used for patches, repairs, inserts, or coatings often perform differently than the original material. Differences in strength, porosity, and permeability can cause differential movement or stresses that damage the original material, especially if the original is softer or more porous. Chemical differences can cause chemical changes in the original material. Impermeable materials can trap moisture in the material.

That said, materials identical to the original are not always available, affordable, or even appropriate. Certain woods may no longer be available or may be restricted from harvesting. Stone quarries may be exhausted or closed. Manufactured metals may no longer be fabricated. In some cases, the original material may be in poor condition because it was never appropriate for the conditions, or it may not be sufficient to meet current building codes or future environmental conditions. In these cases, a material must be selected that has similar aesthetic and performance characteristics.

Before implementing any conservation program, it is important to identify the full range of options for protection, repair, or replacement, consider the impact on performance, historical integrity, and economics. The impact of future changes in the weather, including temperature and humidity must also be considered.

Barriers and Insulation

Modifications may be necessary when the interior or exterior environment has changed. Changes to interior conditions are the result of new expectations for comfort and energy efficiency, modern HVAC systems and tighter windows and wall systems. Exterior weather changes include changes in climate, including new averages, extremes, and fluctuations in temperature, precipitation, and wind, all of which impact how moisture moves through the exterior envelope. These conditions have led to the introduction of barriers to control the movement of air, vapor, and moisture. However, these barriers can inadvertently trap moisture that deteriorates materials, or create conditions hospitable to fungi, mold, or insects. Excessive insulation can reduce drying of porous exterior masonry, leading to moisture or frost damage.

A large variety of insulation materials and water, vapor and air barrier materials are available today. The pros and cons are a subject of much debate and dispute, and beyond the scope of this discussion except to suggest a few key considerations. Any analysis of new insulation or barrier must consider how they will affect the movement of heat and moisture through materials and systems over a range of future conditions that are largely unknown. Will the barrier function as intended under weather conditions in 20 or 50 years? Can it be modified or removed as conditions change?

The difference between theory and practice should also be considered. A membrane is only as good as its installation. Joints, leaks, and other openings are inevitable and will allow moisture to move in unintended ways. Is the membrane continuous at all areas of the exterior envelope? How does it terminate at windows and doors? What is the fallback position? How will the wall perform if and when there is a leak?

Moisture Management

Any strategy of protection or repair requires a thorough understanding of the original design and details, and how they were intended to divert and manage water. Projecting cornices, bands, window and door heads and sills, and water tables were not merely decorative; they were designed to shed water from the building so that it would have less time to run down and soak into the face of a building. Notches and other articulation on the underside of overhangs and sills interrupt the capillary movement of water before it meets the wall, where capillary action can draw excess moisture into cracks and joints. The most common strategies were:

- Direct water away from porous materials. This can be accomplished with roof and wall covering, membranes or coatings that shed water, and overhangs, drip edges, gutters and other details that divert water away from surfaces.
- Provide adequate ventilation to avoid condensation and promote drying. This may require permeable materials, and details such as weeps, or sacrificial coatings.
- Manage wind stress. This can be accomplished by hardening surfaces to resist wind, by diverting wind forces around a building, or allowing it to flow through the building. Air pressure can force water into exterior materials and assemblies.

An historic material or design may require diligent maintenance or improvements to resist a more aggressive environment. More durable or more porous materials, larger gutters or overhangs or changes in elevation may be required to withstand more severe storms and higher flood plains. Protecting materials from additional heat may require creative strategies to shade or locally cool the environment. Strategies from other climates with a history of more severe weather might be adapted and applied as the weather changes. The possibility of adaptations will raise philosophical questions that will force stakeholders to evaluate their priorities and weigh the importance of retaining the original fabric against the desire to maintain other features of the design.

Training

A major obstacle to maintenance, repair and replacement of traditional materials is lack of knowledge and training of future generations of craftsmen. Although there is a growing demand and premium paid for craftsmen with preservation skills, there is a worker shortage attributed to fewer young people entering the field, lack of union representation, the seasonal character of the work, and a mismatch between available workers and project locations. Many organizations that manage historic properties have incorporated training into their mission out of necessity.

Access to consistent curriculum, a process for standard credentials, and access to hands-on apprenticeships are also necessary. Fortunately, there are a growing number of training programs for heritage crafts running the full gamut from self-trained individuals or single apprenticeships to large scale training programs, and a conscious effort to recruit trainees and expand training programs.

Some are affiliated with manufacturers or industry associations; others are independent non-profits or associated with governmental programs. Each has advantages and disadvantages. They all share a commitment to sharing and expanding a better understanding of traditional materials and techniques. Many of these programs are beginning to include adaptation for climate change.

- Manufacturers offer a range of programs from webinars to multi-day workshops such as deGruchy's Limeworks *Craftwork Training Center* and Edison Coatings, Inc.
- Industry organizations like the International Masonry Institute offer webinars, hands-on workshops, and apprenticeships.
- The *Campaign for Historic Trades* is an organization that seeks to register apprenticeships and promote and recruit for the National Park Service.
- The National Park Service's *Historic Preservation Training Center* (HPTC) uses historic preservation projects for hands-on training.
- HistoriCorps is a non-profit that recruits and trains volunteers to work on publicly owned historic project.

- Several universities offer training in historic preservation but not all offer hands-on training in material conservation. Examples of Universities with an emphasis on conservation are the University of Pennsylvania and the American College of the Building Arts.
- There are several trade schools like the North Bennet Street School in Boston that focus on full-time, hands-on training, and others like the University of Oregon's Northwest Preservation Field School that offer shorter hands-on training workshops.
- Several national and international preservation organizations also share research, and conduct seminars, conferences, and workshop to advance knowledge about traditional materials and crafts. Examples are the Association for Preservation Technology, World Heritage, and ICOMOS.
- Local preservation organizations, especially those that own multiple properties, conduct, and share their research. Examples are Historic New England, who maintain their own properties, and share their experience through a series of white papers.

Conclusions

The very fact that a historic building is still standing indicates that it has exhibited flexibility and resiliency throughout its life. There are many lessons to be learned from historic buildings, from the articulation of the building walls to the use of coatings and sacrificial finishes. Few weather conditions are truly new – most are simply shifting north or south. This provides an opportunity to adapt strategies from one historic climate zone to a new condition. However, adaptations are more easily made to new construction. Modifications to historic buildings will require careful consideration of the consequences and alternatives.

There remains significant uncertainty regarding projected weather conditions and about how historic materials and assemblies will respond. Materials, buildings, and policies must be flexible, adaptable, and reversible as conditions change and new information and alternative become available. Strategic planning and decision-making must be flexible and adaptable to unforeseen and changing circumstances.

Bibliography

ACBA. 2023. *American College of the Building Arts*. https://acba.edu/.

Akins. 2013. "Assessment of Heritage at Risk from Environmental Threat." *English Heritage*. November 5. http://www.english-heritage.org.uk/publications/assessment-heritage-at-risk-from-environmental-threat/.

APTI. 2023. *Association for Preservation Technology, Inc, Training and Education*. https://www.apti.org/training-education.

Edison Coatings. 2023. *Courses for Architects, Consultants, Conservators and Restoration Trades*. https://www.edisoncoatings.com/html/calendar.html.

Historic Trades. 2023. *Campaign for Historic Trades*. https://historictrades.org/.

HistoriCorps. 2023. *HistoriCorps Workforce for Saving Places*. https://historicorps.org/.

HNE. 2019. *Climate Change and Gutters*. Property Care White Paper, Historic New England.

———. 2023. *Historic New England, Sharing Our Preservation Approach*. https://www.historicnewengland.org/preservation/for-professionals-students/property-care-white-papers/.

ICOMOS. 2023. *International Committee on Monuments and Structures*. https://www.icomos.org/en.

IMI. n.d. *International Masonry Institute*. https://imiweb.org/.

Limeworks. n.d. *The Craftwork Training Center*. https://www.limeworks.us/the-craftwork-training-center/.

Lstiburek, Joseph. 2006. “BSD-1006: Understanding Vapor Barriers.” *Building Science Corporation Digest*. https://www.buildingscience.com/documents/digests/bsd-106-understanding-vapor-barriers.

———. 2005. *Builder's Guides*. Building Science Press.

Lubeck, Aaron, and Francis Conlin. 2010. “Efficiency and Comfort through Deep Energy Retrofits: Balancing Energy and Moisture Management.” *Journal of Green Building* 5 (3): 3–15.

NBSS. 2023. *North Bennet Street School*. https://nbss.edu/.

NPS HPTC. 2023. *Historic Preservation Training Center*. https://www.nps.gov/orgs/1098/index.htm.

Place Economics. 2022. *Status of Historic Trades in America*. The Campaign for Historic Trades. https://historictrades.org/wp-content/uploads/2022/11/Campaign-for-Historic-Trades-10.31.2022.pdf.

University of Oregon. 2023. *Northwest Preservation Field School*. https://archenvironment.uoregon.edu/hp/field-schools/pacific-northwest-preservation-field-school/2023-pacific-northwest-preservation-field-school.

World Heritage USA. 2023. *United States Committee of the International Council on Monuments and Sites*. https://worldheritageusa.org/mission/.

CHAPTER 2.2

Wood

Introduction to Wood

Wood is one of the most universal, flexible, and varied building material. It is present in all aspects of our life: sequestering carbon, producing food, and providing raw materials for our shelter and comfort. Most people simply identify trees as either hardwood (deciduous) or softwood (evergreen), but trees are amazingly diverse, with different structures, characteristics, strength, chemicals, and life spans. Two trees of the same species will differ because of differences in their local environment and the way they are harvested and used. Each of these characteristics affects how wood is used, and how it performs. This chapter explores how the nature and nurture of wood influences deterioration and durability in response to weather: UV light, temperature, moisture, and related stressors of bacteria, fungi, and insects.

The Past, Present, and Future of Wood

History

Trees evolved within the surrounding ecology and developed a relationship in balance with other species and with the climate. As a result, much of the wood in historic buildings is locally sourced and adapted to the climate and ecology. Lacking convenient transportation, early builders learned how to best use locally available woods. They understood that the quality of wood depended both on the species and on the specific location and growing conditions.

As forests were depleted and transportation by waterways, carts, trains, and ships became available, builders either used less suitable local wood or imported

DOI: 10.4324/9781003160489-7

wood from other climates. Imported wood was often more resistant to native pests but carried foreign pests that flourished in new habitats without natural defenses. Whether native or imported, the qualities of wood influenced design, construction, maintenance and – eventually – philosophies of preservation. Although lumber is no longer alive, the evolution, genetics, growing conditions, cellular structure and chemical composition of the wood continue to influence its durability and resistance to weather and biological stresses.

Wood construction was naturally more prevalent, and survived longer, in cool climates that limited biological growth and dry climates that caused less moisture damage. However, selection of wood species, careful detailing, and ongoing protection and maintenance have allowed wood construction to survive in many climates that are not naturally hospitable.

Current Social, Cultural, and Climatic Impacts

Changes in temperature and moisture affect the growth, quality and availability of local trees, the viability of biological pests that infest both trees and lumber, and the durability of existing wood in structures.

Loss of old growth forests and increased economic demand has led to a reduction in the number of species available. Attention to sustainability and the environmental effects of deforestation have led to restrictions (some legal, some voluntary) on the harvest and use of many species of wood. The increased demand for wood has increased the use of younger trees that are not as durable as old growth wood. Most trees grown in the past century grow more rapidly and are less dense, partly due to changes in the climate, and partly due to forestry practices. Changes in the climate will continue to affect the growth of trees, and the quality and quantity of wood available for restoration and new construction.

Climatic changes also impact existing wood in structures. The potential impacts vary with geographic location and microclimate.

- Higher temperatures and moisture levels are expected to increase the population and range of damaging biological growth, including bacteria, fungi, and insects.
- Longer and more severe rain events are expected to increase the depth and duration of moisture saturation of wood, making it more vulnerable to deterioration.
- Rapid fluctuations in temperature and moisture cause wood to expand and contract, stressing wood fibers and increasing stresses from dissimilar movement of coatings, substrates, or inserts.
- The density of new wood has decreased as the rate of growth and length of growing season have increased.
- Prolonged heat can lead to reduced moisture in wood cells, causing permanent damage.

Understanding Wood

The Character of Wood

Durability is a function of strength, hardness, and flexibility, as well as resistance to pests and weather and is determined by both the nature and nurture of the tree and by milling operations. No two trees of a single species are identical. The nature of a tree reflects the genetics of the species and individual genetic variations. The nurture of the wood includes the unique growing conditions a tree experiences throughout its life, how the tree is harvested, and how the wood is milled, cured, and maintained. As a result, the character and weather resistance of the original wood in a historic building may be very different than a replacement piece of the same species.

Factors that influence durability of Wood		
Feature	**Characteristic**	**Impact on durability or vulnerability**
Order	Hardwood, Softwood or Monocot	Permeability to moisture through cell structure
		Vulnerability to specific fungi or biologicals
Species	Hardness	Resistance to physical stresses
	Extractives	Resistance to fungal or biological attack
Growing Conditions (temperature, moisture, wind, nutrients)	Speed of growth/Density	Resistance to physical stresses
	Reactive wood	Resistance to physical stresses
Harvesting and curing	Strength	Resistance to physical stresses
	Shrinkage/Cracks	Vulnerability to moisture, insect & biological attack

Figure 2.2.1 Factors that influence durability of wood. (Table by Susan Pranger, 2023.)

Nature: The Structure of Trees

In the Linnean System, softwoods are of the subdivision "Gymnosperm" and Order "Conifers" with needles that remain green year-round, and a cellular structure that can survive in colder climates. Hardwoods are Dicotyledons (Dicots) and woody plants like Bamboo and Palm are Monocotyledons (Monocots), both of the subdivision "Angiosperms" with leaves and a cell structure that adapt to seasonal variations in temperature and moisture. Most trees are also known by various local or common names or by lumber industry "trade names" such as "western fir," "southern yellow pine" or "African mahogany" that group species that are functionally interchangeable.

Softwood and hardwood trees are formed by consecutive growth around a pith. Trees have three layers: bark (which is not considered wood), cambium (a thin layer of reproductive tissue), and wood (sapwood and heartwood). In non-tropical trees, an “annual ring” is formed by the difference between the “early wood” grown early in the growing season, and the “latewood” grown later in the season. The contrast in density and color between early and late growth rings will vary depending on the species and the growing conditions. It may not be possible to identify rings in trees in tropical climates where growth is consistent year-round. Palm and Bamboo are used as lumber but are not technically trees. Their stems are tube-like vascular networks of cells that conduct water and nutrients. The stems grow by adding new vascular bundles and parenchyma cells at the edges of the stem.

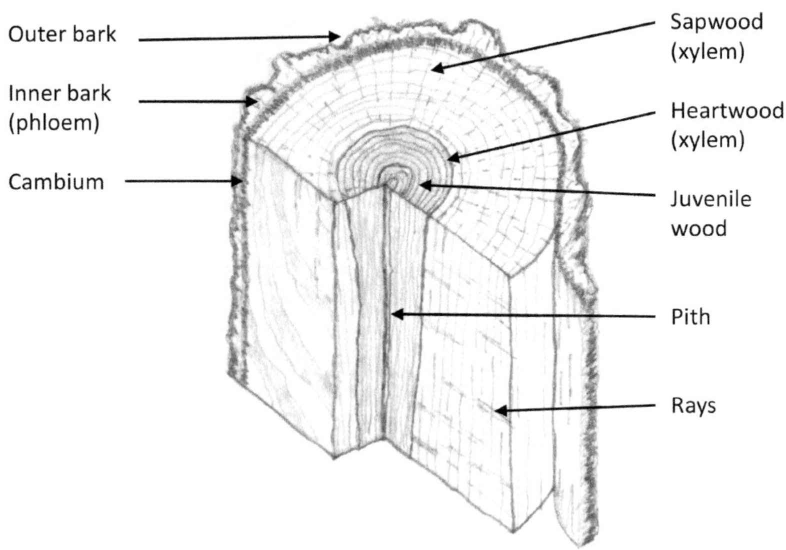

Figure 2.2.2 Cross section of tree. (Artwork by Daniel Thomas, 2023, based on artwork by Vallejos and Grote, 2009.)

All wood is live sapwood in a young tree. New layers are added as the cambium creates cells that vary in size and shape depending on their function. Most cells are elongated and aligned with the vertical growth of the tree. Each cell starts with a wall surrounding a cavity containing a living material, protoplasm. Most cells die within a few days and become tracheid, vessel elements, and fibers that create a physical structure. Parenchyma cells remain alive for years, performing metabolic functions, such as storage of carbohydrates. The shape and organization of these cells are different in hardwoods than in softwoods and determine how the living tree and subsequent wood materials manage water.

In softwood, tracheid cells perform two functions: structural support and water conduction. Early wood has thin walls to maximize water movement, while latewood has thicker walls to maximize strength. Cells in some pines, such as Douglas fir, have membranes in the cells that block the transmission of fluid horizontally through the wood (Groom, 2023). Four specific genera of pine (Pines, Spruces, Larches, and Douglas fir) also have large canals that excrete pitch or resin that seals wounds from boring insects or other damage. "Traumatic Resin Canals" will occasionally appear in response to a trauma in species that don't normally have resin canals.

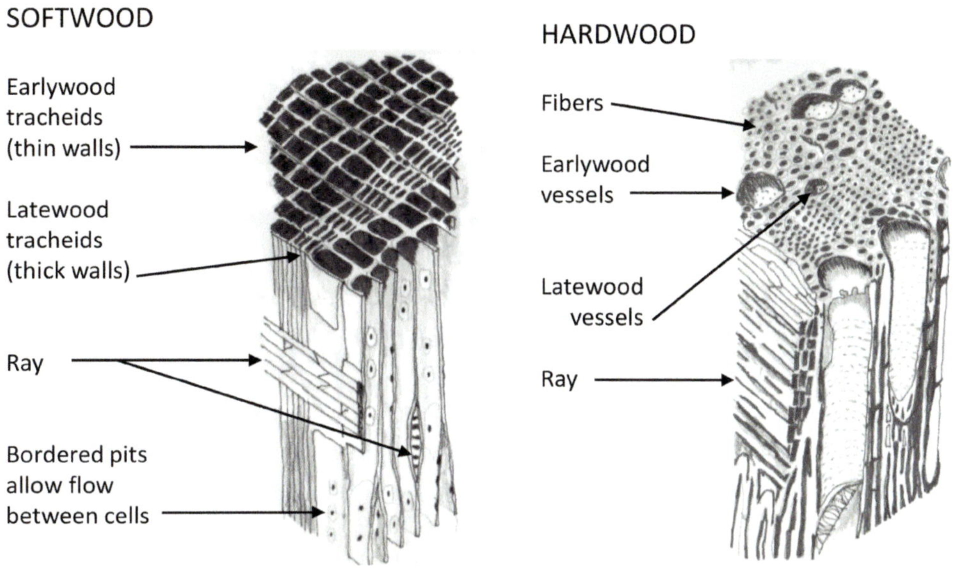

Figure 2.2.3 Water-carrying cells in hardwood and softwood. (Artwork by Daniel Thomas, 2023, based on artwork by an unknown Forest Products Laboratory artist.)

The size and arrangement of cells give each hardwood species a distinctive grain and differences in strength, durability, and decay resistance and can be used to distinguish and identify various hardwood species. Hardwood trees develop more complex cells, including specialized cells that are short with thick walls to provide strength, and vessel cells (xylem) that carry water and other materials. Ray cells are more horizontal and add to the visual character of the wood. Structural cells are shorter, with thicker cell walls. These cells form masses of small-diameter fibers that add strength.

The shape and configuration of vessel cells directly affect how lumber will respond to moisture and humidity. The vessel cells in hardwood are larger than the tracheid cells in softwood, and line up end to end, with porous ends that block air bubbles but conduct water, making hardwoods generally more porous than softwood. Some hardwoods, such as oak, are "ring porous" and produce wide pores during early growth to facilitate water movement to the leaves. Species such as beech are "diffuse porous" with little visible distinction across seasonal growth. Other species may be neither "ring porous" nor "diffuse porous." Species such as white oak are less porous because the pores are blocked by tyloses as the cells transition from sapwood to hardwood.

Heartwood is created as a tree grows and the older sapwood dies. As the sapwood becomes heartwood, the number and size of cells do not change except for swelling or shrinkage as water content changes. However, heartwood cells lose their metabolic functions and become primarily structural. They no longer conduct water, but they accumulate the residue of the (no-longer-living) protoplasts in the cells and accumulate chemical extractives in the cell walls. Extractives vary by species and affect both color and resistance of heartwood to biological attack.

Hardness does not necessarily indicate structural strength, resistance to decay, or durability. However, dense wood is generally more resistant to deterioration.

The Chemistry of Wood

All wood, including Bamboo, is composed of three main constituents (cellulose, hemicellulose, and lignin) plus water and extractives. Variations in the character and proportion of these components contribute to density, stiffness, hardness, and durability of each species, and influence the vulnerability to weather, moisture, and biological threats.

Wood in service has a moisture content of 12% to 18% (weight of water compared to oven-dry weight of wood). Short-term exposure to moisture will usually affect only the surface. Depending on the wood species, it can take weeks of submersion to fully saturate a piece of wood. Expansion along the grain is insignificant but can be as high as 12% to 15% tangentially in some woods when saturated.

As the wood matures, the cellulose increases and hemicellulose and lignin decrease. Cellulose and hemicellulose provide strength and structure but are also hygroscopic and easily absorb and release water, causing expansion and contraction stress. Some types of cellulose and hemicellulose are attractive food sources for insects, fungi and mold, and other microorganisms.

Lignin is a binding agent that connects cellulose and hemicellulose to create cells and connects cells to each other. Softwood will typically contain more lignin and less hemicellulose than hardwood. Bamboo uses lignin to bind the vascular bundles of cells and provide structure.

Extractives are complex chemicals that can be "extracted" by a variety of solvents, including water. Although they are only 1% to 2% of the timber by weight, extractives give each species of wood a unique chemical "fingerprint" and influence the relative water resistance, shrinking and swelling properties, and the resistance to specific insects or fungi. There are hundreds of different extractives; the roles of some are better understood than others.

Extractives vary in color from yellow to orange, red, or amber. If extractives are opaque, it can be difficult to distinguish the heartwood by color. Pitch and resins are extractives that are used by trees to protect damage or wounds. Tannin is also an extractive, typically found in the bark, that can protect the tree from bacteria and fungi by precipitating enzymes and proteins out of the bacteria. After harvest, chemical changes can occur if wood with high tannins comes into contract with iron straps or anchors.

Components of Wood				
Component	**Approximate Percentage**		**Purpose**	**Vulnerability**
	Hardwood	**Softwood**		
Cellulose - a linear polymer	38–49%	40–45%	Microscopic parallel microfibrils that provide strength, elasticity and toughness	Hygroscopic; Food source for insect, fungi, and mold
Hemicellulose – a variety of similar polymeric materials	15–20%	15–20%	Short, branched chains cross link the fibrils, adding strength, flexibility and toughness	Food source for insect, fungi, and mold
Lignin – an aromatic, branched polymer	18–25%	25–35%	Cementing material that binds hemicellulose within and between cells. Hydrophobic lining supporting water movement	Food source for insect, fungi, and mold
Pentosans– polymers of pentoses, five-carbon sugars	19–26%	7–14%	A type of hemicellulose	More sensitive to acids than other hemicellulose)
Water	12–18% by weight under normal conditions		Causes expansion and contraction of cells	Stress from movement; suitability of habitat
Extractives, a residue left in heartwood cells as the parenchyma cells die; varies by species	1–2%, slightly higher in softwood than hardwood, and higher in heartwood than sapwood. The chemical signature varies by species		Influences resistance to water, insects and fungi	Resistance to moisture, insects and fungi varies by species.
Trace amounts of protein, starch, pectin and other materials	Varies throughout the year		Support growth and function	Pectin acts as a binder and may decrease permeability.
Trace amounts of approximately 12 inorganic material including minerals	Varies by species and environmental conditions		Support growth and function	Necessary to growth of tree

Figure 2.2.4 Vulnerability of wood components. (Table by Susan Pranger, 2023, based on content from John Wyllie Simpson and Peter James Horrobin 1970, 136, 139; Robert A. Young 2008; Roger M. Rowell, Roger Pettersen and Mandla A. Tshavalala, 2012.)

Nurture: The Impact of Growing Conditions

Biologists, chemists, and horticulturalists are only beginning to decode how trees communicate through chemicals, bacterial relationships, and fungal networks to access minerals and nutrients in the soil and resist biological threats. Because of these complex relationships, trees grown in diverse natural environments may be stronger than those grown in more sterile monocultures but more vulnerable to changes in the climate.

Sun, moisture, temperature, and wind each affect the growth of a tree and how their wood later responds to weather conditions. Tree rings provide clues to the annual growing conditions. Tight rings reflect slower growth and greater density and are more resistant to insects and decay. Slow growing trees are more dimensionally stable due to the higher percentage of heart wood and extractives. Higher nutrients, especially artificial nutrients, contribute to faster growth, higher turnover, and lower density. Artificial nutrients can also interfere with the natural biological relationships with bacteria in the soil. Research in South America highlights a 10% to 20% difference in the density of species grown in the same climate but with different soils.

Unfortunately, many contemporary tree farming processes stress rapid growth, resulting in trees that have a higher percentage of young "juvenile" sapwood that is less dense, less strong, with and little or no heartwood and more reactive wood. Wood that is grown faster will have thinner cell walls, and less strength. These differences are important when predicting the strength of existing wood or considering new wood for repairs or replacements. The percentage of juvenile wood decreases as the tree grows, and will be substantially less if the trees are harvested after 20 years.

Forest to Building

Durability and weather resistance is also influenced by how the wood is harvested, dried, and milled. Natural growth characteristics such as knots, seasoning cracks and splits, orientation (how the wood is carrying loads relative to the grain of the wood), the environment during harvest (exposure to sunlight, moisture, insects, and other biological factors), storage and milling, and of course, detailing of the design each influence the character and durability of the finished wood product.

Moisture content and temperature both affect the strength of lumber. Reducing moisture content to 12% can double the strength, and oven drying can triple the strength compared to wood in its natural state. Short-term changes in ambient humidity will only affect the surface of the wood, but long-term exposure will gradually change the moisture content. Stresses are created in the wood if the exterior wood dries more quickly than the interior; allowing the wood to dry slowly in a controlled manner can reduce checking.

Strength also decreases as temperature increases. The loss of strength is generally temporary but can become permanent if lumber is exposed to high temperatures for a long time. Humidity and temperature acting together have an even greater impact, as revealed by the ability to bend wood after it is steamed. Higher temperatures, longer heat waves, and rainstorms of longer duration therefore present a greater risk to wood, especially to rapidly grown sapwood.

Grain orientation is often selected for its aesthetics, but also affects strength, where cracks will develop, and how the wood will absorb moisture. The type of saw and techniques used to cut and mill wood will determine how the fibers are cut, the shape of the exposed cells, and how the exposed edges absorb or release moisture. Hand-cut wood tends to follow the grain, cutting between cells. Cells that are ripped open will wick moisture into wood, especially the continuous cells of softwood. Wood milled prior to the mid-1800s on reciprocating saws, identified by parallel striations, have torn fibers that allow more water penetration.

Detailing Wood

Properly detailed joints allow for and sometimes use the inevitable expansion and contraction that occurs seasonally and during weather events. Expansion or shrinkage due to changes in moisture content can cause cupping and other stresses. Increased duration of storms may increase moisture absorption and decrease intermittent drying.

Wood shingles illustrate the relationships between the inherent qualities of a material, fabrication, and weather. Wood shingles can last from 15 to 60 years depending upon the type of wood and the installation. In the 17th and early18th centuries, white pine shingles were hand split from heartwood following the grain of the wood and then dressed with a draw knife or draw-shave to make them even. The heartwood was dense and rich with protective extractives that made them resistant to insect damage. In the 1820s shingles cut by circular saws became popular because they werecheap and readily available. Sawn shingles were considered more weather-tight because they were even, and laid flat with fewer gaps. However, sawing opens the grain more than hand shaving, so machine-made shingles also were more likely to absorb water and swell. Shingles in old barns were installed over battens or spaced sheathing to promote drying after rain events.

Drivers of Deterioration

While most materials deteriorate from chemical degradation or physical processes like erosion, damage to wood is primarily caused by biological decay by fungi and insects. Biological decay is facilitated by increases in temperature and precipitation. The Schaffer Index, updated in 2006 to reflect increases in average temperatures and precipitation in the macroclimate from 1970 to 1971, shows a moderate increase in the probability of wood decay across the United States. Although this index is less accurate than an evaluation of the temperature and humidity at a specific location, it is a useful predictive tool.

The risk of decay is also determined by the local mezoclimate, including the proximity of water bodies, vegetation and shade, and features of the microclimate, including the type of wood and whether the design traps water. The increase of decay in new construction has been attributed to factors that are less applicable to historic construction: the use of commercial lumber that is less resistant to decay and modern details that do not shed water as effectively. Decay attributable to energy efficient construction that reduces outward heat flow that dries exterior materials and the use of less effective but safer preservatives could apply to both new and historic construction.

Temperature

An increase in prolonged heat waves, extreme heat, and daily fluctuations in temperature present a risk to existing wood materials both directly and by increasing biological damage. Fluctuations in temperature can cause fluctuations in humidity that cause expansion or contraction followed by checking, cracking, and creep. Creep reduces the overall strength. Small cracks and checks allow more moisture or biological vectors to penetrate more deeply into the wood.

Wood is generally stronger when dry, but high heat can weaken wood by reducing the moisture content of the wood below a minimum threshold (which varies by species). The reduction in strength is usually temporary but can become permanent after a prolonged period. Temperatures over 100°F break down wood components through "pyrolysis" a process that releases carbon dioxide, water, and traces of minerals. Temperatures in excess of 200°F deteriorate hemicellulose and cellulose and release tar and flammable volatiles. The volatiles and tars ignite at around 350°F degrees.

Moisture

Moisture, in the form of vapor, liquid or ice, is expected to increase either in overall volume or in intensity in most locations. When storms become longer and more intense, there is an increased risk of saturation. Excess moisture damages wood directly by causing swelling that leads to cracks and checks, and by creating conditions suitable for biological attack, and indirectly by increasing the population and habitat of insects. Increased humidity in the air, wood, or soil can increase the risk of damaging fungi or termites.

Wood is hygroscopic – it absorbs and releases moisture. If the surface is not sealed, wood will absorb water through capillary action until the surface layer is saturated. After the wood is saturated, water accumulates on the surface and forms a thin film that prevents further saturation. However, the water can penetrate cracks and joints in concentrated areas and allow moisture, fungi, and insects to penetrate more deeply in the wood. Repeated fluctuations in moisture content can cause repeated changes in dimension, leading to creep, a failure of the bonds between wood fibers, which in turn causes splits and checks on the surface. If the wood is constrained while swelling, the internal structure of the wood and any materials supported by the wood can crush or buckle.

Damage can also occur when existing water management systems, such as gutters, are overwhelmed by the intensity of rain events. An increase in humidity can exacerbate existing problems with inadequate ventilation or inadequate vapor barriers.

Although the old growth wood in many historic buildings is denser, and therefore more resistant to moisture damage than new lumber, excess moisture can still impact wood in a variety of ways:

- Water content can cause shrinkage and swelling, which can lead to seasoning checks and splits.
- Moisture can affect the surface texture of wood.
- High moisture content (greater than 20%) can make wood vulnerable to fungi and mold leading to decay.
- Rain can wash extractives and other water-soluble materials from the wood.
- Submersion in water can limit oxygen, and therefore inhibit certain types of decay.

Wind and Rain

Longer and more intense rainstorms increase the odds that wind driven rain will penetrate wood surfaces more deeply. The extent of damage depends on the distribution of the rain on the façade, and the vulnerability of the façade. The distribution of the rain is impacted by the angle of the rain, the size of raindrops, wind speed, and the intensity and duration of rain. The vulnerability of the wood is determined by the configuration of the building and its surroundings, and by the porosity, cracks, joints, etc. of the wood and any protective coating.

Sunlight

Through a photochemical process, UVB light can disintegrate the cell walls of wood and cause photo oxidation of the lignin, the binder within and between cells. Sunlight can also degrade extractives by making them water soluble and therefore more vulnerable to rain and other moisture. In the absence of other factors, this deterioration is slow, as little as ¼″ to ⅓″ in one hundred years in some western softwoods, but can be as much as ⅔″ to an inch in western red cedar. Disintegrated materials can accumulate on the surface of the wood, temporarily preventing further deterioration, until it is removed by rainfall, wind, or other mechanical processes. Intense rain and wind further accelerate photodegradation by removing the protective layer of debris.

Both visible and UV light can cause photochemical changes to the color of the softwoods and hardwoods. The extent of change varies with temperature, water, and atmosphere. Wood may be bleached, gray, yellow, or brown, depending on the wood species, composition, and extractives. Certain pigments in stains, paints, and coatings can absorb UV rays, protecting wood from damage.

Chemical Damage

Most timber is naturally mildly acidic and not affected by mild acid or salt. The Clean Air Acts in the United States and in England have substantially reduced the risk of polluted air or coal smoke, but salts in marine environments and deicing salts can discolor wood and attack cellulose, breaking surface fibers into sharp filaments of "fuzzy wood." Salt can cause the wood to hold more water, making it more vulnerable to fungi. High alkaline conditions are rare but can occur where lime mortar is applied directly over wood members. Alkalinity can degrade both lignin and hemicellulose, softening and weakening the wood.

Rain washing across very acidic wood can damage metal nails, fittings, and roofing. The chemical reaction not only corrodes the metal, but it can also weaken the adjacent timber. Heat treated or chemically treated wood can also trigger galvanic corrosion of metals.

Fungal and Bacterial Damage

Fungi and bacteria are an essential part of the forest ecology, facilitating communication and the movement of nutrients between plants, bacteria, and the soil. However, they also cause deterioration of wood in nature and in buildings. Warmer, wetter weather increases the risk of biological growth by increasing habitat conducive to fungal growth. A 2007 research project by Noah's Ark Project expects a 50% increase in the risk of rot in wood in northern European climates due to climate change by the end of the century.

To survive, fungi have four requirements,

- Moisture content above 30% of dry wood weight to establish, and above 20% to survive.
- A temperature of 70° to 85°F (40° to 100°F with appropriate humidity).
- At least 20% oxygen
- A source of food – the sugars, proteins, lipids, cellulose, or hemicellulose in wood

Eliminating any one of these requirements, especially moisture content, will deter fungi. However, fungi can go dormant under dry or very cold conditions and recover when conditions are favorable. They can be reintroduced by spores that can be airborne, spread by splashing rain, by contact with infected wood, or vectored though insects. "Dry" fungi can draw moisture from the atmosphere, even when the wood itself is dry.

Fungi is killed at the high temperatures of standard kiln drying but new fungi can reinfect wood when nutrients are forced to the surface during drying. Mold and mildew can be controlled by additives in finish coatings, although there is some doubt as to their long-term effectiveness. Submersion in water can retard the growth of fungi and bacteria by eliminating oxygen. Wood piles can survive for centuries under water but are susceptible to new biological damage if they are exposed by dropping ground water levels caused by construction dewatering or other uses.

Different fungi consume different elements of wood, causing damage that ranges from discoloration to structural failure. Both wet and dry fungi, including brown rot, white rot, soft rot, and black rot, use enzymes to break down and consume the cell wall, feeding on the cellulose. The wood will become brittle, and shrink and crack, creating cube-like sections, eventually destroying the structural integrity of the wood. The depth of cracking is generally less with wet rot than with dry rot.

In living trees, heartwood is more susceptible to decay than sapwood. However, in lumber, the nutrients in sapwood make it more vulnerable to attack than heartwood. Heartwood and sapwood in many spruce and true firs are similar in color and not easily distinguished after harvesting.

Types of Fungus or Rot			
Description	**Location**	**Food Source**	**Appearance**
Cubic brown rot (dry rot) Serpula lacrymans	Found in softwoods Draws moisture from air or other sources	Consumes cellulose and hemicellulose, leaving the brown Lignin	Reduces the wood to easily identifiable chunks
White rot (Wet rot) Coniophora puteana	Found primarily in hardwoods, but occasionally in softwoods Wood must be wet	Consumes cellulose and Lignin	Leaves a white stringy residue, eventually causing the wood to collapse.
Soft rot	Attacks hardwoods that have been subjected to long term wet conditions	Consumes cellulose at the cellular level	Reduces structural capacity while slightly discoloring the wood
Brown or Black rot, or cellar rot (wet rot)	Requires damp conditions at 40 to 50% humidity	Consumes cellulose	Turns the wood black so that it appears to have been burnt
Chlorociboria	Growth conditions are not clear. Some like decayed wood, some harder wood.	Erode wood cell walls	Deposits a green stain in the wood, and will eventually weaken the wood
Sap Stain	Colonizes the cells of the sapwood	Sugars, proteins, lipids in the cells of the sapwood	Causes bluish, grey or black discoloration
Molds	On surface of wood	Sugars, proteins, lipids in the cells of the sapwood	Vegetative cells (hyphae) on the surface of the wood

Figure 2.2.5 Types of fungus and rot. (Table by Susan Pranger, 2023, based on content by Robert A. Young, 2008, Frea Irace, 2011, Anne S. Kaslegard, 2008.)

Dry rot, also called "water syphoning fungus" draws moisture from the air or by extending threads to adjacent sources of moisture. These threads can extend several feet across concrete or masonry, connecting fungi in dry wood to moisture in the soil. Wood-staining fungi known as "sapstains" or "bluestains," consume the materials stored in parenchyma cells of sapwood but do not attack cell walls, and therefore do not affect the strength of the wood.

Mold and mildew are a type of dark gray or black fungus that will remain on the surface of the wood. They require high humidity and cause discoloration but do not structurally damage wood. However, some mold can penetrate cell membranes, making them more permeable and vulnerable to rapid swelling that leads to checking, decay, and uneven absorption of coatings.

Insect Damage

Increasing temperature and humidity and a reduction of lethal winter temperatures have increased the potential habitat for both native and non-native insects while sometimes eliminating beneficial predators. Changes in climate are also increasing the size of pest populations by improving the quality of their habitats, increasing both the number of insects hatching in a cycle, and increased the number of life cycles in a year. Insects hatch earlier, with additional reproductive cycles reported in at least 44 species that historically reproduced only once annually.

Many of these expanding insect populations, including termites, beetles, carpenter ants, carpenter bees and marine borers, are capable of destroying lumber and living trees to consume food or to create a habitat. Both activities compromise the structure of the wood in buildings and allow moisture to penetrate deeper.

Trees communicate through airborne chemicals, effectively warning their fellow trees to proactively deploy their chemical or biological defenses but cannot adapt their natural defense mechanisms quickly enough to respond to new or increased threats. Wood in buildings contains extractives evolved to deter natural pests but are often overwhelmed by increased or repeated attacks.

Subterranean Termites

Subterranean termites are the most widespread and damaging of the three types of termites. Warmer temperatures have already expanded their range and increased their population. The habitats for 12 of 13 termite species studied in 2017 are expected to increase significantly by 2050, including a 15% to 20% expansion of the destructive *Coptotermes formosanus* termite. There is speculation that termites, who create significant methane, contribute to global warming, creating a feedback loop that increases habitats, colonies, and more emissions.

Figure 2.2.6 Douglas fir damaged by a subterranean termite. Note insect galleries along the grain. (Photo by Suzana Radivojevic, 2023.)

Colonies must grow to a certain size before additional winged termites appear, swarm, and spread out to create new colonies. The decrease in winter freezes allows more termites to survive the winter, and warmer temperatures allow existing colonies to grow and expand. Warmer weather has already created an overlap in the mating times of the *C. formosanus* and *C. gestroi*, both native to Florida. These two species are now crossbreeding, creating a hybrid even more robust and destructive.

Temperature and humidity are important factors to the survival, but most termites use wood as a food source, not a habitat. Of the three types of termites, subterranean termites are the hardest to control because they retreat far underground. There are multiple species of subterranean termites, and their existing habitat spreads across the southern United States, Hawaii, and part of Africa, Asia, Australia, and South America. Termites are prevalent in the Southeast US, followed by the East and West Coast, but their range is expected to expand north to Canada's Hudson Bay as warm, humid climate conditions shift north over the next 100 years.

Termite treatment has traditionally included a combination of physical barriers and repeated chemical treatments applied to the soil. Moisture control, heat treatments, physical barriers, or biological treatments are more ecologically sound than the current widespread use of pesticides. Physical barriers separate the termite's home in the soil from the wood that is their food source, but some species build tunnels of soil to bridge the distance. Barriers do not eradicate the termites, and so require ongoing vigilance.

Progress has also been made in the use of fungus to invade and destroy termite colonies. A wide variety of fungal isolates are effective against different species of termites, and do not affect other insects, such as beneficial bees. Applications of fungi have successfully eradicated an entire termite colony, but additional studies are required to determine effective doses and procedures to treat wood in buildings.

Carpenter Ants and Bees

Weather conditions that lead to wet wood will increase the risk of carpenter ant and bee damage. Carpenter ants and bees do not attack wood for food but cause extensive structural damage by drilling into wood to create habitat. Carpenter ants are present across the United States, but cause the most damage in the Pacific northwest where moist or damaged wood provides a constant source of water and proper moisture for eggs. Carpenter bees prefer untreated wood that is not directly exposed to sunlight. Unlike carpenter ants, carpenter bees over-winter in the tunnels previously excavated to lay their eggs, so their damage to wood is usually less extensive. However, the damage can become more extensive if they return year after year to the same wood, extending the tunnels, or if woodpeckers discover the noisy larvae, and peck through the surface wood to eat them.

Eliminating wet wood and sources of water such as condensation and roof or plumbing leaks is key to preventing insect infestation. A variety of chemicals can be used to treat the nests or bait food that ants will carry back to the colony. Nesting can be prevented by keeping the wood painted. Stains and other surface treatments can help but are not as effective because bees do not ingest the wood or absorb the chemicals.

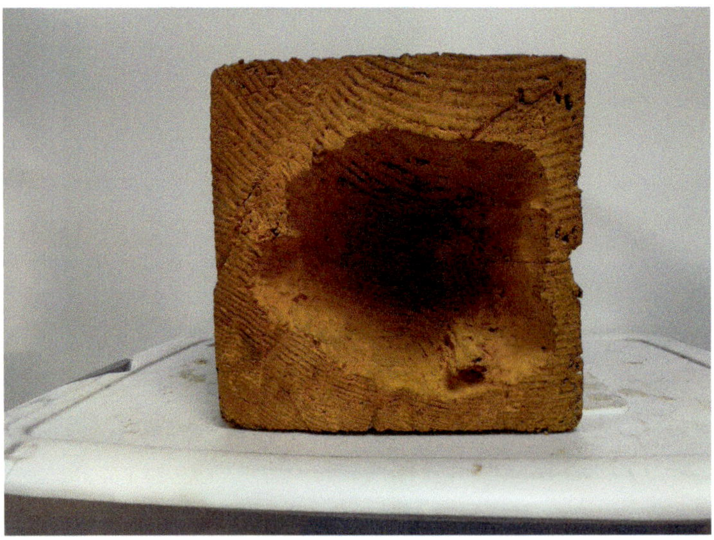

Figure 2.2.7 Carpenter ant damage removed in preparation for epoxy repair, Gilkey Covered Bridge, Oregon. (Photo by Kevin Groom, 2023.)

Wood-Boring Beetles

About 25% of all known species are beetles, including 400,000 species in the order *Coleptera*. Beetles feed on plants and fungi, and break down animal and plant debris, including wood. Wood-boring beetles are usually adapted to a particular tree species, and usually limited by climatic conditions. However, they have caused extensive damage when they are inadvertently transported to new locations and introduced to species that lack natural deterrents.

The decline of oak trees in Europe appears to be due in part to several bark and wood-boring beetles. Elevated temperatures in Norway have accelerated beetle reproduction, and the combination of drought and heat are making Norway spruce more susceptible to beetle damage. The eventual destruction of Norway spruce in these warm dry climates is expected to drive the beetles to less drought-affected spruce forests. A ten-year study of insects in multiple English Heritage properties indicates that even modest temperature increases can alter insect lifecycles, increase the number of eggs, growth rate, and the days of flying (which can enhance dispersal). Beetle infestations have increased in Japan as temperatures have increased.

Powder post beetles (Lycinae) cause more damage to wood in buildings than any insect except termites. Powder post beetles can live out their life cycle indoors, and hence are less susceptible to outdoor temperatures and climate change. As a result, they are found as far south as Georgia and California, and as far north as Canada. They do, however, prefer moisture and damp wood.

Deathwatch post beetles (Anobiidae) attack both hardwood and softwood and are found in North America, the United Kingdom, and Japan. Unlike Lycinae, Anobiidae beetles are found in old or partly decayed wood with moisture levels above 13% or 14%. Infestations can occur from insects flying in from outside or carried in with firewood or other infested wood. The Hadrobregmus gibbicollis (LeConte) member of this family is most prevalent along the coast, where it attacks old fir timbers in barns, bridges, and basements. Weather that contributes to moisture damage would increase the risk of Anobiidae. Death watch beetles have been found within wood members, concealed by sound exterior wood and lacquer finishes in temples in the forests of Japan.

Strategies for Responding to Climate-Based Threats to Wood

Wood has a natural ability to expand and contract with fluctuating temperature and moisture. It also has natural chemical defenses against native biological pests. However, deterioration could be dramatically increased as global changes in the weather increase local UVB, temperature, moisture, and wind, and cause changes to habitat of insects, bacteria, fungi, and chemicals. These impacts can be mitigated by understanding how wood deteriorates, and how man-made strategies can support the inherent resilience of wood. Any defense against weather and its impacts must start with managing water.

Traditional builders and craftsmen acquired and shared their knowledge of wood and wood techniques through trial and error, apprenticeships, family trades, guilds, and local traditions. Modern science has given us a greater understanding of the biological and chemical processes behind this practical knowledge. Strategies for moisture control can be divided into several strategies:

- Select the appropriate wood species.
- Orient grain and tool wood to minimize water absorption.
- Detail to eliminate or shorten exposure to weather.
- Add a coating or surface barrier to protect the wood from moisture, UVB, and biological growth.
- Consider wood-preservative treatments that penetrate the wood to increase its durability or resistance.

Where conditions become more favorable to fungi or insects, it will be increasingly important to control the moisture in wood through coatings, protective detailing, and designs that promote drying of the wood after rain events, and to avoid infecting old wood when introducing new wood in repairs. However, if the underlying cause of deterioration is not addressed, surface coatings, epoxies, and other consolidants can trap moisture and conceal continuing deterioration.

Investigation

Identifying the existing species of wood allows us to understand its performance under different conditions and to select appropriate materials and techniques to repair or protect original wood. The species of wood will define its strength and durability, how it resists moisture, fungal or insect decay or chemical actions. How wood is grown and harvested will also create variations in the performance of the original wood species, and in contemporary wood of the same species.

Maintenance routines should include a regular inspection and evaluation to identify the current condition and any pre-existing damage and identify any changes in the condition of the wood, or in the performance of water management systems. It is critical to determine whether damage is a result of past conditions that are no longer active, or new moisture or biological infestation. Investigation should focus not only the wood itself, but on surrounding water management features including roofs and overhangs, the overall wall construction, and nearby soil conditions. Settling, warping or misalignment of roof or windows can indicate settlement problems or deterioration of the wood itself.

Careful inspection of the surface can identify insect bore holes, mud tubes, frass, or fruiting bodies of fungus that may indicate decay deeper in the wood member. Probing with a sharp pick or awl can identify voids below the surface. In larger timbers, it is important to determine whether the deterioration is limited to exterior sapwood, in which case the underlying heartwood might still be structurally sound. The extent of decay, especially fungal rot, might not be visible on the surface and may require additional investigation or testing. Ideally this investigation should be non-destructive or minimally destructive. Existing coatings might be concealing damage or trapping moisture.

The most practical and useful tests are moisture measurement, remote visual inspection, and resistance drilling. Measuring moisture is useful not only to identify existing decay, but also to identify conditions that might make the wood vulnerable to future attack by fungi or insects. Large timbers can retain moisture, causing interior rot without any visible exterior damage. Remote visual inspection uses specialized tools to view inaccessible conditions or construction within a beam, pocket, or wall, that is not otherwise visible. Infrared thermography, stress wave measurement, ultrasound and digital radioscopy can also provide information but require an expert to accurately interpret the results.

Detailing and Protective Coatings

Rain events of long duration increase opportunities for wood to become saturated and decrease opportunities for wood to dry between rain events, increasing the need for protection and permeable coatings. Overhangs and details can reduce exposure to rain but provide limited protection from extreme events and wind driven rain. Varnish, stains, and paint can reduce, but not prevent, moisture changes in the wood. Coatings must protect the wood from excess moisture but allow the escape of any moisture that inevitably breaches that defense. Moisture trapped behind paint can promote decay as well as break the bond between the paint and the wood.

As climates become warmer and more humid, it will be increasingly important to manage water and paint or seal wood to prevent infestation by beetles. Unfinished crawl spaces or basements should be ventilated and covered with moisture barriers to lower moisture content.

Chemical Treatments

Chemical pesticides and fungicides are a popular way to preserve wood and prevent fungal decay, but should only be used with proper precautions. Surface treatments have limited penetration and life; techniques to penetrate chemicals into the wood with pressure or heat are only available for treating new wood. Solid fumigants that are inserted into wood and release chemicals over time have been successful in preventing insect damage to exposed wood timbers in covered bridges when replaced on a regular schedule.

Preservatives and mildewcides added to paint are popular but have a limited life and must be regularly maintained. Although there are questions about the long-term viability and appropriateness of chemical treatments, they clearly have a role where the alternative is loss of the original material altogether.

Heat Treatments

Heat treatments are an alternative to fumigation and chemical treatment. Heat treatments do not destroy colonies in the ground, but can be kill isolated colonies of beetle, larvae, termites, and eggs by raising the ambient temperature to a level of 140° to 160°F so that the core of the wood is at 120° to 130°F. The required temperatures and duration vary with the density and thickness of the wood and with the type of insect. Usually, the entire structure is treated, but isolated treatment is possible if the location of the infestation is clearly known. There has been extensive research on the use of heat treatment to kill beetles and their larvae at the temples in Nikko, Japan.

Repair and Replacement

Even with the best protection, wood often needs to be repaired or replaced. Replacement materials should match the original where possible but should also consider whether modifications to the material or details are appropriate to better meet the impacts of climate change. The impacts of heavier rain, higher temperatures, UV light, and increased insects should all be considered. The widespread distribution of lumber makes local quarantine procedures cumbersome, but concerned caretakers and contractors could treat or quarantine wood locally before introducing it to a historic structure.

A few strategies specific to wood are as follows:

- Ensure that new lumber introduced for repairs is dry (20% or less moisture content) and free of pests.
- Allow wood to acclimate, or adjust to the ambient humidity, before it is installed (unsealed wood expands and contracts in response to the humidity in the air).
- Protect exposed ends of wood. (Wood will wick water from all surfaces, but especially through the ends.) Run wood siding behind window trim, not butting trim.
- Avoid joints that allow rainwater to enter or be trapped.
- Provide ventilation to allow wood to dry. (Dry wood will last longer and hold paint better.) Install wood siding over furring.
- Reduce or prevent air movement between the exterior and conditioned interiors. (Air carries moisture that can condense on cold surfaces.)

Surrounding conditions should also be considered:

- Avoid contact between the building wood and soil unless the wood is pressure treated.

- Cover exposed soil to prevent moisture that can condense on wood members.
- Ensure that surrounding landscape surfaces drain away from the building.
- Avoid storing stumps, wood debris and other sources of insects, fungus, and biological infection near the building.

Epoxies and Consolidants

Although repair with compatible wood is always preferable, epoxy resins are often used to replace or reinforce missing or decayed wood especially where maximum retention of the original fabric is desired. Resins are easily used in the field, eliminating the need to disassemble components while replacing or consolidating decayed wood.

Two-part epoxy resins have been used for 30 to 40 years and have many strong advantages. They have minimal shrinkage during curing, bind well to wood, resist chemicals including alkalis, and are strong and waterproof. Unfortunately, a comprehensive evaluation of their long-term performance has not been undertaken, and there is concern among professionals regarding their compatibility with wood and the impact of higher temperatures on their long-term performance. While wood will dry out and become more rigid at higher temperatures, epoxies will tend to plasticize, and become more flexible. High temperatures can cause epoxies to creep after they have cured. UV light can degrade epoxy, a risk that may increase with higher UV light levels. Cure time is influenced by temperature; adhesion and uptake vary with temperature, moisture content, extractives or other contaminants, and environmental stability. High moisture content can reduce absorption of epoxies and reduce the bond between the wood and the epoxy.

Wood is weaker and will expand differently than adjacent epoxy patches or epoxy consolidated wood. Repairs must be carefully designed to avoid restricting the movement of joints or creating new stresses in the remaining wood. Moisture can be trapped in porous unconsolidated wood where it meets impervious epoxy fill or consolidants. These stresses may increase with more extreme and rapid fluctuation in daily temperature and humidity, and extended heat, rain, or drought events.

Conclusions

While wood is the most ephemeral of building materials, when well-maintained under the proper conditions it can survive for centuries. However, the continuing increase in temperatures and higher humidity will increase thermal, and moisture driven expansion and contraction, and increase the risk of fungal growth and insect infestations. Strategies must be developed to actively monitor humidity, temperature, wood decay activity and insect populations, and to proactively reduce exposure to moisture and maximize opportunities for wood to dry.

Bibliography

Allen, Edward, and Joseph Iano. 2013. *Fundamentals of Building Construction: Materials and Methods*. New York.Wiley.

Andrady, Anthony L, Hamid S. Hamid, and Ayako Torika. 2003. "Effects of Climate Change and UV-B on Materials." *Photochemical and Photobiological Sciences* 2 (1): 68–72.

Anthony, Ronald W. 2010. "Practice Points 10, Nondestructive Evaluation: Wood." *APT Bulletin, The Journal of Preservation Technology* 41, no. 4 (2010): 1–6. http://www.jstor.org/stable/41000035.

Anthony, Ronald W. 2007. "Practice Points Number 3: Basics of Wood Inspection." *APT Bulletin, The Journal of Preservation Technology* (The Association for Preservation Technology International) XXXVIII, No. 2–3. https://www.apti.org/assets/docs/03-Anthony.pdf.

Anthony, R. W. (2018). *Guidelines to Restoring Structural Integrity of Covered Bridge Members*. Forest Service, Forest Service Laboratory. USDA.

APA. 2018. "Chemical Resistance." (APA The Engineered Wood Association). https://www.performancepanels.com/chemical-resistance#:~:text=The%20principal%20classes%20of%20chemicals,nitric%20acid%20may%20cause%20ignition.

Bowen, R. L. D.D.S. 1956. "Use of Epoxy Resins in Restorative Materials." *Journal of Dental Research* 35 (3): 360–369. https://doi.org/10.1177/00220345560350030501.

Brimblecombe, Peter, and Caroline Truth Brimblecombe. 2013. "Trends in Insect Catch at Historic Properties." *Journal of Cultural Heritage* 16 (2), 127–133 https://doi.org/10.1016/j.culher.2014.05.005

Brimblecombe, Peter, and Mikiko Hayashi. 2018. "Pressures from Long Term Environmental Change at the Shrines and Temples of Nikko." *Heritage Science* 6, 27. https://doi.org/10.1186/s40494-018-0186-1.

Buczkosski, G, and C Bertelsmeier. 2017. "Invasive Termites in a Changing Climate: A Global Perspective." *Ecology and Evolution* 7:974–985. https://doi.org/10.1002/ece3.2674.

Carll, Charles G. 2009. *Decay Hazard (Scheffer) Index Values Calculated from 1971–2000 Climate Normal Data*. General Technical Report, Forest Products Laboratory, USDA. https://www.fpl.fs.usda.gov/documnts/fplgtr/fpl_gtr179.pdf.

Carmeliet, Jan, and Bert. Blocken 2004. "Rainwater Runoff for Evaluating Water Leakage Risks in Building Envelopes." *Proceedings of the Conference Performance of Exterior Envelopes of Whole Buildings IX*. Clearwater Beach, Florida. https://web.ornl.gov/sci/buildings/conf-archive/2004%20B9%20papers/196_Carmeliet.pdf.

Cassar, May. 2023. *NOAH's ARK Project*. https://www.ucl.ac.uk/bartlett/heritage/research/projects/project-archive/noahs-ark-project.

Chouvenc, T, and N-Y Su. 2012. "When Subterranean Termites Challenge the Rules of Fungal Epizootics." *PLoS One* 7(3): e34484. https://doi:10.1371/journal.pone.0034484

Cleary, Ryan. 2014. *Considering the Use of Epoxies in the Repair of Historic Structural Timber*. Thesis, University of Pennsylvania.

DeLucia, Evan H, Paul D Nabity, Jorge A. Zavala, and Mary R Bernbaum. 2012. "Climate Change: Restting Plant-Insect Interactions." *Plant Physiology* 160 (4): 1677–1685.

Dost, William A., and Elmer E. Botsai. 1990. *Wood Detailing for Performance*. Mill Valley, CA: GRDA Publications.

English Heritage. 2012. *Timber*. New York: Routledge.

Evans, Theodore A, Brian T. Forschler, and Kenneth Grace. 2013. "Biology of Invasive Termites: A Worldwide Review." *Annual Review of Entomology* 58 (1): 455–474. https://doi.org/10.1146/annurev-ento-120811-153554.

Forest Products Laboratory. 2021. *Wood handbook-Wood as an engineering material.* General Technical Report FPL-GTR-190, USDA, Forest Service. https://www.fpl.fs.usda.gov/documnts/fplgtr/fplgtr282/fpl_gtr282.pdf.

Forest Service. 1986. "Wood Decay in Houses." *Home and Garden Bulletin Number 73* (United States Department of Agriculture).

Garratts. 2008. "Wet Rot" Garratt's Damp and Timver." https://garrattsdamp.com/service/wet-rot-treatment/.

Garvin, James L. 2001. *A Building History of Northern New England.* Lebanon, NH: University Press of New England.

Grattan, D. W., and R. L. Barclay. 1988. "A Study of Gap Fillers for Wooden Objects." *Studies in Conservation* 33 (2).

Groom, Kevin, interview by Susan Pranger. 2023. *Linn County Road Department* (March 4).

Harvey, Donald W. Jr, and Michael F Schuller. 2010. "Nondestructive Evaluation: Structural Performance of Masonry." *APT Bulletin, The Journal of Preservation Technology* (Association of Preservation Technology International) XLI, No. 2–3.

Hoadley, R. Bruce. 1990. *Identifying Wood.* Newton, Connecticut: Taunton Press.

Irace, Frea. 2011. *Factors affecting the strength of wood.* https://ezinearticles.com/?Factors-Affecting-The-Strength-Of-Wood&id=5908591.

Editor 2018. *Janka hardness.* August 29. https://www.sizes.com/units/janka.htm.

Kaslegard, Anne S. 2010. *Climate Change and Cultural Heritage in the Nordic Countries.* Nordic Council of Ministers.

Kass, Adelaide, F. F. Wangaard, and H. A Schroeder. 2007. "Chemical Degratdation of Wood: The relationship between Strength Retention and Pentosan Content." *Wood and Fiber Science*, June 19. https://wfs.swst.org/index.php/wfs/article/view/1942.

Kherais, M., A. Csebfalvi, and A. Len. 2020. "The Climate Impact on Timber Structures." *International Journal of Optimization in Civil Engineering* 11 (1). https://www.researchgate.net/publication/350373502.

Kozlowski, Roman. 2007. "Climate Induced Damage of Wood: Numerical Modeling and Direct Tracing." *Experts' Roundtable on Sustainable Climate Management Strategies.* Tenerife, Spain: The Getty Conservation Institute. https://www.getty.edu/conservation/our_projects/science/climate/paper_kozlowski.pdf.

Kreber, Bernhard. n.d. "Understanding Wood Discoloration Helps Maximize Wood Profits."

Layne, Elizabeth. 2017. *Are Palm Trees Moncots?* https://sciencing.com/are-palm-trees-monocots-12523232.html.

Lebo, Stan T., Paul A. Cooper, and Patricia K. Lebow. 2006. "Study Design Considerations in Evaluating Environmental Impacts." In *Environmental Impacts of Treated Wood*, by Timothy G. Townsend and Helena Solo-Gabriele. Boca Raton, Florida: Taylor and Francis. https://doi.org/10.1201/9781420006216.ch5.

Lewitin, Joseph. 2019. *Bamboo Flooring Janke Hardness Scale.* http://www.bamboo-flooring-facts.com/bamboo_flooring_janka_hardness.html.

Lovett, Gary M., Marissa Weiss, Andrew M. Liebhold, Thomas P. Holmes, Brian Leung, Kathy Fallon Lambert, and David A. Orwig, et al. 2016. "Nonnative Forest Insects and

Pathogens in the United States: Impacts and Policy Options." *Ecological Applications* (Wiley on Behalf of Ecological Society of America) 26 (5): 1437–1455. https://www.jstor.org/stable/24818127.

Lubeck, Aaron, and Francis Conlin. n.d. "Efficiency and Comfort through Deep Energy Retrofits: Balancing Energy and Moisture Management." *Journal of Green Building* 5 (3): 3–15.

Maeglin. 1987. *Applying the latest research to hardwood trees.* https://www.fpl.fs.fed.us/documnts/pdf1987/maegl87a.pdf.

Moore, Gregory F., Donald E. Kline, and Paul R. Blankenhorn. 1983. "Impregnation of Wood With a High Viscosity Epoxy Resin." *Wood and Fiber Science* (3): 223–234.

Myles, Timothy. 2004. "Advancing Termite Control in Canada." *Research Bulletin #23* (Center for Urban and Community Studies, University of Toronto). https://www.researchgate.net/publication/290379396_Advancing_termite_control_in_Canada.

Nishioka, Tsunekazu, and Jiro Kohara. 1978. *The Building of Horyu--ji.* Translated by Michael Brase. Japan Publishing Industry Foundation for Culture.

Park, Sharon C. 1989. *Preservation Brief 19, The Repair and Replacement of Historic Wooden Shingle Roofs.* National Park Service, Government Printing Office. Wahington DC. https://www.nps.gov/tps/how-to-preserve/briefs/19-wooden-shingle-roofs.htm.

Peterson, Chris J. 2010. "Termites and climate change: Here, there and everywhere?" *Earth Magazine*, February 16. https://www.earthmagazine.org/article/termites-and-climate-change-here-there-and-everywhere.

Potter, Michael F. n.d. *Carpenter Ants.* University of Kentucky Cooperative Extension Service.

Potter, Michael F. n.d. *ENTFact-611 Carpenter Bees.* University of Kentucky.

Potter, Michael F. n.d. "ENTFact-616 Powderpost Beetles." *UK Entomology.* http://www2.ca.uky.edu/entomology/entfacts/ef616.asp.

Pretzsch, Hans, Peter Biber, Gerhard Schutze, Julia Kemmerer, and Enno Uhl. 2018. "Wood density reduced while wood volume growth accelerated in Central European forests since 1870." *Forest Ecology and Management.* https://doi.org/10.1016/j.foreco.2018.07.045.

Quarles, William. 2006. "Thermal Pest Eradication in Structure." *IPM Practitioner*, May/June.

Radivojevic, Suzana. 2016. "Effects of Climate Change on the Long Term Deterioration Trends of Wood in Cultural Heritage." *A Century of Design in the Parks, Preserving the Built Environment in National and State Parks.* Santa Fe, New Mexica.

Rath, Andrew. 2010. "The Use of Entomapathogenic Fungi for Control of Termites." *Biocontrol Science and Technology 10*, Taylor and Francis. 563–581. https://doi.org/10.1080/095831500750016370.

Rowell, Roger M., Roger Pettersen, and Mandla A. Tshabalala. 2012. *Handbook of Wood Chemistry and Wood Composites.* Routledge Handbooks.

Sabbioni, Christina, Roger-Alexandre Lefevre, and Peter Brimblecombe. 2009. "Vulnerability of cultural heritage to climate change." *Pollution Atmospheric.*

Salle, A, L.-M. Nageleisen, and F. Lieutier. 2014. "Bark and Wood Boring Insects Involved in Oak Declines in Europe." *Forest Ecology and Management* 328: 79–93.

Seybold, V.R., and S. J Lewis. 2010. "Wood Boring Beetles in Homes." *UC IPM Statewide Integrated Pest Management Program.* http://www.ipm.ucdavis.edu/PMG/PESTNOTES/pn7418.html.

Simpson, J.W., and P. J. Horrobin. 1970. *The Weathering and Performance of Building Materials*. Aylesbury, Bucks: Medical and Technical Publishing Co.

Stork, Nigel E., James McBroom, Clarie Gely, and Andrew J. Hamilton. 2015. "New Approaches Narrow Global Species Estimates for Beetles, Insects, and Terrestrial Arthropods." *Proceedings of the National Academy of Science of the United States of America* 112 (24): 7519–7523. https://doi.org/10.1073/pnas.1502408112.

Suszkiw, Jan, and Erin K Peabody. 2007. "Fungal Foam Seeks and Destroys Termites." *Agricultural Research*, September.

Temperli, Christian, Harald Bugmann, and Che Elkin. 2013. "Cross-Scale Interactions Among Bark Beetles, Climate Change, and Wind Disturbances: a Landscape Modeling Approach." *Ecological Monographs* (Wiley on Behalf of the Ecological Society of America) 83, No. 3: 383–402. https://www.jstor.org/stable/23596922.

The Nature Conservatory. 2015. *Don't Move Firewood: Invasive Species*. https://www.dontmovefirewood.org/.

U.S. Forest Service. n.d. *Tannins*. https://www.fs.fed.us/wildflowers/ethnobotany/tannins.shtml.

USDA; APHIS. 2010. *Risk Assessment of the Movement of FIrewood within the United States*. USDA.

Vallejos, Christobal, and Walter Grote. 2009. "Wood Moisture Content Measurement at 2.45 GHz." *Microwave and Optoelectronics Conference (IMOC)*. https://doi.org/10.1109/IMOC.2009.5427593.

Waldvogel, Michael, and Patricia Alder. 2018. *Carpenter Bees*. NC State Extension.

Wohlleben, Peter. 2015. *The Hidden Life of Trees*. Random House.

Young, Robert A. 2008. *Historic Preservation Technology*. Hoboken, New Jersey: John Wiley & Sons, Inc.

CHAPTER 2.3

Stone

Introduction to Stone

Stone is a symbol of permanence that is constantly changing, being destroyed, and re-created by nature and by man. Industrial processes of the past two centuries have accelerated deterioration by adding chemicals that react with stone. Changes in the climate are creating weather patterns that accelerate the mechanical and chemical erosion of stone, threatening both the natural environment in which we live and the historic buildings we treasure.

The Past, Present, and Future of Stone

History

Humans manipulated rock to build shelters and defensive structures when they established permanent communities using locally available stone. As civilizations developed, improvements in transportation, infrastructure, and technology allowed stone to be cut more easily, moved from deeper quarries, and transported longer distances. Builders became selective, choosing a stone for its aesthetics and durability. The discovery of lime and natural cement led to more weather resistant masonry construction. Stone building traditions and techniques were spread by migration or conquest and adapted to local materials.

Rubble foundations and rough-cut stone were used as early as 1,000 BCE in the far east. More sophisticated techniques were developed around the globe to dry-build with dressed stones and tight joints, or to wet-build with clay or lime grout and plasters. Although many were deteriorated by the action of rain, wind and sun, many ancient stone structures remain today.

DOI: 10.4324/9781003160489-8

Adobe or stone rubble with mud mortar was used for domestic dwellings, but important ceremonial or religious buildings were constructed of stone or covered with tightly joined stone veneer by the Egyptians and Greeks (1300 to 1200 BCE) and Pre-Columbian cultures of the Aztec, Mayan, and Incan (1200 to 900 BCE). Large, sophisticated communities of stone were built in North America by the Anasazi Indians in 1500 BCE and the Zuni and Hopi in 850 CE.

Figure 2.3.1 Chaco stone wall c. 800 CE, Una Vida, Chaco Canyon, New Mexico. (Photo by Daniel Thomas, 2021.)

By 300 BCE, the Roman Empire had acquired stone masonry skills from the Egyptians and Greeks, and refined formulas for lime mortars and cement. They shared this expertise as they expanded as far north as Hadrian's Wall in England (122 BCE) and as far east as Hagia Sophia in Constantinople, Turkey (537 CE). The collapse of the Roman Empire in the 5th century led to a decline in masonry construction throughout Europe, although earlier structures were deconstructed for their mortar, brick, and stone. By the late 8th century, religious, civic, and military buildings were again pushing the limits of stone construction with pre-Romanesque styles. Large, dressed ashlar blocks and flamboyant details of the 14th-century Romanesque were followed by tall, thin, composite Gothic walls by the end of the 16th century.

Migrant communities were attracted to locations with familiar natural resources and climates, and more likely to build with stone if they came from an area where those techniques were familiar. When materials and skills were available, stone was preferred as a strong, durable material, able to withstand weather and hostile assaults. However, the need for economic and social stability often dictated the use of timber, which was cheap and fast, and required less labor.

Stone building was rare in North Americas in the 16th and 17th centuries due to a lack of experience and limited access to lime. Dutch, French, and Scottish settlers maintained brought traditions of stone construction to the Hudson River Valley in New York, New Jersey, and along the St. Lawrence River in Canada from the mid-17th to the early 19th century. Settlers on the American frontier were more focused on survival; stone was limited to foundations to protect wood structures from ground water, and the occasional commercial establishment or grand home.

Spanish influence proceeded more rapidly across the American south and southwest due to the influence and expansion goals of the Spanish King and Catholic Church. Most missions were constructed of timber and adobe, but presidios, fortifications, and some mission church complexes in Florida, Texas, and California were constructed of local sandstone or limestone, usually with a protective layer of adobe mud or lime plaster.

Masonry construction thrived in 18th-century North America as timber was depleted, lime became available, and the economy, transportation, and social conditions improved. Catastrophic fires in European and North American cities strengthened the demand for fireproof construction. Limestone and granite became a symbol of solidity and permanence in public and commercial buildings, while residential buildings were constructed with less expensive sandstone and brownstone.

The Industrial Revolution brought innovative technology and transportation to the stone industry. Waterpower and frame saws made it economical to work large blocks of stone and provide pecking, batting, rustication (rockface, punched, or hammered) finishes.

Current Social, Cultural, and Climatic Impact

Coal burning during the Industrial Age increased acid fumes, carbon in the air and acid rain that accelerated chemical weathering of stone. Damage decreased during the 20th century as coal use declined and industries began to scrub their fumes, but manufacturing continues to add chemical and biological agents to the atmosphere and oceans. Stone deterioration is being accelerated by increased UVB light, temperature, rainfall, and carbon dioxide. Widespread forest fires and volcanic action also contribute particles and chemicals that damage stone.

Modern agricultural processes, the production of cement, and burning of fossil fuels continue to release carbon dioxide into the atmosphere and directly and indirectly damaging stone. The transportation of stone products around the world adds to the carbon "footprint." Even stone that is sourced locally is often shipped overseas for less expensive cutting, shaping, and finishing.

Understanding Stone

The Rock Cycle

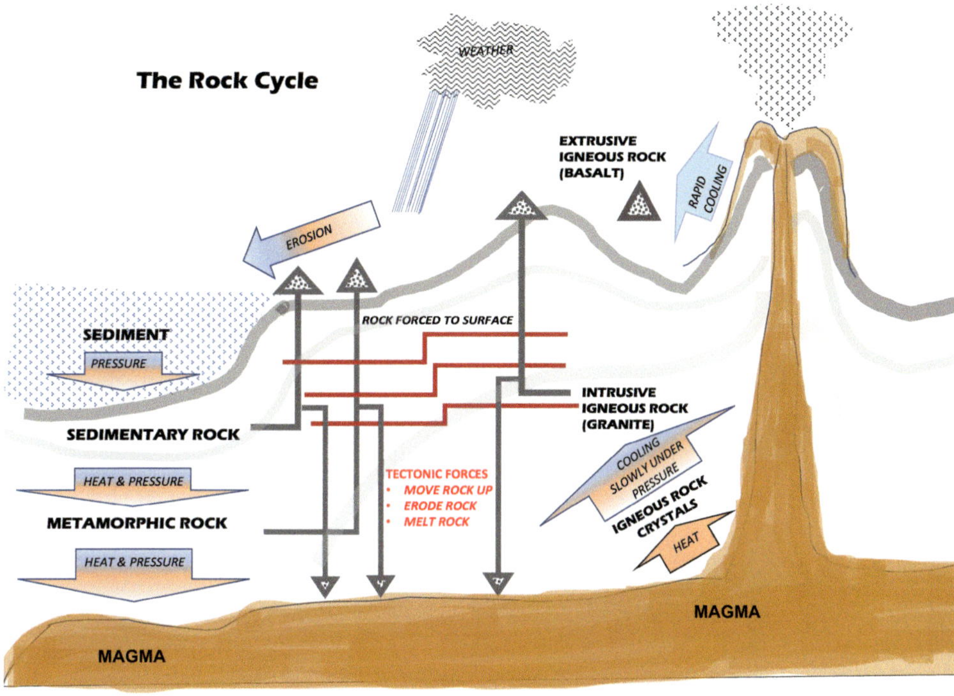

Figure 2.3.2 The rock cycle. (Diagram by Susan Pranger, 2023.)

The rock cycle is similar to the carbon cycle. Composites of minerals transform from magma and lava into igneous rock that is worn into aggregates that form sedimentary stones. Sedimentary stone that is subjected to tectonic pressure or heat is chemically changed to metamorphic stone. Igneous, sedimentary, and metamorphic rocks are eroded into sediment or returned to magma by tectonic or volcanic movement, completing the cycle. These geological processes determine the cohesion, strength, and porosity of stones with igneous considered the hardest and least porous, followed by metamorphic. Sedimentary stones are the least strong and most porous. Most stones exist somewhere on a continuum between these three categories or are a mix of other rocks held in a sedimentary binder.

Igneous and metamorphic rock is formed as minerals heated by tectonic plate or volcanic action cool. If the magma cools slowly underground under pressure, it will form dense, coarsely grained intrusive (plutonic) rocks with large crystals rocks, such as granite. Molten rock that erupts through to the surface as lava will expand and cool quickly to form extrusive volcanic rock that is fine grained with small or no crystals, such as basalt.

Over time, the grinding of tectonic plates, and erosion from weathering and other biological or chemical changes erode rock to create smaller aggregates and sediments. These aggregates may be identical to the original rock or a blend of aggregates and organic matter that have been transported and mixed by glaciers, rivers, or the wind. The aggregates and organics accumulate and are compressed under low temperature and pressure to form sedimentary rock, often in flat areas that were once covered by oceans or vast lakes. Limestone was formed from organic matter in the oceans, shale from undecayed organic plants, while brownstones and sandstones were formed by compression of non-organic mineral sediments. Metamorphic rocks such as marble, dolomite, or slate are formed when sediments or sedimentary stones are subjected to pressures and temperatures that are high enough to cause chemical changes, but not high enough to return the rocks to a molten state.

The Chemistry of Stone

The strength and cohesiveness of a stone is determined by geological processes and the specific combination of elements and minerals. Chemical and electrical forces, along with environmental pressure, temperature, humidity, and acidity, combine, separate, and recombine elements and minerals to form and destroy different types of stone. The ability of a stone to resist these forces corresponds with its ability to resist manipulation by man and the forces of weather.

Different stones are made from an almost infinite combination of the 115 basic elements. Most stones are a mix of different minerals, the most common being feldspar, mica, quartz, and calcite. Binders are generally based on calcium, clay, or other fine particles.

Feldspar is a group of minerals that include alumina and silica, including aluminum silicates of soda, potassium, or lime. It is the most common type of mineral on earth, and is present in most rocks, soil, clay, or sediment. Feldspar is recognized by its tendency to cleave, or split at an angle close to 90 degrees, and by a pearly, translucent, and vitreous luster on the split faces. Feldspars resist acid and are relatively hard (6 out of 10 on the Mohs scale) but erode easily because of their tendency to cleave or split. Up to 60% of granites, monzonites and syenite are orthoclase or plagioclase feldspar. Orthoclase feldspar is light pink or orange with a symmetrical cube-like crystal. Plagioclase feldspar is clear or gray with an asymmetric crystal.

Mica is a mineral group composed of phyllosilicate minerals, which cleave into very thin sheets. It is most common in igneous and metamorphic rocks, especially granites, pegmatites, and schists, but also occasionally as flakes in sedimentary rocks.

Quartz contains silicon and oxygen and is present in almost all colors and types of rock across the globe. It is harder than most rocks, does not split easily, and resists both mechanical and chemical weathering. It is abundant on mountaintops and is the main component of most beach, river, and desert sands.

Calcite, also known as calcium carbonate ($CaCO_3$), is the main component of limestone and marble, and also found in other sedimentary, metamorphic, and igneous rock including travertine and tufa, and in chalk and caliche soils. Calcite is formed in nature by the compression of shell, coral, algae, or other debris at the bottom of shallow oceans, or by leeching out of limestone. Man-made calcite is made from ground limestone or precipitated out of mixture of carbon dioxide and calcium hydroxide. Calcite is vulnerable to mechanical weathering because it cleaves in three directions and is very water soluble.

Acid or acid rain converts calcite to lime and carbon dioxide.

Traces of ferrous iron are found in many sedimentary, metamorphic, and igneous stones. Veins of iron, pyrites, or black shale bituminous are generally too fine to cause staining or weathering, but iron contaminates can oxidize during weathering. Iron oxide can block the reaction between water, salts, and the other components in stone.

Porosity and Permeability

The durability of stone depends strongly on porosity and permeability, which is determined by the size of particles and how they are compressed. Water can change from vapor to liquid to ice as it moves through the pores of a stone. It can evaporate, condense, or expand in response to temperature, relative humidity, solar radiation, and hydrostatic or wind pressure. Moisture in the air can condense on cold stone as temperatures drop and be drawn into porous stones through capillary action though the smallest of pores.

The extent of damage from moisture movement is determined by how well the pores can resist the migration of water, and by how well the binder agents in stone can resist pressure from expansion, compression, and chemical change. Porous stones have a larger internal surface area with more opportunities for water to interact with the minerals and binders. If the binder is soft, water or vapor moving through the pores can dissolve and carry away calcium and clay binders, causing the stone to disintegrate. However, if the stone is both porous and permeable, the stone will both absorb and release moisture quickly, limiting damage from water and chemicals.

The Movement of Moisture

Salts or minerals that originate in a stone, atmosphere, or ground water are carried by moisture through connected pores and cracks in stone. As the water evaporates, or the temperature drops, salts will precipitate within the pores (crypto florescence) or on the surface (efflorescence) of the stone. A stone with small pores may dry out more slowly, which can either carry salt to the surface, or allow it to precipitate and accumulate within the pores.

In some soft, porous limestones, moisture within the stone migrates toward the surface when quarried stone is exposed to air, dissolving silica or calcite binders and redepositing them in pores near or on the surface as it evaporates. While this can "case harden" the surface, it can also create a plane of weakness between the hardened surface and the softer underlying layers. Trapped ice, water pressure, or salts can accumulate behind the hardened layer exerting pressure that eventually causes spalling or delamination.

Variations in sun and wind exposure can accelerate and cause uneven hardening of the surface. Areas that are not as hard will erode first, causing pockets that honeycomb the surface. Salts and pollutants can accumulate in pockets sheltered from the rain and accelerate deterioration. After construction, moisture can continue to move binders within and between adjacent stones, mortar, or concrete.

Stone as a Building Material

There are local variations, even within a single quarry, and yet the qualities of igneous, metamorphic, and sedimentary stones are remarkably consistent across the globe, resulting in similar qualities and construction practices for each type of stone across time and location.

Igneous Stone

For building purposes, gabbro, basalt, pegmatite, some schist, gneiss, diabase, and diorite granite are often classified as granite, although geologically they have different compositions. Granites are used for everything from aggregate and rough-cut curbstones to highly polished building facing, pavers, or tiles.

The igneous character of granite makes it more challenging to quarry, shape, and polish than metamorphic and sedimentary stones, but small pores and a uniform grain make it resistant to impact, rain, and chemicals. Granite is less porous than sedimentary and metamorphic stones, and less vulnerable to water infiltration, wet-dry, and freeze-thaw cycles and preferred for foundations, coastal locations and other harsh or wet conditions. However, some granites have small but permeable pores that absorb water by capillary action and retain the moisture longer.

Sedimentary Rock

Sedimentary rock is formed when layers of organic or inorganic sediment are compressed under directional pressure and low heat. They are relatively soft and porous compared to the metamorphic and igneous rocks which are subject to higher temperatures and pressures. Nonetheless, sedimentary sandstone and limestone are most often used for construction because they are readily available in many parts of the world and are easy to quarry and shape. Most, but not all, have a moderate pore structure that is durable enough to resist weathering. However, limestones and sandstones vary in quality and are vulnerable to weathering and erosion if not correctly installed and protected.

Sedimentary rock particles fall into the following categories, but may also be a mix:

- Clastic (sandstones, some shale), formed from minerals and weathered rocks.
- Organic (limestone, calcareous sandstones, siltstone, some shale), formed from skeletal debris accumulated on the ocean floor, or undecayed organic matter trapped in mud.
- Chemical (usually limestones), precipitated from man-made or natural solutions.

The type of binder has the most impact on durability. Sandstones with siliceous binders are considered the strongest. Limestones and calcareous sandstone, bound with calcite, and ferruginous sandstone, bound with iron oxide, are less durable, and argillaceous (clay) sandstones are the least durable.

- Silica binder forms a durable quartzite sandstone (clastic) that does not react with acids.
- Calcite and dolomite binders are found in limestones and calcareous or dolomitic sandstone. Calcite, and to a lesser degree, dolomite, are readily dissolved by acids. Limestone is generally more porous overall than sandstone because the binding calcite material is also very porous.
- Iron minerals form a ferruginous sandstone that is colorful, but not very durable due to the tendency of iron to oxidize.
- Fine clay cements are found in argillaceous sandstone, will absorb and release water, and is subject to expansion and frost damage.

Sandstone can have a high permeability but a low porosity, which means that it is saturated quickly without absorbing a lot of water and vulnerable to damage from swelling of clay in the binder or salt crystallization in the pores. Wet-dry cycles increase damage from repeated cycles of expansion and contraction. The sandstones at Iglesia de San Mateo in Tarifa, Spain, absorbed only 4% water before becoming saturated. The clay was evenly distributed on the gray sandstone but concentrated along the bedding planes in the brown sandstone. This increased capillary action along the bedding plane and made the brown sandstone more susceptible to cycles of wetting and drying and damaging expansion-contraction.

Figure 2.3.3 Spalled sandstone, Iglesia de San Mateo, Tarifa, Spain. (Photo by Daniel Thomas, 2022.)

Clastic sedimentary rocks include breccia, conglomerates, and caliche. Breccia is formed by angular fragments of other rocks, bound together by smaller aggregates and minerals. Conglomerate is similar to breccia, but the rocks are rounded. Caliche is a combination of rock fragments and other sediment bound together by $CaCO_3$ or other minerals as either a loose soil or a highly consolidated rock.

Shale is made up of clay particles (illite, kaolinite, or smectite) or other similarly tiny grains of quartz, chert, or feldspar, cemented together by even tinier particles of silt. A black color usually indicates undecayed organic matter from an environment that lacked oxygen – such as mud. Red, brown, or yellow shales contain iron oxide or iron hydroxide, hematite, or limonite minerals from oxygen-rich environments.

Sandstone is formed from small-, sand-sized particles (up to 90% quartz, with some weathered feldspar) bound by still smaller clay or silt particles, and can be coarsely grained, or fine grained with a consistent color and texture similar to limestone. Durability depends on the relative proportion of quartz, feldspar, and other grains of rock (lithics) and on the type of mineral cementing those grains to each other. Quartz particles are very hard and durable, but are not as tightly bound which makes them vulnerable to abrasion and weathering. Colorful inclusions of different minerals can add character but also potentially weak fault lines.

Organic sedimentary rock includes light-colored chalk, diatomite, and most dolomites and limestones, which are composed of calcite ($CaCO_3$), and dark-colored shale and coal, which are formed from undecayed organic plant matter compressed in mud.

Calcite is created when plankton and other organisms in warm shallow oceans combine carbon dioxide ions with bicarbonate ions to form shells or skeletons. Shells and sediment are eventually compressed into organic limestone, dolomites, and similar materials. Deposits that combine with magnesium form a rock somewhere on the continuum between limestone, dolomitic limestone, and dolomite. Limestone and dolomite are very similar in color and texture, but dolomite is slightly harder and slightly more resistant to acid. A soft, gray calcareous limestone found in many parts of the world has a close compact grain that approached marble, often include fragments of shells. It is easily quarried, but "case hardens" after exposure. There can be weak pockets where large fossils prevent compaction of binder within the shell.

Limestone with large pores will absorb and release water quickly, but those with small pores absorb water through capillary action and retain water longer, which gives the water more time to dissolve calcite binders, or cause frost damage. Limestone that is poorly compacted, contains highly soluble minerals, or has a high clay content will be less durable. Geologically older stones are more compacted and less porous as pressure converts small crystals to fewer larger crystals.

Chemical sedimentary rocks are precipitated from a solution. Naturally occurring examples are chert (flint), iron ore, rock salt, and some limestones. Chert was used in early toolmaking because it breaks in a conchoidal fracture pattern, producing sharp pointed tools or weapons. Limestone is generally produced from organic sediment, but it can also be formed by the precipitation of $CaCO_3$ from lake or ocean water. Stalagmites and stalactites in caves are an example of chemical limestone formed by calcium deposited by dripping water.

Tufa is formed when carbonate precipitates out of ambient temperature water. Tuff is a porous sedimentary rock formed of more than 75% volcanic ash.

Metamorphic Rock

Metamorphic rock, including shist, slate, dolomite, and marble, is created when sufficient pressure and heat change the mineralogy, texture, and chemical composition of sedimentary particles or rocks. Their strength, chemical resistance, and workability depend upon the intensity and duration of pressure and heat and the types of aggregate and binder.

Metamorphic rock can be foliated with visible layers, or non-foliated with a more three-dimensional quartz texture. They are more likely than igneous rocks to have veins or inclusions of other minerals between layers, or changes in material from one layer to the next. Foliated rocks such as gneiss, phyllite, schist, and slate have a distinctive layered appearance formed by directional pressure on layers of different composition. The mineral content and degree of heat and pressure will determine whether a foliate rock can resist splitting along its layers. Non-foliated rocks such as hornfels, marble, novaculite, quartzite, and skarn are more uniform in composition and less likely to split. A layered quality can be useful when quarrying or working metamorphic rock but can cause spalling and delamination if not properly bedded with weight carried perpendicular to the layers. Bedding layers that are exposed vertically can be vulnerable to delamination from freeze-thaw action; the edges of horizontal bedding layers can erode unevenly.

Marble and dolomitic marble are formed when pressure or heat chemically changes the $CaCO_3$ in limestone to a dense and almost impervious material. During this metamorphic process, the structure of the original fossils and sediment structure is lost, and the calcite recrystallizes into interlocking calcite crystals that grow as the process continues. Most marbles are non-foliated, with a homogenous quartz structure. Although foliated marbles experience more directional pressure than non-foliated marbles, both can be brittle with traces of veins and bedding layers that can split or exfoliate under stress. Veins can be more or less resistant to weathering than their parent limestone, and subject to sugaring or loss of their binder, especially in polluted or acidic conditions.

Marbles are classified according to their workability, uniformity, flaws, and imperfections. Only Class A is suitable for long-term exterior use. Most marbles are hard, in the range of 3 to 5 on the Mohs rating system but porous marbles are vulnerable to efflorescent or freeze-thaw damage in wet climates. Marble is susceptible to chemicals in the air, especially acid which dissolves the $CaCO_3$ binders. Marble with a hard aggregate but soft binder can be uneven in strength and difficult to polish.

Sedimentary shale is transformed by a similar metamorphic transformation but will be foliated into layers. With time and pressure, shale is transformed first into slate, then phyllite, then schist. The last step is gneiss, where the mica is recrystallized into granular mineral with alternating light and dark bands of interlocking feldspar and quartz. The pressure and heat transforms clay in shale into flat, hard, and very thin, layers of mica and other metamorphic components that can be mechanically split (cleaved) into smooth planes to form slate or shist roof shingles or pavers.

Slate is primarily composed of quartz, mica, and chlorite, which are not very chemically reactive, and various amounts of rutile, zircon, monazite, tourmaline, organic matter, iron sulfides, and carbonates. Although all slates are non-porous, variations in the type of minerals, intrusions, and the amount of pressure exerted during its creation make some slates more brittle and durable than others. Slates with iron sulfides of pyrite and pyrrhotite, and carbonates are not stable under thermal expansion. Oxidation of iron sulfides can cause discoloration. Carbonates in slate can convert to gypsum, causing discoloration, expansion, or disintegration of the slate. Increased rain and temperature from climate change could lead to increased oxidation and damage to stones with impurities.

Stone Construction

Bedding

The importance of the bedding direction of sedimentary and metamorphic stone cannot be overstated. Bedding refers to the way that layers (beds) of sedimentary rock are oriented, first as they are formed and compressed under directional pressure, and later as they are set in masonry construction. Less dense bedding layers are formed by a change in material or pressure, and dictate the maximum size block that can be quarried intact.

Sedimentary and metamorphic stones are strongest when oriented in the same direction as found in the quarry, or with compressive loads perpendicular to the layers. Cornices and other projecting pieces are stronger with the layers turned vertically, and arches should be laid so that the bedding layers are parallel to the radius of the arch. Exposed bedding layers are more likely to delaminate from the pressure of water, ice, or salts forming behind the surface. However, the face of horizontal bedding layers of softer stones may wear unevenly.

Figure 2.3.4 Marble bedding, Danby Quarry, Vermont. (Photo by Susan Pranger, 2021.)

Laying Masonry

Stone is strong in compression but not in tension, which originally limited construction to vertical walls and short spans. The simplest stone walls were laid one or two stones deep with a thick layer of mortar. Compound or "coffin" walls were built with a core of earth or stone rubble and were between 2 and 15 feet thick depending on their height and the skill of the masons. Thick walls prevented moisture from penetrating to the interior. Projecting details, including overhangs, reglets, projecting lintels, caps, and water tables were also added to shed water before it could penetrate the wall.

Mortars, Plasters, and Washes

Porous sedimentary stones were protected in wet climates by a sacrificial layer of mud or lime, based on the premise that it was easier to replace the damaged coating periodically than to repair the underlying adobe, brick, or stone. Both coatings were flexible and breathable, protecting the underlying stone while allowing moisture and salts to escape.

The introduction of Ordinary Portland Cement (OPC) was both a boon and a curse to masonry construction. OPC was hard, impermeable, and could be mixed and applied by unskilled labor. However, OPC coatings did not expand and contract with moisture or temperature and were more likely to trap moisture in the wall. Differential movement led to cracks in both the mortar and in the adjacent masonry.

Drivers of Deterioration

Climate-Related Damage

Stone begins to change as soon as it is exposed to the atmosphere. It can deteriorate slowly or quickly through a complex interaction of biological forces, temperature, solar radiation, precipitation, cloud cover, relative humidity, fog, and wind speed.

Mechanical deterioration is caused by movement, expansion, or abrasion that physically disrupts and breaks the bonds between particles. Chemical reactions change the chemical composition of the stone, including dissolving the materials that bind components, or cause the accumulation of salt or water. The extent of damage depends on the character, structure, and cohesion of the stone. Deterioration can be accelerated by the stone's bedding direction relative to external loads, lack of compatibility with surrounding materials, exposure to weather and other conditions that influence the movement of air, water, and chemicals. Materials that are dissimilar in hardness, water absorption, thermal expansion, or chemical composition can cause mechanical stresses, trap moisture or salt, and encourage chemical changes. Climate and weather can cause a singular event of deterioration but will more often trigger a cycle of deterioration.

External Drivers	Type of Damage to Stone						
	Disaggregate	**Crack**	**Delaminate or Flake**	**Deposit Salt**	**Change chemical structure**	**Stain**	**Erode or Pitt**
Humidity/ Vapor	✓	✓	✓	✓	✓	✓	
Rain	✓			✓	✓	✓	✓
Ice	✓	✓	✓				
Rising/Falling damp	✓	✓	✓	✓	✓	✓	
Wet/Dry cycles	✓	✓	✓	✓	✓	✓	
Freeze/Thaw cycles	✓	✓	✓				
Excess load or impact, Vibration	✓	✓	✓				
Wind			✓	✓			✓
Solar /UV	✓	✓	✓	✓	✓		
Heat/ Temperature	✓	✓	✓	✓	✓		✓
Biological Growth		✓	✓			✓	
Salts	✓	✓	✓	✓		✓	✓
CO_2	✓		✓		✓	✓	✓
Acid	✓		✓		✓	✓	✓
Extreme Heat	✓	✓	✓	✓	✓	✓	
Earthquakes	✓	✓	✓				
Volcanic activity	✓	✓		✓	✓		

Figure 2.3.5 Table of external drivers of damage. (Table by Susan Pranger, 2023.)

Disaggregation is a breakdown of the binder, which can be dissolved or washed away by water or chemicals, or mechanically broken by stresses from salt crystallization, swelling clay, thermal expansion, wet-dry cycles, or freeze-thaw cycles. Repeated cycles of expansion and contraction weaken, then break, the bonds between particles. Disaggregation of the binder on the surface causes erosion, pitting, or flaking as the aggregates separate from the surface of the stone.

Delamination and flaking occur when the bonds are broken along the bedding layers in sedimentary and some metamorphic stones. It can be caused by incorrect bedding, an inherent weakness in the microstructure, erosion of the binders, or pressure from localized deposits of salt crystals or swelling clay.

Efflorescence on the surface is evidence of water and salts moving through the stone, unsightly but not always damaging. However, surface efflorescence can indicate that binders have been dissolved, or that damaging salt deposits (crypto fluorescence) are accumulating under the surface or within the stone where they can cause disaggregation, flaking, or delamination.

Staining is caused by materials deposited by rainwater or created by chemical changes within the stone. Rainwater and other moisture traveling through stone can deposit ferrous ions or copper patina from nearby anchors or metalwork or hydrate and oxidize ferrous iron and copper minerals that naturally occur in the stone.

Rust jacking occurs when expanding rust can exerts force on the stone and causes cracking, delamination, and loss of entire chunks of stone. Ferrous inserts, anchors, and reinforcement can expand as much as seven times its volume when exposed to water. A damaged stone in an otherwise sound wall could indicate a rusting hidden anchor. Anchors in traditional construction were often bedded in lead or in brimstone, however the smallest void in that protection can allow water to infiltrate and cause corrosion.

Water

Increases in humidity and rain, storms of longer duration, and frequent wet-dry cycles, especially when combined with tighter envelopes and impervious coatings, interfere with the ability of stone walls to dry and increase the odds that the stone will become fully saturated. Changes in humidity can increase the movement of water, chemicals, and soluble salts through stone and cause clay particles to swell and shrink. High humidity and fog increase the capillary action that draws moisture and related chemicals into the pores of stone and increases the potential for condensation, disaggregation of binders, and corrosion of ferrous materials. Condensation and rain deposit dirt, pollutants, and other chemicals that adhere to the surface or react with the chemicals in the stone or mortar.

Penetrating dampness and rising damp are both more likely to occur in porous sedimentary stones than less porous metamorphic or igneous stones. However, water can also migrate through mortar joints or more porous core materials and then gradually infiltrating pores and micro cracks and pores in less porous stone. Long-term saturation from rising or penetrating damp can lead to swelling of clay, disintegrate clay or calcium binders in sedimentary stones, and cause sugaring (separation of the crystals) of metamorphic stones. Coatings intended to exclude moisture from a wall can inadvertently trap dampness and dissolved salts that enter a wall assembly through cracks and rising or falling damp. The gradual accumulation of penetrating dampness can be more damaging than rainfall because it penetrates deeper into the wall and acts over a longer duration.

Ground water can absorb mineral salts from the earth and carry them up through capillary action into porous stone and into the mortar joints of all stone walls. When the water evaporates, the salts or chemicals are deposited in the pores or on the exterior surface of the stone. Salt deposits can attract water, creating a cycle that brings more water and more salt into the stone. This accumulation of water and crystalized salts can exert pressure on the structure of the stones, causing pores to expand and binders to swell, eventually causing layers of stone to crack, delaminate, spall, or flake apart.

Salt carried by rising damp is deposited based on their solubility. Salts that are not very soluble crystalize near the base of the wall, whereas highly soluble salts crystalize at higher elevation, creating efflorescence near the ground or high under the roof line. Salts may travel higher in the wall adjacent to pervious joints. The damage from rising damp is most severe near the ground, where moisture will attempt to escape horizontally along the path of least resistance. If water is prevented from evaporating horizontally, it will continue to migrate up the wall.

The damage from wet-dry and freeze-thaw cycles is exponential as each cycle opens cracks in the stone and enlarges pores, allowing more moisture to enter, more opportunities for moisture to linger, and salts and other damage to accumulate. Wet-dry cycles can also move unstable minerals in sandstone to the surface of the stone where they bind with dirt and pollutants on the surface, creating a crust on the surface of the stone. This process can block pores in the stone and create differential movement between the surface and underlying layers, eventually causing delamination.

Heat and Solar Radiation

Heat plays an important role in maintaining the equilibrium of moisture within stone. It can drive moisture out of stones and adjacent mortars and plasters. However, rising temperatures can also contribute to the movement of salts within or on the surface of the stone by changing the relative humidity and increasing the capacity of the air and water to hold salts in solution. Falling temperatures reduce solubility, causing salts to precipitate.

Diurnal temperature swings and the evaporative cooling effect of rainfall can cause sudden temperature changes on stone surfaces. Sunlight creates additional stress by heating exposed surfaces more quickly than concealed or shaded areas, causing differential expansion or movement of moisture.

Heat can expand trapped moisture, exerting powerful forces that can break the bonds between molecules of stone and cause disintegration or spalling. Thermal cycles can cause micro-cracks in the calcite grains of limestones and marbles and in the feldspar and biotite in granites. Water molecules can expand ten times more than granite; an increase of 30°F can exert 5,600 pounds of pressure on capillary walls of a dark granite with small pores. Microcracking can lead to sugaring, scaling, and flaking, and create pathways for chemical and biological damage.

High temperatures increase the likelihood and rate of chemical reactions that can change the composition of clay and increase acidic reactions in limestone. UVB light can increase the acidity in the atmosphere by triggering a series of reactions that convert sulfur to sulfuric acid, especially in the presence of ozone, oxidants and organic compounds found in smokestacks and other industrial urban environments. Sulfuric acid dissolves the $CaCO_3$ (calcite) in limestones and marbles and causes harmful gypsum crusts to form on the surface of the stone.

Chemical Damage

Carbonate dissolution – Karstification – is projected to account for 96% to 99.5% of the surface erosion of carbonate stones over the next century. Stone can be deteriorated by chemicals found in the air or water, in dissimilar materials, or within the stone itself. Dissolved minerals can migrate from an adjacent dissimilar stone, such as the migration of alkaline calcite from limestone or marble into granite. Ferrous minerals in stone can migrate with rising damp, leaving a light brown band on the face of the stone that is distinct at the bottom and gradually fades as it rises. Both acidic and alkaline chemicals can react with the minerals in stone to dissolve binders or create salt deposits. The extent of chemical deterioration is determined by the density of chemicals and the natural durability of the stone, and is generally increased with higher temperatures and humidity.

Salt crystals attract moisture, creating a cycle whereby salts in the stone attract more moisture, which brings more salt, which attracts more moisture, eventually causing the salt crystals to grow beyond the capacity of the pores. Low temperatures cause salts to precipitate. Rapid changes cause greater damage, as the fluids do not have time to dissipate through the pores in response to pressure.

Water becomes alkaline when it contains more OH^- ions than H+ ions. The alkalinity of puddles, streams, and lakes fluctuates as ions are added by rainwater and dry fallout. Alkaloid materials can migrate from alkaline water, cement mortar, concrete, or coatings to react with ferrous materials to create stains of ferric hydroxide or dissolve silica cement in quartz sandstone causing case hardening or flaking.

Unpolluted water is naturally slightly acidic due to its tendency to absorb carbon dioxide. Although carbonic acid is relatively weak, it combines with calcium and magnesium carbonates to create soluble bicarbonates that are more easily washed out of the stone, causing deterioration, pitting, and erosion of the stone surface of limestone and marbles. As water dries, the bicarbonates may be redeposited on the surface resulting in a powdery surface. Water can also simply wash away soluble clay binders, disintegrating sedimentary stones with high feldspar or silt content.

Acid also affects the binders in granite, by leaching Sodium (Na) and Calcium (Ca) salts from the boundaries of the harder grains. Acid also reacts with the feldspar to create kaolinite (white clay), and can react with micas to produce kaolinite, iron, potassium, and magnesium. The change from feldspar to kaolinite reduces the hardness of the binder from 6–6.5 on the Mohs scale to 2–2.5.

Sulfate Damage

Sulfates are mineral salts that occur naturally in stone and can form as pyrites (iron disulfide) on the surface of newly quarried stone. More destructive are the atmospheric pollutants sulfur dioxide (SO_2), sulfur trioxide (SO_3), and hydrogen sulfide

(H_2S), which combine with water to form various sulfurous and sulfuric acids. These acids can be deposited on stone as dry aerosols (minute particles suspended in the atmosphere) or wet solutions (acid rain). Corrosive acid fog can form when humidity is high and can be particularly damaging because it can penetrate the surface of the stone. London's foggy environment and early industrial era coal burning combined to provide the ideal environment to generate a black gypsum crust on its historic buildings.

Acids combine with $CaCO_3$ (calcite) in limestones and marbles to form a relatively insoluble calcium sulfite that reacts quickly with oxygen to form a much more soluble calcium sulfate dihydrate (gypsum). Gypsum is washed out of limestone by rain, leaving a clean but eroded surface, but will remain on surfaces protected from the rain, gradually trapping particulates in the air and forming a dark crust. This crust blisters and cracks over time, exposing the eroded limestone, and beginning another cycle of deterioration. In magnesium limestone, the reaction with acids produces magnesium sulfate (Epsom salt) which is even more soluble and causes dramatic erosion and cavernous decay.

Sulfites do not react with silica binders, but will react with calcite, iron oxide, or clay binders in sandstone. Unlike the gypsum crusts that accumulate in protected areas of limestone, crusts of dirt and acids accumulate on rain-washed surfaces of sandstone. The locations of dirt crusts on washed or unwashed surfaces can help identify whether the stone is limestone or sandstone.

Chloride Damage

Chlorides are initially mildly base (alkaline) but are easily converted to hydrochloric acid which dissolves carbonate rock. Chlorides occur naturally in desert dust and ocean spray in the atmosphere and can be deposited directly on the surface of stone or be absorbed by surface water, filter into ground water, and rise into buildings through capillary action. Although natural dusts and salts are more widespread, man-made salts from deicing salts and industrial pollution can be more damaging as the salt causes thermal shock by rapidly melting ice under the surface of the stone. Calcium chloride is very soluble and will keep porous stones wet and deteriorating for a longer period. Chlorides can cause severe corrosion of both natural impurities and man-made inserts of ferrous metals.

Sodium chlorides ($NaCl$) found in sea spray can accelerate the swelling of certain clays found in sedimentary stones when capillary action draws the salt spray into the stone, and the salt attracts more moisture. Cracks form when the wall dries and the clay shrinks, increasing water infiltration in each wet-dry cycle.

Electrochemical Reactions

Electrochemical reactions can cause corrosion of stones that contain minerals with iron, including ferrous-ferric oxides and ferrous-ferric silicate minerals. Galvanic action occurs when certain catalysts, including pyrite, marcasite, hematite, and magnetite, are exposed to an electrolyte such as water.

Biological Damage

Climate changes that increase humidity and moisture have the potential to promote biological growth and increase the related damage, especially in porous stones that retain moisture over a longer period. Diurnal temperature swings can cause condensation that promotes the growth of mold, fungi, algae, and bacteria.

Bacteria can excrete metabolic acids that etch or dissolve the stones and organic chelating agents that separate heavy metals into metals that are more water soluble. Aerobic organisms produce carbon dioxide that increases the carbonic acid that dissolves the stone and produces salts. Bird droppings and nesting materials are broken down by bacteria that release sulfuric acids. The bacteria's mycelium enters the stone, transporting acids that dissolve stone and depositing salt crystals.

Fungi require a source of nutrition on the surface, but once established can penetrate deeply into the stone. Fungi produce organic and inorganic acids and chelating agents that dissolve and bind to calcium, silicon, iron, manganese, and magnesium so that they become water soluble and washed out of the stone.

Algae require an appropriate surface, sufficient temperature and humidity or water to be active but prefer low light levels. Photosynthetic organisms like microalgae can produce a phototropic, cyanobacterial film that can protect the surface of the stone, but most algae, lichen (a symbiotic partnership of fungi and algae) and mosses deteriorate stone by retaining water, or by producing organic acids and chelating agents. Some lichen and some algae penetrate the stone by dissolving carbonates in calcium rich limestones and dolomites or by extracting metallic cations.

Expanding roots of trees and other plants split rocks or trap moisture, providing opportunities for more biological growth and additional damage from water, acids, and salts. Moss causes less damage because they have rhizomes, not roots, and require a layer of soil to grow, however they are can penetrate and deteriorate mortar joints.

Wind

Even at lower speeds, wind and windborne particles can erode stone by scrubbing aggregates or binders from the exposed surface. Wind can increase evaporation, concentrating deposits of salt crystals. Wind creates pressure on the windward side of a wall and suction on the leeward side causing changes in pressure that can drive vapor, rain, and pollutants through stone. Wind can cause rapid drying of the stone, especially at corners where wind speed tends to be higher, causing crypto-florescence, salt deposits below the surface.

Earthquakes

Buildings avoid damage from the lateral movement of earthquakes by one of two mechanisms: either they are flexible, plastic enough to absorb the movement without failure, or they are rigid, strong enough to maintain their integrity. Massive masonry walls can withstand earthquakes because the mass is able to absorb the movement without failing or because flexible lime mortars can absorb some movement. Mechanical anchors that connect stones but allow some movement can allow a stone wall to absorb the vibration from an earthquake. However, the use of rigid mortars makes some assemblies less flexible and plastic, and unable to absorb seismic activity.

When the mortar and connections between the stones are rigid, or the stone is restrained by an external frame, the stress is transferred to the stones and can damage the stone or the bonds between stone and mortar. These cracks might not cause immediate failure of the wall, but they create a path for future weather-related damage. Subsequent deterioration in turn leaves the wall more vulnerable to future earthquake damage.

Strategies for Responding to Climate-Based Threats to Stone

Strategies for responding to climate threats that are similar for many different materials have been discussed in a previous chapter. Following are a few strategies specific to stone.

Inspection and Testing

Visual inspections are useful in identifying the type of stone, which will provide clues as to the type of deterioration most likely to occur under current and future conditions. However, visual inspections of stone have limitations; a sound surface can conceal damage directly below the surface or deeper within the stone. The presence of efflorescence, erosion, pitting, and cracks may suggest deeper damage.

There are many non-destructive tests to investigate the density of the stone or locate hidden flaws or anchors. Variations in density, heat, or moisture content can provide clues to subsurface conditions. For example, X-ray diffraction (XRD) was used to determine the minerology of a stone and clay binders at San Mateo church in Tarifa, Spain.

Destructive testing of stone should be avoided if the stone cannot be replaced. Even if identical stone is available, changes in the surface from sun and other weathering will often make a repair stand out. It may be possible to test a similar stone from the same quarry, a hidden face or fragments that have already broken away.

Figure 2.3.6 Selection of Rock Samples. (Photo and labels by Susan Pranger, 2023 using samples and descriptions from the "Classroom Rock Collection" by GeoScience Industries.)

Maintenance and Cleaning

Stone buildings need regular maintenance of mortar joints and traditional plaster coatings and washes. Attempts to reduce maintenance by using impervious paint or cementitious coatings have proven to be more damaging than no maintenance at all.

Removal of dirt or stains may be desired for aesthetic reasons, to remove previously applied coatings or remove harmful accumulations of salts, chemicals, or biological materials. Standard preservation practice is to use the least aggressive method to achieve the desired results without damaging the stone by introducing excessive water, chemicals, or abrasion. An experienced consultant and contractor should determine the appropriate materials, concentrations, and techniques. Testing should verify that the cleaning process will not damage or erode the original fabric.

A gentle water wash is often sufficient to remove dirt or gypsum crusts that have accumulated in areas protected by the rain. However, excess water can dissolve salts and redeposit them within the stone. Acidic and alkaline cleaners each have their uses should be used with care. Acidic cleaners react with the calcite in limestones and marbles, while alkaline cleaners can damage granite. Neutralizing agents may be required to balance the pH of the stone after it has been cleaned.

Inert poultices are often a good strategy for removing water-soluble salts on the surface of the stone. Sequestering agents chemically react with metallic stains caused by residue from copper or iron flashing or anchors, or from the corrosion of natural minerals in the stone so that they become soluble and can be removed by a poultice. Chemical poultices are used to remove contaminants that are not water soluble but can mobilize new staining material if used incorrectly, and can leave chemical residues if not properly neutralized.

Stone Repair

A minimal approach to repair of individual stones is often best, depending on the preservation goals. Eroded or cracked stone can be retained if it is stable and not admitting water. However, modifications maybe required to ensure the future survival of the overall structure.

Stone repair can take many forms:

- Coatings, that sit on the surface of the stone, and provide some protection or aesthetic purpose.
- Consolidants, that are absorbed into the existing material, to consolidate or strengthen.
- Patches, that use a mortar or other synthetic mixture that replaces missing material.
- Dutchman, that uses a matching stone to replace a part of a damaged stone.
- Replacement, that replaces an entire stone with a new stone or synthetic replacement.

Consolidants should only be considered where the material is continuing to deteriorate, and only after rigorous testing to determine the impact on the material's performance and appearance. When patching stone, it is best to use aggregates that duplicate those in the original stone, and to ensure that the final product is compatible with original material.

Where appropriate, deteriorated stone should be replaced with a dutchman of the same material and size, if not from the same quarry, then with a stone of comparable appearance (color and texture), chemical composition, strength, and porosity. Exceptions may be made where the stone is no longer available, or not available in the desired size, or when the original stone is not able to withstand current conditions. Weather conditions may dictate that alternative materials be considered where the original stone has proven to be inadequate and where the stone is in a particularly vulnerable location, that is, foundations, tops of walls, water tables, or other projecting stones.

Cement-based, synthetic or fiberglass substitute materials are often considered due to cost, ability to match detail, decreased weight or increased durability but may have different porosity, permeability, thermal expansion and strength, perform differently than the original, and create different conditions within an assembly. The pros and cons of these products are beyond the scope of this book, except to suggest a few criteria:

- The role of the weight in the structural performance of a wall should be considered before using lightweight replacements.
- Synthetic materials may react differently to cleaning, weather and age than adjacent natural stone.. Subtle difference in texture can cause rain to clean or discolor a synthetic repair differently than its natural neighbor. UV light can fade some synthetic materials.
- Replacements that are not inert might contain materials that migrate to, or absorb materials from, the adjacent stone or mortar.
- Synthetic materials can trap moisture or physically restrain original materials with different porosity or strength.

Mortar, Plaster, and Coatings

Traditional lime washes and plasters can protect porous stone while allowing vapor and salts to escape. There has been a movement to replace previous applications of impervious cementitious mortars and plaster with lime-based materials, especially in historically significant buildings, the increased cost and necessary skills required for lime coatings means that cementitious materials continue to be used.

Increases in rain volume, duration, and intensity will require increased maintenance of lime materials, but also are likely to increase failures of cementitious coatings.

Epoxy, Consolidants, Resins, Repellants

An ever-increasing number of chemical mixtures have been developed and marketed to seal, waterproof, consolidate or patch stone, and other masonry materials. The most common products are alkali silicates, alkoxysilanes, and a variety of synthetic resins, but acrylic co-polymers, epoxies, polyurethanes, and microcrystalline waxes have also been used. Given that technology and research continue to evolve, it is not feasible to provide a list of available products, let alone their pros and cons. Qualified professionals should be consulted when selecting and applying materials.

The use of fillers, consolidants and coatings may be appropriate where the alternative is total loss of the existing material. Modern consolidants have been successfully used, especially on sandstones that have lost their calcareous or argillaceous binders. However, extreme caution and appropriate research and testing are recommended. The following criteria should be considered:

- Consolidants can affect the porosity of stone and subsequent migration of salts. Materials that block moisture movement can cause more damage than they prevent.
- Transparent coatings can change the appearance of the stone. Pigments may fade or darken over time; pigments, emulsions, and solvents can migrate into and stain adjacent stone.
- Compatibility with the substrate is critical, and should consider strength, density, porosity, permeability, and chemical composition, as well as color and texture. Patches that are stronger than the supporting stone can stress and crack the original stone.
- The new material will only be as strong as the material it bonds to. If the substrate deteriorates or is stressed, the bond to the new material will fail.
- Treatments should be reversible or retreatable.

Interest in water repellants is increasing with the severity and duration of rain events but should be considered only after every effort is made to reduce exposure to rain through maintenance of roof, gutters, and other details to divert the rain. Repellants should block excess moisture from absorbent stone without trapping moisture. Most repellents are transparent but can darken stone or add a sheen. Some are sticky and will attract dirt. Water repellants may wear off after a few years or weather unevenly creating a blotchy appearance or concentrating capillary action at untreated areas.

Research into consolidation continues. There are promising developments in the use of "Bacterial carbonatogensis" or "biomineralization" to protect and repair limestone. Bacterial cultures, along with a nutritional solution, are injected or spread onto the limestone where it naturally produces $CaCO_3$. The mixture is reapplied to build up the desired thickness. While this process is currently prohibitively labor intensive and expensive for most current projects, it offers a window into the future.

Conclusions

The best strategy for protecting stone from an increasingly aggressive climate is to establish an equally aggressive maintenance program. Stone, for all its monumental qualities, is inconsistent and vulnerable, especially to chemical attack enabled by water, wind, and sun.

Bibliography

Abouzeid, Adam, David Channon, and Phil Sever. 2007. *Bird Damage to Historic Buildings*. Cathedral Communications Limited. https://www.buildingconservation.com/articles/birddamage/birddamage.htm.

Adham, A.K.M, and Akira Kobayashi. 2010. "The Effect of Global Warming on the Dissolution of Limestone." *Journal of Rainwater Catchment Systems* 15 (2): 1–9.

Ali, Harith E., Suhail A. Khattab, and Muzahim Al-Mukhtar. 2015. "*The Effect of Bio-deterioration by Bird Droppings on the Degradation of Stone Built.*" *Engineering Geology for Society and Territory* 8: 515–520.

Atchison, Bob. 2021. *Columns and Capitals of Hagia Sophia*. https://www.pallasweb.com/deesis/columns-and-capitals-of-hagia-sophia.html.

Bartoli, Flavia, Neil Thomas William Ellwood, Laura Bruno, Simona Ceschin, Lorenza Rugnini, and Giulia Caneva. 2019. "Ecological and Taxonomic Characterisation of Trentepohlia umbrina." *Annals of Microbiology* (Springer) 69 (Supplement 10): 1059–1070. https://doi.org/10.1007/s13213-019-01472-9.

Basu, sudenshna' Orr, Scott Allan, and Yasemin D. Aktas. 2020. "A Geological Perspective on Climate Change and Building Stone Deterioration in London: Implications for Urban Stone-Built Heritage Research and Management." *Atmosphere*. https://www.mdpi.com/2073-4433/11/8/788/htm.

Boynton, Robert S. 1985. *Chemistry and Technology of Limestone*. Hoboken: Wiley-Interscience.

Carmeliet, Jan, and Bert Blocken. 2004. "Rainwater Runoff for Evaluating Water Leakage Risks in Building Envelopes." *Oak Ridge National Laboratory*. http://web.ornl.gov/sci/buildings/2012/2004%20B9%20papers/196_Carmeliet.pdf.

Cartwright, Mark. 2018. *Byzantine Architecture*. https://www.worldhistory.org/Byzantine_Architecture/.

Encyclopedia Iranica Foundation. 2021. *Construction Materials and Techniques in Persian Architecture*. https://iranicaonline.org/articles/construction-materials-and-techniques-in-persian-architecture.

English Heritage. 2012. *Stone*. Edited by Alison Henry David Odgers. Surrey: Ashgate Publishing.

Feiereisel, Amy. 2018. *North Country at Work: How the Scottish Came to Hammond*. June 18. https://www.northcountrypublicradio.org/news/story/36343/20180612/north-country-at-work-how-the-scottish-came-to-hammond.

Friedman, Donald. 2010. *Historical Building Construction*. Second. New York: W. W. Norton & Company, Inc.

GeoScience Industries. 2019. *Classroom Rock Collection. GeoScience Industries, Fort Collins, CO.*

Griffin, P. S., N. Indictor, and R. J. Koestler. 1991. "The Biodeterioration of Stone: a Review of Deterioration Mechanisms, Conservation Case Histories, and Treatment." *International Biodeterioration* (Elsevier Science Publishers Ltd, England): 187–207. https://repository.si.edu/bitstream/handle/10088/43127/mci32104.pdf?sequence=1&isAllowed=y.

Gwilt, Joseph. 1842. *The Encyclopedia of Architecture*. London: Crown Publishers.

Hagia Sophia. 2021. https://www.hagiasophia.com/.

Hall, Mathew Robert. 2017. *The Mechanisms of Moisture Ingress and Migration in Rammed Earth Walls*. Thesis, Ann Arbor: ProQuest.

HNE. 2019. *Climate Change and Gutters*. Property Care White Papers, Portland Maine: Historic New England. https://s3.amazonaws.com/hne-assets/wp-content/uploads/2019/03/06161642/Gutters-and-Downspouts-03-Climate-Change-2019-v5.pdf.

U.S. Department of Labor. n.d. *Hydrogen Sulfide*. https://www.osha.gov/hydrogen-sulfide.

IMA-NA. 2021. *What Is Calcium Carbonate?* https://www.ima-na.org/page/what_is_calcium_carb.

Jenkins, Joseph C. 2016. *The Slate Roof Bible*. White River Junction: Chelsea Green Publishing.

Kielmas, Maria. 2018. *Weathering Process of Granite*. https://sciencing.com/info-8604681-weathering-process-granite.html.

King, Hobart. 2021. *Igneous Rock; Minerals:* https://geology.com/rocks/gabbro.shtml.

Klemm, A., and D. Wiggins. 2016. "Sustainability of Natural Stone as a Construction Material." Chapter 12 in *Sustainability of Construction Materials*, edited by Jamal M. Khatib, 283–308. Woodhead Publishing. https://doi.org/10.1016/B978-0-08-100370-1.00012-3.

Lee, Seung Yeop, Soo Jin Kim, and Min Hoon Baik. 2007. "Chemical Weathering of Granite Under Acid Rainfall Environment, Korea." *Environmental Geology* (Springer) 55: 853–862.

Lstiburek, Joseph. 2014. *BSI-011 Carpillarity- Small Sacrifices*. Buidling Science. www.buildingscience.com/documents/insights/bsi-011-Carpillarity-SmallSacrifices.

Lumencandela. n.d. *The Mughal Period*. https://courses.lumenlearning.com/boundless-arthistory/chapter/the-mughal-period/.

McDowell, Katherine Ann. 1997. *Characterizations and Conditions Assessment of the Sacristy Window Mission San Jose y San Miguel de Aguayo San Antonio, Texas*. Master's Thesis, Philadelphia, PA: University of Pennsylvania, Scholarly Commons. https://repository.upenn.edu/hp_theses/377.

Moore, John T. 2011. *Chemistry for Dummies*. Indianapolis: Wiley Publishing, Inc.

Morrison, Hugh. 1952. *Early American Architecture*. New York: Dover Publications.

Perazzo, Peggy B. Perazzo and George. 2021. *Structures and Monuments in Which Florida Stone was Used*. https://quarriesandbeyond.org/states/fl/florida.html#other.

Ressler, Stephen. 2013. *From Quarry to Temple - Building in Stone*. Produced by The Great courses. Kanopy. https://boston.kanopy.com/video/quarry-temple-building-stone.

Ringenbach, Paul T. 2014. *San Antonio Missions, Nomination to the World Heritage List by the USA*. Los Compadres de San Antonio Missions National Historic Park, San Antonio: UNESCO. https://whc.unesco.org/en/list/1466/documents/.

Roemer, Ferdinand. 1995 *Roemer's Texas 1845-1847 Eakin Press* ISBN 10: 1571680403 ISBN 13: 9781571680402

Sarpun, I. H., Unal, R., & Tuncel, S. (2006). Grain Size Determination in Marbles by Ultrasonic Techniques. European Conference for Non-Destructive Testing. Berlin.

Sebastian, Eduardo, Giuseppe Cutrone, David Benavente, Lucia Linares Fernandex, Kerstin Elert, and Carlos Rodriguez-Navarro. 2008. "Swelling Damage in Clay-Rich Sandstones Used in the Church of San Mateo in Tarifa, Spain." *Journal of Cultural Heritage* (Science Direct) 66–76. https://doi.org/10.1016/j.culher.2007.09.002.

Sellards, E.H. 2014. "Limestone, Mineral Industries and Resources of Florida." Sixth Annual Report of the Florida State Geological Survey. https://quarriesandbeyond.org/states/fl/florida.html#other.

Short, Nicholas M. 2006. *Some Basic Concepts Underlying the Science of Geology*. February 20. https://web.archive.org/web/20060220043153/http://rst.gsfc.nasa.gov/Sect2/Sect2_1a.html.

Siegesmund, S., K. Ullemeyer, T. Weiss, and E. K. Tschegg. 2000. "Physical Weathering of Marbles Caused by Anisotropic Thermal Expansion." *International Journal Earth Sciences*, June: 170–182. https://doi.org/10.1007/s005310050324.

Stubbs, John H. 2009. *Time Honored, A Global View of Architectural Conservation*. Hoboken: John Wiley & Sons, Inc.

UNESCO. n.d. *Chaco Culture*. https://whc.unesco.org/en/list/353/.

Vitruvius. n.d. *The Ten Books of Architecture, Book*. Vols. Book II, Chapter VII.

Vivian, Gwenn, and Bruce Hilpert. 2012. *The Chaco Handbook: An Encyclopedic Guide*. Salt Lake City: The University of Utah Press.

Weaver, Martin E, and F. G. Matero. 1997. *Conserving Buildings: A Manual of Techniques and Materials*. New York: John Wiley & Sons, Inc.

Weber, Tomas. 2021. "Could bacteria help save historic buildings from environmental damage?" *www.dw.com*. June 14. https://www.dw.com/en/could-bacteria-help-save-historic-buildings-from-environmental-damage/a-57767479.

Wikipedia. 2021. *Tufa*. May 20. https://en.wikipedia.org/wiki/Tufa.

Winkler, E. M. 1994. *Stone in Architecture, Properties, Durability*. New York: Springer-Verlag.

Woolfitt, Catherine, and Graham Abrey. 2000. *Poultices, The True or Plain Poultice and the Cleaning and Desalination of Historic Masonry and Sculpture*. https://www.buildingconservation.com/articles/poultices/poultice.htm.

World History Encyclopedia. 2021. *Roman Architecture Timeline*. https://www.worldhistory.org/timeline/Roman_Architecture/.

Yao, Lei, Xiaolong Fan, Chao Yan, Theo Kurtén, Kaspar R. Daellenbach, Chang Li, Yonghong Wang, Yishuo Guo, Lubna Dada, Matti P. Rissanen, Jing Cai, Yee Jun Tham, Qiaozhi Zha, Shaojun Zhang, Wei Du, Miao Yu, Feixue Zheng, Ying Zhou, Jenni Kontkanen, and C Tommy. 2020. "Unprecedented Ambient Sulfur Trioxide (SO_3) Detection: Possible." *Environmental Science & Technology Letters* 7 (11): 809–818. https://doi.org/10.1021/acs.estlett.0c00615.

CHAPTER 2.4

Earth, Clay, Brick, and Terracotta

Introduction to Earth, Brick, and Terracotta

The terms "loam," "earth," and "mud" all apply to soils used for earth construction. The structures made from earth are as diverse as the earth itself: rammed earthworks, mud and wattle walls, sun dried cob, adobe blocks, fired clay bricks, and terra cotta are only a few. The strengths and vulnerabilities of sun dried earth, brick, and terracotta in the face of weather have led to traditions of repair and maintenance that extend the life of an otherwise fragile building material.

David Easton aptly stated that "Soil is so diverse, its properties and reactions so widely variable, that to fully understand its uses and limitations, and to build with it successfully, takes years of study and experience. Civil engineers understand density and moisture content. Agronomists comprehend porosity and tilth. Geologists speak of composition and classification.... Earth builders must know it all. Take your time."

The Past, Present, and Future of Earth, Brick, and Terracotta

History

Over the past 11,000 years, many techniques were developed to build structures from soil. Although earth construction is most often associated with the hot dry climates of the American Southwest and the Middle East, it has also been used in cold, wet climates of Canada and Europe, and in the hot, wet climates of South America and Africa. Brick has been produced for 6,000 years wherever firewood was also available.

DOI: 10.4324/9781003160489-9

The earliest builders used locally available soil but were quick to understand that clay had special qualities that allowed it to hold its shape when mixed with gravel, sand, silt, and fibers. Evidence of rammed earth structures date to 7,000 BCE, mud bricks and adobe date to at least 6,000 years BCE. Earth construction found in Mexico, Central America, and South America date to 1,500 BCE, including homes by Pueblo Indians that are still standing today.

Mud and wattle and rammed earth require were used during the Medieval period in Europe and spread to cities as migration created a demand for construction less expensive than stone and less flammable than wood. Earth construction was popular in the United States and Canada through the 19th century for homesteads, farm buildings and high style buildings, but fell from favor during the Industrial Revolution due to competition from cheaper materials and the misconception that earth construction was inferior. Rammed earth and adobe experienced a resurgence in the 20th and 21st century in industrial countries including Canada, the United States, Germany, and Australia because of its low environmental impact and a growing desire to be self-sufficient. Earth construction continues to be used in developing countries because it uses local materials and does not require skilled labor.

Clay bricks have been fired to increase strength and durability since about 3,500 BCE and was well established as a craft by the time of Nebuchadnezzar in 604–562 BCE. Brick traditions spread from Mesopotamia to Persia, Indian, China, Egypt, Asia Minor, Greece, and Italy. Romans spread the use of brick throughout Europe. Techniques for adobe and fired brick were spread across the Americas by the Spanish in in the 16th century and the English in the 17th century.

Terra cotta, a fired cast clay product, was introduced in the late 19th century as a less expensive and more lightweight alternative to carved stone and later became a popular cladding for the steel skeleton structure of early skyscrapers. Production declined during and after World War II but saw a small revival with demand from the preservation movement in the 1970s.

Clay mortars were used well into the 1800s until lime mortars became more available. Both clay and lime mortars were replaced by natural cement and Ordinary Portland Cement (OPC) mortars in the early 20th century.

Current Social, Cultural, and Climatic Impacts

Due to ease of construction and readily available materials, earthen houses are still used by one-third to one-half of the world's population, primarily in developing countries. Similar techniques are popular in the United States and other wealthy countries in recognition of their sustainability.

Misperceptions about its durability, a lack of knowledge and standards, and the lack of applicable building codes are obstacles to proper maintenance and new construction. However, research and education have accelerated in the last few decades and several countries including Germany, Australia, and Spain have issued building codes for building with earth. The CRATerre in France and the Building Research Laboratory in Germany promote earth construction through educational projects and workshops. Most of these programs focus on new construction, however the research will inevitably lead to better maintenance of existing earth buildings.

Still, natural disasters and recurring rain events have become more frequent and severe, requiring more frequent maintenance and repairs, and raising doubts about the safety and durability of earthen construction. Gradual deterioration, the need for constant maintenance and the traumatic loss of buildings reinforces the perception that earth houses are inferior structures that cannot withstand rain and storms. Uncertainty about the climate can influence attitudes more than the actual climate conditions.

Earth, brick, and terracotta construction is already adapted to a wide range of climatic conditions, and uniquely suited to manage the projected increases in temperature and humidity. Brick and terra cotta continue to be perceived as strong and long-lasting materials, but historic bricks can be damaged by the introduction of modern coatings or systems intended to reduce maintenance or improve comfort and efficiency. However, changing weather patterns have the potential to create new problems and aggravate existing ones. The impact of climate on fired and unfired earth materials will be even more varied because of the variability of materials and techniques. Many structures depend on a traditional cycle of repair and renewal that no longer exists or is no longer sufficient to address more aggressive weather conditions.

Understanding Earth and Clay

The Geology of Soil

In nature, soil includes a layer of topsoil over layers of subsoil, over bedrock. Healthy topsoil is a mix of organic and inorganic particles that hosts a thriving ecology of plants, animals, bacteria, fungi, air, and water, essential to the health of our planet. However, topsoil is not a good building material because it constantly changes in character and volume as organic matter grows, dies, and decay. Unlike topsoil, subsoil is a relatively stable mix of inorganic particles of clay, silt, sand, gravel, and minerals.

The strength, porosity, and permeability of subsoil are determined by the types of inorganic particles and whether they are homogenous or were blended with particles of different sizes or minerals. Soil particles are classified by their size and mineral type as granular (sand and gravel) or fines (clay and silt). Larger boulders, cobbles, or fragments may also be present. Materials may be sharp edged if formed by erosion or rounded if washed by water.

Gravel can be fine, medium, or coarse, ranging in size from 2 to 80 mm. It can be a uniform in character, weathered in place from local bedrock, or it can be a mix of minerals and rocks transported by glacier or water. Sand particles are usually fine to coarse, 0.06 to 2 mm quartz, but may also contain feldspar, calcite, mica, and small amounts of metallic oxides and sulfates. Silt is made of finely ground quartz and feldspar minerals in the 0.002 to 0.006 mm range, slightly larger than fine clay, but smaller than sand. Water affects the workability, cohesiveness, and durability of soils.

Water is driven through the pores of the material either by diffusion (changes in moisture concentration) or by suction (capillary action). When soil is in a solid or plastic state, a mix of water and air coexist in the voids between particles and bind the soil through suction. Earth soils with a large proportion of silt and clay have narrow pores that can create a powerful capillary action sufficient to overcome gravity and draw moisture up from the ground below. In the liquid state, the pores in the soil are saturated with water, and can push the particles apart, for a total loss of cohesion.

Clay has unique qualities that facilitate both chemical and mechanical bonding of clay to itself and to other materials in soil. The more clay in a mix, the more cohesive it will be. However, a mix with too much clay will shrink as it dries, while a mix with too little clay will lack workability and strength. Cohesion also requires a careful balance of particle types and sizes and appropriate mixing and compaction. Coarse, granular materials like gravel, sand, or silt will not bind together without the addition of cohesive materials like clay. Although silt particles are similar in size to clay, they lack clay's unique ability to bond with water and other materials. A soil with too much silt will absorb excess water and be susceptible to damage from weathering and freeze-thaw action. Sandy soils are cohesive over a small range of water content and lose their strength as water is added.

Qualities of Clay

Although the term "clay" is often applied to any mineral smaller in size than 0.002 mm in diameter, clay particles are minerals of the group phyllosilicate, formed by the weathering of the hydrous aluminous silicates, such as feldspar. Crystals of clay do not break down into other minerals and have a hexagonal lamellar structure and high surface area that allow the crystals to bind with themselves and other materials. The surface area of the lamellar crystalline structure determines the internal forces of cohesion, binding force, compressive strength, and tensile strength, as well as the amount of expansion in water.

There are three types of water in clays and soils: bound water, absorbed water, and capillary water. Bound water is chemically bound to the particles of minerals, but released at temperatures over 400°C to 900°C. Absorbed water is electrically bound to clay minerals, but released at temperatures above 105°C. Capillary (pore) water moves through the pores of the material, and is released at temperatures above 105°C.

Clay expands and contracts as water moves through it, drawn in by capillary action or changes in relative humidity. Even when clay appears dry, water molecules remain in the tiny pores between the clay crystals, forming a membrane that bridges between clay crystals and draws the crystals together. This strong attraction creates a cohesion that allows clay to deform without cracking. However too much water will fill the voids, push the crystals apart and destroy the cohesion.

As water evaporates, the crystals of clay compact and the voids between particles shrink. This contraction can cause shrinkage cracks. The extent of expansion and shrinkage depends on the type and quantity of clay, the distribution of silt and sand, and the amount of water. Cracking is generally managed by using the minimum amount of water required to maintain plasticity, or by the addition of sand or fibers to increase workability without adding water. This balance of water and clay is determined by the size, shape, and number of voids in the soil, which is determined by the size, shape, proportion, and distribution of clay, silt, and gravel particles, and by how the mix is compacted.

The three main categories of clay are kaolinite, smectite, and illite. They each have different lamellar structures determined by their mineral content and how they were geologically formed that make them appropriate for different uses. Smectite clays, including montmorillonite and bentonite, are "swelling" or "expansive" clays that absorb water within the building blocks of the crystal, causing the crystals themselves to swell. They have more compressive strength than other clays, but will shrink and crack as they dry, which will reduce water resistance of the final product. Swelling class is usually not suitable for building construction. Kaolinite is the least expansive but also less cohesive. Illite swells more than kaolinite, but less than smectites.

Earth as a Building Material

A building material must be flexible (plastic) enough to shape, cohesive enough to hold that shape, strong enough to resist compression and gravity, and stable enough to remain intact and in place as it cures. The final product must withstand local weather conditions. Clay-rich loams maintain their strength and shape indefinitely if protected from excessive moisture, excessive drying, and erosion. Clay mortars adhere well to other clay substrates unless the swelling properties are different, which is rarely a problem if they are from the same source. Silt, sand, and gravel all lack the cohesive binding qualities of clay but help reduce shrinkage by reducing the amount of clay. A mix of particle sizes allows the mix to be compressed to the density required for compressive strength while maintaining the voids required for movement of water.

The consistency, density, strength, and durability of earth materials, and the rate of capillary action, will vary with the proportion of organics, clay, silt, sand, gravel, and water but also heavily depend on the techniques used to mix, compact, place, and cure (dry or fire) those materials.

Unfired earth mixes are not resistant to excess water because they absorb moisture more quickly than other building materials, however they can absorb humidity without becoming wet or losing stability. The speed of capillary action depends upon the construction method and the mix of materials.

Unfired earth products:

- are reusable and can be recycled indefinitely.
- can absorb humidity without becoming wet or unstable.
- have a low moisture equilibrium that can conserve timber elements by keeping them dry.
- can absorb and bind pollutants in water and might absorb pollutants in the air.

Brick and terracotta vary in porosity and permeability depending on the type of clay, how they have been fired, and whether they are glazed. The hotter and more consistent the firing process, the harder and more impervious the material will become. Absorption can vary from 2% in highly fired clinker brick to 25% in bricks with high calcium carbonate. Surface glazes can make the surface less porous, but moisture can still enter the material through cracks in the glaze, joints or from the unglazed surfaces. Fired brick and terracotta are more water resistant, but are still porous and vulnerable to frost, freeze-thaw cycles, or salt damage.

Fired earth products:

- retain their chemical composition after grinding.
- are more durable in wet conditions.
- can be used as pozzolanic material in lime and cement mixes.

Inexpensive, locally available plant and animal fibers were traditionally added to "wet" mixes to improve workability. Fibers do not improve strength but can distribute tension as the mix dries and shrinks, so that cracks are small and evenly distributed. Fiber can also reduce surface erosion and freeze-thaw damage. Porous plant fibers like straw, reeds, seaweed and cork, and light weight aggregates like pumice, lava, expanded clay or perlite, were added to clay mixtures to increase its porosity and insulating quality. Cellulose from manure enhances binding and cellulose fibers act as reinforcement. In addition to clay and carbonites, lime, iron oxides, salts, water, and organic material, such as animal manure, milk, whey, urine, bull's blood, were historically added to bond or stabilize earth mixtures.

Portland cement (OPC) and natural hydraulic lime are added today to stabilize clay and soil mixes, and reduce cracking, especially in rammed earth, but too little or too much can weaken the mix. Added lime or OPC should cement the larger particles together without filling the pores. Lime works best to bind Montmorillonite clay, while cement is more effective with kaolinite. A small amount of lime can make the wall weaker than no lime at all by increasing the optimal water content and lowering density. Pozzolana, a natural volcanic cement, can be added with lime to achieve additional strength and reduce the amount of lime.

A small amount of cement increases water absorption and may improve resistance to surface abrasion and rain. Tests in England confirmed that adding 3% cement (by mass) to rammed earth increased the porosity by dispersing the clay particles and increasing the size of the pores. Lower percentages (2% to 4%) interfere with the binding action of clay and can reduce the compressive strength while percentages higher than 6% reduce porosity to its original level. Australia and United States typically use about 6% in modern rammed earth construction. Cement, unlike lime, will start to set up as soon as it is added so the mix should be used immediately. A delay of several hours can reduce the compressive strength by as much as 50%. Mixes with cement should be protected from sun and wind or sprinkled with water to slow the drying process.

Methods of Earth Construction

Unfired earth construction includes "dry" and "wet" methods. Dry methods use less water and clay, and are compacted by ramming. Wet methods have higher water and clay content and use a variety of methods to compact and place the material. Fired adobe, brick, and terracotta begin with a "wet" method before being fired. Each method requires a different mix and compaction method to achieve the best balance of workability, binding strength, and compressive strength.

Compaction increases density and reduces water in the voids, increasing the suction and attraction between particles. The workability of the mix and the amount of compaction will depend on the moisture content and on the size and grading of particles. Too little water will cause excessive friction between the particles while too much water will saturate the voids. The percentage of other materials – silt, sand, and gravel – will depend on the specific method of construction.

Comparison of Clay Content in Wet and Dry Construction Methods				
Construction Method	**Moisture content**	**Clay Content**	**Linear Shrinkage**	**Density**
Wet methods	18 to 30%	10 to 30%	3 to 12%	1500 to 2000 kg/cubic meter (Cob wall depending on sand and gravel)
Dry method (Rammed Earth, Compressed soil blocks)	12 to 16%	7 to 15%	0.4 to 2%	1900 to 2200 kg/cubic meter
Fired Brick or Terra Cotta (wet)	7 to 30%	20 to 30%	2 to 4% at drying and 2.5 to 4% at firing	2000 kg/cubic meter

Figure 2.4.1 Comparison of clay content in wet and dry construction methods. (Table by Susan Pranger, 2023.)

Dry Method: Rammed Construction

The "dry" method is used for rammed construction or compressed soil blocks. Walls are built up in layers compacted in a formwork and briefly dried to accommodate shrinkage before adding the next layer. The earth is traditionally rammed by hand, a technique that continues today wherever labor is available. Electric or pneumatic ramming was introduced in the mid-20th century. A similar mix is compacted in forms to create earth blocks that are laid with a mud or clay mortar. Rammed earth shrinks much less than wet methods because of the reduced clay and moisture. The walls are more monolithic and more densely compacted than other methods, which results in fewer joints and a longer life.

"Wet" Methods: COB, Mud and Wattle, Adobe, Bricks

The "wet" method is used for all other types of earth construction, including extruded or bagged loam, cob, mud and wattle, mud or clay bricks or blocks, and adobe. The method of working wet loam varies widely based on local skills and traditions. The stronger the force used to throw loam into forms or frames, the better the compaction and strength after drying. Sudden impact is more effective than gradual pressure in reducing pore size and aligning the crystals of clay.

Clay particles are loosened by additional kneading, slicing, treading animals, or manual stamping. In colder climates, the raw material is allowed to freeze and thaw over the winter. After mixing with other ingredients, the clay mix is compacted by stamping or ramming with hand or machine tools or throwing the mix against a wall or into a form. Although these methods are still common today, large-scale operations will use mixing machines with rollers or paddles.

Shrinkage is a major issue with all wet methods. Shrinkage is controlled first by minimizing water in the mix, second by adding fibers, and third by reducing the length of the building elements (blocks or larger elements) and finally by increasing the drying time. Letting the wet mix cure for 12 to 48 hours before shaping can also improve the binding forces.

Adobe Blocks

Traditional adobe blocks are made from a wet but stiff and workable mix that is shaped by hand or thrown and pressed into formwork and then dried in the sun or lightly fired. Given their handmade nature, adobe blocks come in many sizes and shapes, limited only by what can be easily handled by a single worker, or at most, two men. Adobe blocks will stick to each other without mortar if they are soaked in water before they are stacked, however mortar also allows more adjustment and alignment of the courses.

Drying time was controlled by turning blocks upright, shading from direct sunlight and wind so that the loam mix dries slowly and evenly. External loading of "green" blocks, blocks that have dried enough to stop shrinking but are still wet, can cause compaction or settlement. Mechanical presses developed in the 18th century used a drier mix that eliminated the need to dry the unfired adobe blocks before stacking. However, unless cement was added, these pressed blocks were not as strong as the hand thrown adobe.

Figure 2.4.2 Adobe block wall with lime plaster, San José de Tumacácori, Arizona. (Photo by Daniel Thomas, 2021.)

Cob

Cob is a stiff wet mix shaped by hand into its final configuration without the intermediate step of making blocks or bricks. The large mass will result in shrinkage cracks, which can be limited by reducing clay and water and increasing fibers or can be filled with wet mix after the cob dries. Expansion joints can control the location of shrinkage cracks.

A cob mix must be compacted by ramming, beating, pressing, or throwing before or while placing. A variety of traditional methods are still used in different countries. In India, a pasty mix is carried to the site and poured in layers that dry in the sun before the next layer is added. In Ghana, wet balls are formed, and then stacked and pressed into place to form circular walls that are later smoothed with plaster. In Yemen, clods are forcefully thrown in place and then beaten into place with a wood trowel. England has used a 2-man technique since the 15th century: one person forms large clods "as large as two fists" and throws the clods to a man standing on the base of the wall, who catches it with a pitchfork, throws the clog forcefully on the wall, and compacts the wall with his feet. The walls are sliced to finish the sides. In Germany, the straw loam is stacked with a pitchfork and compacted using feet or rams in layers; the sides and smoothed with a spade after drying.

Figure 2.4.3 Reconstruction of Pueblo cob structure, Coronado Historical Site, New Mexico. (Photo by Daniel Thomas, 2021.)

Wattle and Daub

In wattle and daub construction, wet loam is thrown onto a woven framework of small branches and twigs. If larger trees or lumber were available, a more substantial timber frame was infilled with compressed clods of wet loam or mud or a wattle and daub mesh. Both methods relied on a loam coating free of cracks to prevent deterioration of the underlying wood members although more substantial wood members might be left exposed.

Timber-framed construction has proven to be very resistant to earthquakes because the flexible wood framework absorbs vibrations. Failure of the mud infill does not compromise the integrity of the wood structure, and the lightweight construction limits damage when failure does occur.

Earth Roofs

Although most common in dry areas, such as the American southwest, earth roofs were also used in wet areas with seasonal maintenance. Flat and domed roofs were built with a primary structure of tree trunks or bamboo, infilled with a dense lattice of twigs and branches or leaves, and covered with a loam mix. Vaulted or domed roofs have also been constructed with adobe blocks, primarily in religious buildings in Europe but also in homes and other buildings in southern Europe, Asia, and Africa.

The top layer was carefully smoothed to eliminate cracks. In wet areas, loam roofs were sealed with a salty coating that dries slowly, reducing cracking. The coating swells when wet, closing cracks (similar to the way that swelling clays are used for waterproofing, or wood shingles on roofs swell to block rain). These roofs required maintenance – a new coating of clay or a sprinkling of additional salt or a wash of salt water as rain eventually washed away the salt.

Brick

The heat of firing permanently changes the chemical properties of clay in fired brick and terracotta. Absorbed water is first driven out of the clay in the "smoking period." As the temperature continues to rise, bound water is driven from the clay molecules, and mineral impurities are oxidized. At the highest heat the clay is vitrified, converting crystalline silicates to nanocrystalline glass. Unlike unfired clay or mud brick, fired brick cannot be returned to its previous plastic state by adding water, although the bonds between the inert fired particles can still be deteriorated or fractured by water, ice, fire, or salt crystals. The type of clay and duration and temperature of firing determine the extent of chemical change and the strength of the final product.

Fired brick is usually shaped by a wet method similar to those used for adobe blocks: by hand, in forms or extruded. Sand or water struck bricks are made from a sloppy mix that is cast into molds that have been coated with sand or water to prevent sticking. Wire cut brick uses a stiffer mix of ground clay or shale that is extruded through an auger machine into the cross section of a brick, and then wire cut to create individual bricks. The dry method uses a clay mixture that is nearly dry, and then forced under pressure into molds, similar to rammed earth construction. Both methods produce a green brick that must be dried before firing.

Bricks were hand stacked and fired in a large "clamp" or "stove kiln" until the late 19th century. Burning brick was a skilled art that required that brick and firewood were distributed evenly throughout the kiln. The iron content of the clay, consistency of the clay/sand mixture, the temperature of the fire, the proximity of the brick to the fire, and the cooling process were all variables that affected the finished color, hardness, strength, and porosity of the brick. Bricks became more uniform in size and consistency with the introduction of brick making machines in the 1830s. Soil was crushed, mixed, and extruded by rollers, sliced by wires, and dried in an oven before firing. Today this is accomplished by a computer-operated assembly line.

Terracotta

Terracotta is made of clay and grog mixed with sand and water, cast into molds, dried, and fired. Grog is refractory clay (a heat-resistant clay with a higher percentage of alumina) that acts like aggregate to control shrinkage and cracking. The lack of large aggregates gives terracotta a fine, dense, texture appropriate for casting ornamental work, roof and floor tiles, interior non-load bearing partitions and fire proofing. However, without the larger aggregate, terracotta is not as strong as brick, and is not load bearing.

The traditional method is to hand press the mix into a mold, but it can also be slipcast or ram-pressed into molds, extruded, or hand sculped. It can be fired unglazed or coated with clay slurry for a matt finish. Adding fluxes or coloring agents to the clay slurry will create a glass like vitreous glaze during firing. Exterior terra cotta was initially keyed into masonry walls and backfilled with brick and mortar and temporarily supported with simple bent bars as the backup mortar cured. Later more elaborate structural systems were used to provide gravity support, lateral support, or resist overturning of cantilevered pieces. Terracotta cladding for steel structures was left unfilled to reduce weight and stress and was secured with embedded ferrous fasteners and anchors. Stresses from differential movement caused cracks in the terracotta or mortar, admitted water into undrained cavities. Rusting anchors and freeze-thaw damage increase damage to the terracotta.

Figure 2.4.4 Broken terracotta cornice at Michigan Central Station, Detroit, Michigan. (Photo by Angela Wyrembelski, 2018.)

Mortars

Unfired adobe and fired brick or terracotta rely on mortar joints to adhere blocks to each other and to fill gaps and imperfections. Fired bricks and terracotta also rely on the joints to release moisture from the wall, and to accommodate thermal movement. Adobe walls and early brick walls were laid with a mortar mix of clay, sand, and water, sometimes with a small amount of lime to stiffen the mortar. OPC mortars are popular because they require less expertise to mix and set easily but they trap moisture and transfer the stress of thermal or moisture expansion into the adjacent clay masonry. As a result, the use of lime mortars or natural cement mortars is becoming more prevalent in restoration work.

Figure 2.4.5 Mortar damage to brick face at San José de Tumacacori. (Photo by Daniel Thomas, 2022.)

Plasters and Coatings

There is a long tradition of using coatings to fill cracks, protect the substrate, and provide a smooth appearance on masonry walls. Clay and lime were both considered sacrificial coatings that bore the brunt of the exterior weather, could absorb moisture and salt from the substrate, and be easily replaced or renewed. The desire to reduce maintenance and prevent water intrusion led to the use of OPC coatings in the early 20th century, and to elastomeric and other impervious coatings in the mid-20th century. It was not long before it became clear that these impervious coatings would trap water that could delaminate the coating and worst case, damage the substrate.

Loam plasters contained sand, silt, and just enough clay (5% to 12%) to bind the ingredients and make the mixture adhesive. Performance is linked not only to the proportion of clay, silt and water, but also to the grain sizes and distribution of the sand and the type of clay. The shrinkage is controlled by using sufficient sand and by including fibers of animal or human hair, coconut, or sisal, cut straw or hay. In cold climates, loam plaster must have a low vapor resistance, to allow moisture to migrate to the exterior, and must be flexible enough to accommodate thermal and moisture expansion. Loam construction should be protected from rain with a combination of a roof overhang, base plinth, and a good surface coating. The finished plaster can be stabilized with wash of a pervious lime or lime casein finish.

Clay plaster mixes must be properly worked and compacted. Substrates should be free of loose material, dry enough that it has stopped shrinking, but damp enough that the plaster can bond. Slapping the plaster onto the substrate improves penetration and binding. A rough texture is required on both the substrate and plaster for a good physical bond. Plastered edges should be rounded or protected with wood or brick corners to minimize damage.

Drivers of Deterioration

The main vulnerabilities of earth construction, including early brick and terracotta, are:

- the lack of standardization.
- the tendency to expand when exposed to moisture and shrink when drying.
- the lack of water resistance.
- a vulnerability to erosion, and cycles of wet-thaw, freeze-thaw, and salts

Changes in climate and weather will increase these threats. Heavier rains, more frequent storms, storms of longer duration and intensity will increase opportunities for clay masonry to absorb water and decrease opportunities for it to dry. Frequent temperature swings, cycles of wetting and drying, and freeze-thaw cycles increase expansion and contraction, water saturation and erosion, especially when impervious coatings trap moisture. Although firing improves the durability of brick and terracotta, their rigid structure and smaller pores are less resistant to frost, salt crystallization, and rust jacking.

Water

After their initial drying, unfired clay products are stable if they are protected from rain, frost, and ground water. Loam and hand-made adobe have large pores that can absorb water vapor without swelling and are generally frost resistant, although contact with water will cause them to swell and weaken, and repeated wet-dry or freeze-thaw cycles will break the bonds between clay particles.

Droplets of wind-driven rain will shatter on the surface of well-constructed, dense loam. Only small droplets will be absorbed into the pores, and additional penetration is prevented when the surface layer becomes saturated. However, prolonged rain, large pores, cracks, or other defects in the wall exponentially increase the likelihood that capillary or hydrostatic action will draw moisture deeper into the wall.

Heavy precipitation can increase surface water and rising damp, increasing settlement, erosion and cracks at the base of a wall. Rising damp can carry ground water and mineral salts by capillary action into both unfired loam and fired brick walls. When the water evaporates, salts are deposited in the pores or on the surface of the material. The salt tends to attract more moisture and accumulates in the pores, swelling and exerting pressure on the pore structure of fired masonry. Damage is most severe near the ground where rising water attempts to escape. If impervious coatings prevent evaporation, the salt-laden moisture will continue to migrate up the wall.

Many earth structures are in dry climates with extreme diurnal temperature swings that cause moisture to condense on cold surfaces. The evaporative cooling effect of rainfall can also cause sudden temperature changes to materials and lead to condensation. The high capillary action of unfired clay and permeable fired bricks normally carries excess moisture to the surface rather than condensing it within the wall, however, condensation is causing more damage than was previously seen due to both environmental and cultural changes. Airtight exteriors and heavy use increase humidity in kitchen and bathrooms, and higher interior temperatures increase the absolute humidity, even when the relative humidity is the same. Condensation freezing and expanding within the pores is usually accommodated in the large pores of unfired loam but can permanently damage the small rigid pores of brick and terracotta.

As areas experience heavier rains, longer and more frequent rains, materials are more likely to reach critical levels of saturation and less likely to dry out between events. At the same time, as areas get warmer, they may experience more cycles of freeze-thaw, where previously the construction may have stayed frozen for longer periods of time. Each cycle of freeze-thaw expansion increases the damage to the wall, opening larger and larger cracks.

Heat and Solar Radiation

Changes in ambient temperature, solar radiation, and mechanical heating and cooling can cause thermal changes and differential movement in materials. Changes in temperature drive the movement and expansion of moisture and salts. The combination of high temperatures and dry air can drive capillary and bound moisture from unfired clay and cause new shrinkage cracks. Repeated cycles can break bonds within fired and unfired materials and increase infiltration of rain and moisture.

Differential thermal movement causes stresses that crack masonry, joints, and coatings. Concrete and cement expand three times as much as loam or earth construction, and twice as much as lime mortar.

Coefficient of Linear Expansion	
Heavy loam	From 0.0043 to 0.0052 mm/m·K;
Mud brick masonry	Up to 0.0062 mm/m·K
Sandy mud mortar	up to 0.007 mm/m·K.
Soft lime mortar	0.005 mm/m·K
Strong cement mortar	0.010 mm/m·K
Concrete	0.010 mm/m·K

Figure 2.4.6 Coefficient of linear expansion. (Table by Susan Pranger, 2023.)

Efflorescence and Chemical Damage

Chemical damage to earth, clay, brick, and terracotta materials can generally be attributed to salt crystallization within the material or on the surface. While efflorescence on the surface of the masonry is unsightly, it can indicate a more serious problem – deposits of crypto florescence salts within the walls.

Acid rain can also cause efflorescence in damp climates by dissolving mortar joints to form calcium sulfate crystals that wash down the face of the brick.

Although typically attributed to poor workmanship, efflorescence is due to a complex combination of factors and is therefore difficult to diagnose and to remedy. Efflorescence is impacted by the porosity of the material, the lime content of the mortar, and the quality of the mortar joints. However, exposure to moisture during and after construction, and fluctuations in temperature and moisture can contribute to efflorescence. Salt forms at the border between saturation and evaporation and is initially deposited on the surface. However, the border between saturated and dried material gradually moves into the wall so that salt is deposited in the pores. Crypto florescence, the accumulation of salts within the pores attracts more moisture, which attracts more salt.

Wind

Heavy winds can cause changes in pressure, increase abrasion, and increase moisture penetration or increase drying. Changes in pressure between one side of a structure and another, and between the exterior and interior can drive or suck moisture, salts, and chemicals into a wall. Heavy winds combined with low humidity and high temperatures can damage unfired materials by removing moisture, causing shrinkage and cracking. Wind, especially when carrying sand, pebbles, or other debris, can abrade materials or damage coatings and unfired or soft fired materials.

Earthquakes

Earth and brick walls that have been damaged or cracked by more severe or prolonged rain events are more likely to suffer diagonal cracks, displacement or collapse from earthquake vibration. A study of the damage from the Northridge earthquake in California in 1994 found that adobe buildings that were well maintained suffered minimal damage compared to poorly maintained buildings with pre-existing water or earthquake damage.

Most earthquake damage to earth and brick construction is caused by horizontal movement that opens cracks that make the materials vulnerable to other types of deterioration. Failures are often due to other design flaws such as a sloping site, the structural system, or the layout of openings, rather than to an inherent weakness of the material.

Earth construction is common in the earthquake zones of the Equatorial and Pacific rings. Although they have a reputation of failing, earth buildings have a combination of mass and flexibility that can accommodate earthquake movement. Traditional construction techniques resist earthquake forces by using flat sites and plinths to separate earth walls from the ground, and curved walls or corners, vaults and domed roofs that brace walls and absorb movement. Interlocking blocks and ring beams, and vertical or horizontal wood reinforcement has been used in the original construction or added to reinforce existing structures.

Most earth and brick structures rely on the mass and cohesion of the wall, the interlocking of blocks and the rigidity of the overall structure to resist structural failure. The thickness of most earth construction helps it to absorb lateral movement and maintain stability and bearing capacity even when severely cracked. The thicker the wall, the more likely it is to withstand earthquake movement. Rammed earth houses in Argentina with walls that are 24 to 40 inches thick have withstood earthquakes for centuries.

Unfired earth structures within a wood framework are often ductile, flexible, and elastic enough to deform and absorb kinetic energy without failure. Although comparatively thin, both mud and wattle construction and timber framing infilled with loam, adobe, or cob are flexible enough to absorb movement without failure of the structure. Loss of infill does not lead to failure of the structure and can be easily repaired. Unreinforced brick and terracotta structures are more rigid and more fragile, relying on flexible clay or lime mortar joints to absorb movement.

Biological Damage

Earth construction is not hospitable to mold and fungus due to low acidity and low moisture content. Fungus prefers a pH 4.5 to 6.5, which is more acidic than the pH of 7 to 8.5 of earth (earth dug in industrial areas may be more acidic due to acid rain). Fungal growth needs 20% humidity, which is much higher than the 0.4% to 6% equilibrium moisture content of unfired earth. Earth materials can absorb interior humidity, keeping it below the level required to support mold and fungi in enclosed spaces. However, mold will grow on damp, shaded exterior surfaces of fired and unfired earth.

Unfired loam that contains lightweight straw will support fungal growth during construction and while it dries, especially in humid climates that prolong the drying period. Spores can be reactivated if the finished wall is later penetrated by rain, condensation, or surface water.

Mice and insects only survive in earth walls if there are gaps or voids or a high organic contact. Small insects can live in and attack low-density additives such as lightweight straw.

Figure 2.4.7 Lichen on brick, Spencer-Peirce-Little House. (Photo by Susan Pranger, 2023.)

Strategies for Responding to Climate-Based Threats

Managing moisture is the most important strategy to prevent damage from almost all drivers of deterioration. The risks of increased rain, humidity, wind, and UV are best managed with regular maintenance and protection.

Investigation

A visual inspection can identify missing sections, surface deterioration, or severe cracking, water damage or staining. With experience, the sound of tapping with a ball peen hammer can provide clues to the density of a brick or terracotta, or the presence of voids in a wall. The extent of decay, especially fungal rot, might not be visible on the surface and may require additional investigation or testing to determine density. Investigation of wood buried in exterior masonry walls can present special challenges both due to lack of access and to the potential for moisture or other damage to be concealed.

Unlike loam construction, brick and terra cotta cannot be broken into its component materials. However, it is possible to do both laboratory and field tests to identify its qualities of porosity and strength. Load testing can determine the amount of weight that an individual brick or wall can support. There are endless field and laboratory tests to determine the composition and character of unfired earth materials.

Stereo-photogrammetry and other three-dimensional imaging systems have been used to evaluate erosion on earth walls over time. The technology has the potential to evaluate the condition of existing buildings, and to aid in future repairs.

Protection of Earth Walls and Surfaces

Earth structures were traditionally made weather resistant through a mix of strategies that include construction details, the composition of the wall itself, and the treatment of the exterior wall surface. Water management strategies are especially important to protect porous earth materials. Of particular importance are strategies to prevent wicking of water into the material.

In wet locations, it is important to protect porous fired and unfired earth from rising damp (moisture wicking up from the ground below) as well as from rain and runoff or flooding at the ground level. Ideally, earth walls are built on a plinth of impervious material, traditionally stone, to separate it from the ground and protect it from splashing rain.

The exterior surfaces of unfired earth can be strengthened by adding a sacrificial coating, or by compacting the surface. Traditionally this was done by forcefully rubbing the surface with a trowel or smooth stones to consolidate and polish the surface. Oily or latex plant juice from sisal, agave, bananas, and *Euphorbia herea* combined with lime, double boiled linseed oil, or starch and molasses have each been effective in stabilizing walls, but can reduce vapor diffusion.

Veneers of shingles, planks, glazed tiles, or fired brick are frequently added to protect unfired or lightly fired earth materials. Ideally, an air cavity would collect and drain away any water that does infiltrate.

Maintenance, Repair, and Adaptation

Preventive maintenance that limits water intrusion has the most potential to protect both unfired and fired earth and clay construction. Lack of maintenance can result in cracks and flaws in the wall or gaps in mortar or plaster coatings that make the materials more vulnerable to weather and earthquake damage. Repairs should be as true to the original construction where possible. However, adaptation or the use of new materials or techniques may be appropriate where the original construction is flawed and failing, or no longer able to withstand the new weather conditions.

- Avoid adding coatings or paints that trap moisture in the wall.
- Avoid adding insulation and vapor barriers that trap moisture or interrupt drying.
- Consider sacrificial coatings of clay or lime plaster to absorb salts or weather.
- Consider traditional or modern additives to increase the durability of coatings or mortars.

It can be difficult to patch or repair unfired earth construction because wet patches may not bond well with dried construction. Where required, substitute materials should be similar to the original in appearance, flexibility, porosity, and strength. Patching brick and terracotta generally involves replacing individual units with new units that are similar in color, texture, and porosity using salvaged material or new materials fabricated to replicate the original. While it is tempting to use materials or additives that are stronger and more water resistant than the original, great care must be taken to avoid creating or transferring stresses to the original material. Fiberglass is a popular and lightweight substitute for complex terracotta shapes but can fade with UV exposure and repel dirt so that it often looks cleaner than adjacent terracotta. Concrete substitutes are often heavier and more rigid.

The structural loading should be considered. Heavy masonry parapets and cornices sometime serve a structural function of stabilizing arches below. A lightweight substitute may not perform in the same way.

Treatments and Coatings

In theory, weather-resistant paint can protect earth surfaces from rain, but in practice is likely to crack in response to movement of the underlying masonry. Paint can also trap moisture within the wall, causing swelling and deterioration from within. Effective paint barriers should repel water but allow trapped water vapor to escape. Flexible, permeable coatings of clay, lime, or linseed oil are preferred.

Many modern materials have been found to stabilize loam. Synthetic resins, paraffins, synthetic waxes, and latex have each been found to stabilize loam, but are expensive, vulnerable to UV degradation, and create a vapor barrier. Silicone-based moisture retarding admixtures have been used for rammed earth in Australia for new construction.

New coatings include water repellants that impregnate the loam surface, including silane and siloxanes, polysiloxanes (silicone resins), silconates, acrylic resins, silicate ester and silicates with hydrophobising additives. Because different products react differently chemically with the minerals in the loam, any coating should be tested. Coatings can reduce water uptake by up to 90%, but also reduce vapor diffusion. Modern "permeable" coatings have "micro-pores" that allow passage of water vapor and have the potential to protect surfaces without trapping moisture but should be tested and used with caution.

Earthquake Resistance

Extensive research has been conducted to develop strategies and building codes to retrofit historic adobe and unreinforced brick buildings for anticipated earthquake movement. Strategies to reinforce low strength adobe by introducing stiff members of steel or concrete, or coatings of cement have had mixed results, depending on the severity of the earthquake and the bond between rigid new members and flexible existing earth components.

A more successful strategy is to improve overall structural continuity, taking advantage of the inherent stability of mass adobe walls. Horizontal straps or cables that restrain rather than prevent movement of thick walls have effectively absorbed seismic energy and prevented collapse. While not invisible, straps are reversible and require less destruction of the original fabric. Vertical rods inserted with epoxy grout in the cores of adobe walls are more invasive but has some success in delaying or stopping the spread of cracks. Improved anchorage and connections between walls and roof or floor can prevent overturning.

. Fired brick and terracotta are inherently more rigid, and therefore more likely to successfully integrate new rigid seismic reinforcement than more flexible abode structures.

Modifications should be designed by consultants with a thorough understanding of the material properties.

Conclusions

Climate change presents risks to both earth construction and fired brick and terracotta, primarily because of the increase in severe rain and wind-driven rain but also due to changes in temperature that affect drying, humidity, wet-dry and freeze-thaw cycles. Earth and clay-based construction is especially vulnerable to the accompanying risks of rising damp, salt, and crypto florescence. These risks can be compounded by changes in performance caused by the addition of coatings, vapor barriers, insulation, and mechanical systems designed to reduce maintenance and improve comfort and energy performance. Although changes in the frequency and severity of earthquakes have not been confirmed, cracks caused by sun, wind, and rain can make this construction more vulnerable to damage from the movements caused by earthquake, and less able to withstand or absorb the movement.

Damage from the weather has traditionally been managed by maintenance – the regular replacement of coatings or mortar joints. Increased weathering will require more regular maintenance and consideration of additives to traditional materials to increase their durability.

Bibliography

Alex, Dorothee. 2018. "Recognition of Heritage in Danger: Rammed-Earth Architecture in Lyon City, France." *IOP Conference Series: Earth and Environmental Science*. Ho Chi Minh City: IOP Publishing. https://doi.org/10.1088/1755-1315/143/1/012054.

American Face Brick Association. 1922. *The Story of Brick; the Permanence, Beauty and Economy of the Face Brick House*. Chicago: American Face Brick Association.

APT. 2020. "*Conserving Earth and Timber Vernacular Architecture*" *workshop*. October 2. *APT & National Trust 2020 Joint Conference*. Edmonton.

Avrami, Erica, and Hubert Guillaud. 2008. "Terra Literature Review." Edited by Erica Avrami, Hubert Guillaud and Mary Hardy. Los Angeles: The Getty Conservation Institute.

Bosman, Gerhard, and Diaan van der Westhuizen. 2014. *The effects of climatic conditions on*. Review Article/Oorsigartikels. https://www.ajol.info/index.php/actas/article/download/109757/99506.

Boynton, Robert S., and Kenneth A Gutschick. 1966, reprinted 1990. *Masonry Mortar Technical Notes #4: Efflorescence of Masonry*. National Lime Association.

Brick Industry Association. 2006. *Technical Notes on Brick Construction 9*. https://www.gobrick.com/docs/default-source/read-research-documents/technicalnotes/9-manufacturing-of-brick.pdf?sfvrsn=0.

Brickarchitecture. 2017. *The history of Bricks and brickmaking*. https://brickarchitecture.com/about-brick/why-brick/the-history-of-bricks-brickmaking

Bui, Q. B., J. C. Morel, and B. V. Benkatarama Reddy. 2009. "Durability of rammed earth walls exposed for 20 years to natural weathering." *Building and Environment*.

Canada's Historic Places. n.d. "Sandards and Guidelines."

Carmeliet, Jan, and Bert Blocken. 2004. "Rainwater Runoff for Evaluating Water Leakage Risks in Building Envelopes." *Oak Ridge National Laboratory*. http://web.ornl.gov/sci/buildings/2012/2004%20B9%20papers/196_Carmeliet.pdf.

Ciancio, Daniela. 2011. "Use of Rammed Earth in Aboriginal Remove Communities of Wetern Australia." *International Workshop on Rammed Earth Materials and Sustatinable Structures & Hakka Tulou Forum 2011*.

Ciancio, Daniela, and Michael Boulter. 2012. "Stabilized Rammed Earth: A Case Study in Western Australia." *Engineering Sustainability*. https://doi.org/10.1680/ensu.10.00003.

CRAterre. 2021. *Earthen Architecture around the World*. http://craterre.org/?new_lang=en_GB.

Easton, David. 1998. "Using Air to Build Earth Walls." *Environmental Building News* 5 (5): 11–12.

Elder, Jeffrey L. 2009. "Efflorescence. Can it be Controlled." *Brick Architecture and Landscape, Summer 2009*. http://www.interstatebrick.com/TechData/Efflorescence%203%202009.pdf.

English Heritage. 2015. *Earth, Brick and Terracotta*. Edited by Iain McCaig, Clara Willett, Sophie Godfraind, John Stewart, Alison Henry. Vols. Earth, Brick & Terracotta. Surrey: Ashgate Publishing.

Freidman, Donald. 2010. *Historical Building Construction*. New York: W. W. Norton & Company.

Garvin, James L. 2001. *A Building History of Northern New England*. Lebanon: University Press of New England.

Gerns, Edward, and Rachel Will. n.d. *Number 15, Architectural Terra Cotta*. APT Practice Points, Association of Preservation Technology. https://www.apti.org/assets/docs/PracticePoints_15_v8_A_Layout2sm.pdf.

Ginell, William S., and E. Leroy Tolles. 2000. "Seismic Stabilization of Historic Adobe Structures." *Journal of the American Institute for Conservation* 39 (No. 1): 147–163. https://www.jstor.org/stable/3179971.

Hall, Mathew Robert. 2017. *The Mechanisms of Moisture Ingress and Migration in Rammed Earth Walls*. Thesis, Ann Arbor: ProQuest.

Historic England. 2015. *Earth, Brick and Terracotta*. Edited by Ianin McCaig, Clara Willett, Sophie Godfraind, John Stewart Alison Henry. Vols. Earth, Brick & Terracotta. Surrey: Ashgate Publishing.

HNE. 2019. *Climate Change and Gutters*. Property Care White Papers, Portland Maine: Historic New England. https://s3.amazonaws.com/hne-assets/wp-content/uploads/2019/03/06161642/Gutters-and-Downspouts-03-Climate-Change-2019-v5.pdf.

Horvath, Devina. 2020. "Adobe Structures in Earthquake Country: Balancing the Preservatoon of Traditional Buildng Methods with Safety in Southern California."

King, Hobart M., PhD, RPG. n.d. *Caliche*. https://geology.com/rocks/caliche.shtml.

Kleier, Gary. 2000–2002. *Brick Structures; Oh What Damage We Do Inflict*. http://www.oldlouisville.com/circa1900/brick-structures.htm.

Leick, Gwendolyn. 1988. *A Dictionary of Ancient Near Eastern Architecture*. London: Routledge.

Lstiburek, Joseph. 2014. *BSI-011 Carpillarity- Small Sacrifices*. Buidling Science. www.buildingscience.com/documents/insights/bsi-011-Carpillarity-SmallSacrifices.

MCAA. n.d. *History of Masonry*. http://www.masoncontractors.org/aboutmasonry/historyofmasonry/.

Minke, Gernot. 2007. "Building with Earth - 30 Years of research and development at the University of Kassel." *CESB 07 PRAGUE Conference*. https://www.irbnet.de/daten/iconda/CIB7808.pdf.

Minke, Gernot. 2006. *Building with Earth, Design and Technology of a Sustainable Architecture*. Basel, Birkhauser.

Minke, Gernot. 2001. *Construction Manual for Earthquake-Resistant Houses Built of Earth.* Eschborn: GATE-BASIN (Building Advisory Service and Information Network. https://www.preventionweb.net/files/5230_ManualMinke%5B1%5D.pdf.

Nabokov, Peter, and Robert Easton. 2011. *Native American Traditional Architecture.* Edited by David J. Wishart. University of Nebraska. http://plainshumanities.unl.edu/encyclopedia/doc/egp.arc.035.

Niroumand, Hamed, M.F.M. Zain, and Maslina Jamil. 2013. "Assessing the Critical Parameters on Earth Architecture and Earth Buildings as a Vernacular and Sustainable Architecture in Various Countries." *2nd Cyprus International Conference on Educational Research.* Elsevier Ltd.

NPS. 2019. "Measuring the Effects of Rainstorm Intensity on Adobe Walls." U.S. National Park Service. https://www.nps.gov/articles/sodn_adobe_test_walls.htm.

Pranger, Susan. 2014. "Moisture Mischief: Using WÃ¤rme und Feuchte instationÃ¤r (WUFI) to Predict Its Next Move." *New England Real Estate Journal.* October 21. https://nerej.com/78526

n.d. *Rammed Earth.* https://en.wikipedia.org/wiki/Rammed_earth.

Rezende, Prof. Dr. Marco Antonio Penida. 2020. "Understanding vernacular earth and timber architecture, Western Canada and South America." *APT & National Trust 2020 Joint Conference.* Edmonton.

Ritchie, T. 1955. "Studies of Efflorescence on Experimental Brickwork Piers and Produced on Ceramic Wicks by Masonry Mortars." *Journal of the American Ceramic Society 38*: 10.

Tschoepe, Ray. n.d. *The Short Course on Historic Mortar.* http://www.oldhousejournal.com/magazine/1484.

Virta, Robert L. n.d. "Clay." *Chemical Industry Applications of Industrial Minerals and Metals.*

Woods, P.E., and Amy Lamb. 2021. "Historic Mortar Replication and Pointing." *BSA Historic Resources webinar.* June 10.

CHAPTER 2.5

Lime

Introduction to Lime

Lime is a common name for calcium carbonate ($CaCO_3$), a mineral compound found in limestone, seashells, skeletons, and lime-rich soils. Limestone was originally quarried and used as masonry. However, man quickly learned that limestone and seashells could be processed and reconstituted into mortars, interior plasters, exterior renders, and washes with chemical properties similar to the original stone or added to loam to add strength, stability, or plasticity.

Historic lime materials were designed to protect adjacent masonry and wood in several ways:

- Render coatings protect masonry and wood from rain, humidity, heat, chemicals, and abrasion.
- Render coatings absorb salts that are carried by rising damp in masonry walls.
- Render coating are sacrificial, protecting underlying substrates.
- Washes wear rather than peel, simplifying maintenance.
- Mortars are soft and flexible and can absorb movement from expansion or earthquakes.
- Lime is a natural herbicide and insecticide, inhospitable to molds, fungi, and insects due to its pH between 9 and 10.

The durability and other qualities of lime-based products will depend on many factors:

- The materials (the original source of lime and other aggregates and additives).
- The process used to refine the lime (firing, slaking, and re-carbonation).

DOI: 10.4324/9781003160489-10

- The techniques used to mix, apply, cure, and maintain lime-based products.
- The environmental conditions throughout its life.

Lime-rich products absorb moisture but are also capable of regulating and releasing that moisture. Excessive moisture, frequent wet-dry or freeze-thaw cycles, and the accumulation of salts can cause damage but, overall, lime mortars and renders are characterized by good workability, low shrinkage, flexibility, adequate compressive strength, and good resistance to salt and frost damage.

Lime products undergo a series of chemical reactions that are very sensitive to moisture, heat, and imperfections in the raw materials. Compared to many other historic materials, a high level of expertise is required to produce and maintain durable lime products. Traditional builders understood and worked with these qualities.

Efforts in the 20th century to use harder, less permeable materials to block moisture and reduce maintenance have largely failed. Hard and impervious mortars and coatings transfer stress to adjacent materials and trap moisture. Manufacturers, craftsmen, conservators, and architects today are developing a new understanding of lime and lime production, in an effort to restore and protect historic materials as they age, and experience more severe weather impacts due to climate change.

Impacts – Past and Future

History

Lime mortars and renders have been made from limestone or calcium-rich shells for more than 8,000 years. Builders quickly recognized that limestone from different regions had different qualities that made them appropriate for different uses. The process of firing limestone to create lime, and then combining it with water, carbon dioxide (CO_2), and aggregates, has been fairly standard throughout history. Techniques were documented by Cato (234–149 BC) and by Vitruvius (27 BC–14 AD) in *The Ten Books on Architecture.*

Use of traditional lime decreased with the introduction of artificial cements in the early 1900s but had a limited but continuing resurgence beginning with the preservation movement of the 1970s. The desire to use historic techniques and materials is supported by the realization that impervious cement products were damaging to traditional masonry and wood substrates. However, the use of lime products has been hampered by a lack of education and training in the nuances of traditional lime practices.

As conservators, contractors, and artisans learn from historical records, testing, research, and each other, there have been more successful restorations of lime renders and mortars. Increased demand is expanding the production of lime products in continental Europe, and lime quarries in the United Kingdom are being reopened.

Current Social, Cultural, and Climatic Impacts

The restoration and use of lime illustrate a range of preservation attitudes, as shown in the variety of conservation approaches at Missions of San Antonio, Tumacácori and San Xavier del Bac.

- Restoration with original materials to original condition.
- Restoration limited to extant materials.
- Restoration with modern additives to improve durability.

The desire to preserve original lime renders, especially where shaped or painted, has led to efforts to reinforce, modify, or reattach original fabric that is weak or delaminated. The increased stress of weather conditions will increase demand to modify historic materials so they are more durable. Many areas are predicted to receive more intense rainfall, which increases the danger that masonry walls will be wet longer, and that lime will be dissolved and leach out of unprotected mortars and renders. This process would weaken the mortar and leave efflorescence on the surface. Non- and low-hydraulic mortars will be particularly susceptible. However, attempts to block moisture with impervious coatings have proven to be counterproductive. As a result, it is important to understand how and why lime weathers, and how to protect it by working with its natural qualities.

There is also movement to preserve historic sites in their current "ruined" condition, which often leaves earth and stone structures without their original protective lime coating. It is questionable whether this approach is sustainable over the long term, especially with increased climate change impacts.

Lime and cement are both energy intensive to produce but lime is arguably the more sustainable material. Lime coatings better protect the underlying materials, and can be more easily restored, removed, and reused. During the 1939 demolition of the Empire State building and Chrysler building, lime mortar could be removed so the masonry could be salvaged. Salvaged masonry with cement mortar is not easily separated and is ground up for use as fill or road aggregate. Lime-based materials are less likely to retain moisture and promote biological growth than alternative gypsum and Portland cement-based materials.

Understanding Lime

Calcium Carbonate ($CaCO_3$) is an important component in natural limestone and shells, and in man-made renders, mortars, artificial stone, and other artifacts. Products made from pure $CaCO_3$ limestone are considered "non-hydraulic" because it requires carbon dioxide (CO_2) to cure and will not cure when fully submerged in water. Hydraulic lime, natural cement, and artificial cement are all lime-based materials that also contain "reactive silicates," clay, and other impurities that chemically cure even when fully submerged. Non-hydraulic and hydraulic limes will cure and perform differently based on the percentage of carbonates and reactive silicates.

Non-Hydraulic (Pure) Lime

Non-hydraulic lime is produced from relatively "pure" or high calcium sources such as chalk limestone or seashells. The lime cycle describes the chemical changes as pure $CaCO_3$ is converted from pure limestone into other lime-based materials when heat, water, or CO_2 are added or released during calcination (burning), slaking (hydration), and carbonation.

Other names for non-hydraulic lime are air lime, fat lime, rich lime, pure lime, high calcium lime, or simply calcium lime. Non-hydraulic lime is frequently called putty lime by conservators since most non-hydraulic lime is currently used in putty form. Powdered non-hydraulic lime is sometimes called "hydrated lime" even though putty lime is also hydrated.

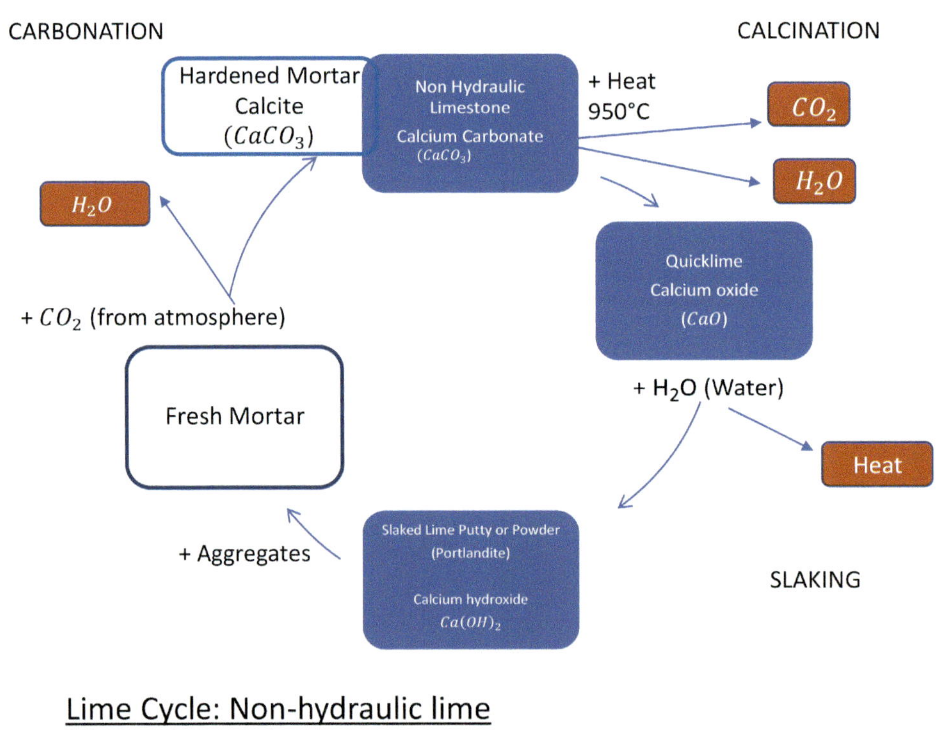

Figure 2.5.1 Lime cycle with non-hydraulic lime. (Diagram by Susan Pranger, 2023.)

During calcination, calcium carbonate ($CaCO_3$) in limestone or shell is first burnt in a fire or kiln. Temperatures exceeding 800°C trigger a chemical reaction that releases CO_2 and H_2O and converts $CaCO_3$ to calcium oxide (CaO), also called quicklime or burnt lime, and releases CO_2 gas. The fired lime is then mechanically ground, sifting out any unburned lumps of limestone or shell.

$$CaCO_3 + Heat = CaO + H_2O + CO_2$$

Magnesium lime, now called dolomitic lime, is produced from Dolomite, a sedimentary stone that is a mix of calcium carbonate and magnesium carbonate. Magnesium calcinates at lower temperatures (510° to 750°C) and begins to "overburn" at temperatures over 900°C. Overburned material will slake very slowly, and so magnesium lime is traditionally slaked for a long time to ensure complete slaking of any overburned lime. Dolomitic lime has a reputation for having properties similar to a slightly hydraulic lime.

The resulting lime powder is very unstable and must be slaked to prevent it from reacting with water and carbon dioxide in the atmosphere. Adding water to quicklime causes a rapid chemical reaction that releases heat and converts the quicklime to the mineral Portlandite, calcium hydroxide ($Ca(OH)_2$).

$$CaO + H_2O = Heat + Ca(OH)_2$$

"Slaking" and "hydration" are terms that have historically been used interchangeably. However, slaking usually refers to a process that uses excess water to produce a lime putty while "dry slaking," "sand slaking," or "hydration" are processes to produce a dry hydrated lime. Both hydraulic and non-hydraulic quicklime can be slaked to produce lime putty or hydrated to produce a dry lime powder.

"Air slaking" refers to the gradual unintentional hydration of quicklime left exposed to the atmosphere. Air slaking can occur at relative humidity as low as 15% and will reduce the binding capacity and strength of the quicklime.

Slaked lime putty should have about 30% excess or "free" water surrounding the chemically combined water and quicklime. This excess water results in higher rates of carbonation, higher plasticity, higher strength, and better water retention. Pure water is preferred for slaking, to avoid sulfates of chloride ions that can weaken the putty. The mix can be agitated or vigorously stirred to distribute the heat, water, and lime particles, accelerate the slaking rate, and produce a more complete hydration.

Traditionally, slaked lime was covered with a film of water and sealed in wood or brick vats for at least a few months, but preferably a few years, to allow all the quicklime to chemically change to slaked lime putty. The brick or wood enclosures were placed underground to prevent exposure to the atmosphere that would dry the mixture. Putty that is slaked for a shorter duration should be screened to remove any unslaked particles; otherwise, the particles might expand within the render as the particles belatedly hydrate.

An alternative method is to hydrate quicklime with the least amount of water required to complete the reaction; the resulting "dry" slaked lime can be dried, ground, and stored as a powder. However, dry lime should be used within six months, or it will carbonate by reacting with water and CO_2 in the air.

Dry slaking with sand was developed in response to the demands of expanded construction and industrialization in the 19th century. The burnt lime (quicklime) was quickly slaked by covering it with the wet sand. The sand had to provide enough water to hydrate the CaO, but not enough to trigger the hydration of the reactive minerals. The hydrated material was then sifted to remove unslaked lumps. The process was quick, had fewer steps, and production could be adjusted daily to match the demand of bricklayers and plasterers.

Sand, aggregates, and water were added to dry lime to create a wet render or mortar. "Hot mortar" for immediate use was produced by adding additional water to the dry slaked mixture to deliberately trigger hydration of the reactive minerals. The heat encouraged rapid evaporation of excess water, an advantage when working in cold conditions. It also increased the bond to non-porous hard stones like granite and increased the pozzolanic quality of the sand.

The final step is passive carbonation. The slaked putty or dry slaked lime is calcium hydroxide: ($Ca(OH)_2$). The $Ca(OH)_2$ combines with CO_2 from the air and converts back to $CaCO_3$, with qualities similar to the original limestone.

Carbonation is a two-step process. First, the CO_2 in the atmosphere must be diffused throughout the pores of the lime, where it combines with pore water to create a mild carbonic acid. The carbonic acid then dissolves the lime ($Ca(OH)_2$) releasing calcium ions that combine with the carbonic acid to create $CaCO_3$. Ideally, this reaction continues until all of the lime is converted to $CaCO_3$.

$$H_2O + CO_2 = H_2CO_3$$

$$Ca(OH)_2 + H_2CO_3 = CaCO_3 + H_2O$$

The proper hydration and mixing of lime render require skill and experience. Pure lime cannot carbonate when it is fully saturated with water or under water, hence the name, non-hydraulic. Carbon dioxide (CO_2) cannot convert to acid if there is too little water but cannot enter the pores if they are filled with water or blocked by surface water. Therefore, humidity should be carefully controlled both during the initial mixing and the subsequent curing. Humidity should be around 60%, no less than 40% and no more than 80%, to facilitate diffusion of CO_2.

The lime will take several months to set, and several years to reach full strength because carbonation occurs from the outside in and is very slow. CO_2 will penetrate more deeply and quickly into porous mixes with high aggregate ratios, and more slowly in less porous mixes with a high binder ratio. Even with adequate humidity, carbonation will eventually stop if diffusion of CO_2 is blocked by calcite crystals on the surface or in the pores. Non-hydraulic lime may never fully carbonate or reach its final strength, especially deep in a wall.

	Non-Hydraulic			Hydraulic			Natural Cement	Artificial Cement
Source	"Pure" Limestone or Shell		Dolomite	"Impure" Limestone or Dolomite			"Impure" Limestone	Clay or Slag + Pure Limestone
Burning Temp	About 95°C		510°C to 750°C	Above 1000°C			Above 1200°C	Above 1260°C
Hydraulic Quality	None			Semi	Moderate	Eminently	Hydraulic	Hydraulic
Clay*	Less than 2%	Less than 6%	Less than 10%	6 to 12%	12 to 18%	18 to 25%	More than 25%	35%
Slaking	Very fast	Fast	Very slow	Slow	Slow	Very slow	Very slow	
Expansion	Significant	Double	Varies	Slight	Slight	Slight	Slight	
*Clay is primarily comprised of silica and alumina, with some iron and magnesium								

Figure 2.5.2 Relationship between hydraulic qualities and clay content. (Table by Susan Pranger, 2023.)

Hydraulic Lime and Natural Cement

Most limestone is not pure and contains clay and other impurities that chemically react during calcination, hydration, and curing. The temperature of firing and the amount of clay will determine the percentage of non-hydraulic carbonated lime binder and the percentage of chemically bonded hydraulic components. In artificial cement, almost all calcium carbonate ($CaCO_3$) is converted into hydraulic components.

Natural hydraulic lime is produced from lime or dolomite with naturally high percentages of clay and other minerals. Clay is primarily comprised of silica and alumina, with some iron and magnesium. Silica and alumina melt at around 1,200°C to form reactive silicate and aluminate by combining with some of the calcium oxide (CaO) produced by burning calcite in the lime. At 1,260°C, the silica, alumina, and calcium oxide combine to form different silicates and aluminates, which hydrate to form even harder cements. Impurities in the limestone can also produce slag that can clog pores and makes the surface impervious to water.

The resulting mix of quicklime, reactive silicates and aluminates, inert silica and uncombined clay must be carefully slaked (hydrated) to form a hydrated but dry mix of calcium hydroxide ($Ca(OH)_2$) and reactive particles without triggering a chemical set of the reactive materials. Reactive particles slow the slaking process and reduce expansion.

Water and aggregates are added at the onset of the carbonation process. The silicates and aluminates chemically bond relatively quickly with the water and calcium hydroxide to form hydrated calcium silicates and aluminates, while any remaining calcium hydroxide is carbonated in the usual slow process to form calcite. The chemical curing of silicates does not require carbon dioxide and can occur under water. As hydraulic lime cures, it becomes stronger, but less permeable.

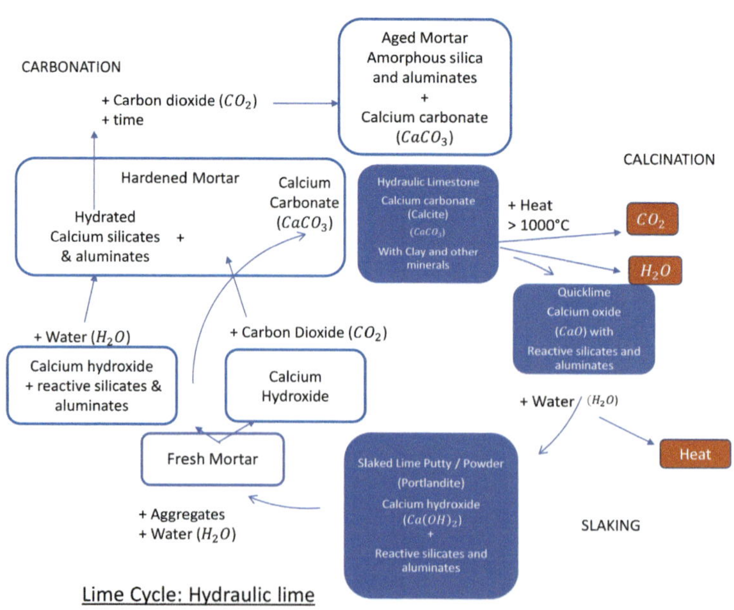

Figure 2.5.3 Lime cycle with hydraulic lime. (Diagram by Susan Pranger, 2023.)

Natural cement is a highly hydraulic binder, produced by burning impure limestone that contains more than 25% clay at temperatures higher than 1,200°C. Most of the calcium oxide produced by burning the calcite in the limestone will combine with the silica and alumina of the clay to produce the mineral "belite" (dicalcium silicate), leaving very little calcium oxide (CaO) to convert to calcium hydroxide ($Ca(OH)_2$).

Higher temperatures, above 1,260°C, will form the mineral "alite" (tricalcium silicate) which will hydrate more quickly and create harder cement with a denser crystal matrix. However, over burning the lime, at around 1,400°C, will produce a darker lime with less chemical reactivity, greater density, and reduced porosity – as little as 8% to 12% compared to a porosity of up to 54% in white, soft burned lime.

Natural cement is ground and stored dry, without slaking and will set up in 15 to 30 minutes when hydrated. Natural cement was popular for engineering work, especially tidal or underground work. It was also popular for exterior render, stone patching and pointing, because it adhered well, were strong, and shrank very little. Natural cement is being reintroduced to restoration work on projects like the Oregon lighthouses because it is more weather resistant than less hydraulic limes but more flexible and porous than artificial cements.

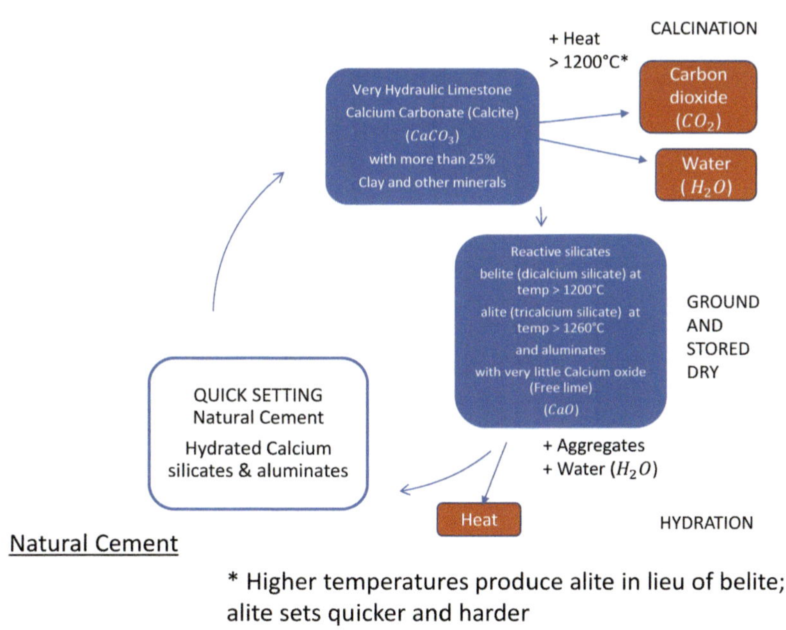

Figure 2.5.4 Lime cycle with natural cement. (Diagram by Susan Pranger, 2023.)

Mortar and Render Mixes

The durability of lime renders and mortars will depend on the type of lime, and the aggregates, fibers, and other additives, as well as the methods and rate of agitation, the temperature, and humidity.

Non- or lightly hydraulic lime mortars are inherently capable of mild self-healing. Lime particles in a lime mortar or render can dissolve and redeposit, filling minute hairline cracks, while artificial cement mortars do not have this capacity. Lime mortar is also flexible and can absorb slight movements and stresses (unlike cement mortars which are very rigid). Lime mortars are relatively porous, allowing moisture to escape through the mortar from adjacent masonry. This flexibility and porosity make lime mortars and renders somewhat resistant to frost and crystal damage after they have cured (releasing excess water). However, uncured non-hydraulic lime render has a high-water content and is easily damaged by the expansion of salt or ice crystals. Hydraulic and pozzolanic lime products are generally more durable and suitable for exposed locations such as chimneys, foundations, quoins, copings, or the top of walls.

Masonry cement is created by adding either white cement or ordinary Portland cement to a combination of limestone, hydrated lime, and air entrainers. Masonry cement and other artificial cement mortars are stronger and less porous than lime-based mortars and have a higher coefficient of thermal expansion than lime mortars and most brick and stone. Cement mortars are more likely to contain calcium sulfates and sodium salts that can leach from the mortar into adjacent stone or effloresce.

Premixed dry masonry cement mixed with water on site are popular today for both new construction and restoration. Their use with historic masonry peaked from 1940 to 1970 but continues today even though they are generally too strong and too impervious for restoration work.

A good mortar or render requires the appropriate type and proportion of sand and other aggregates, which add color and texture and improve compressive strength and porosity. Aggregates reduce the shrinkage by reducing the amount of binder and ensure more complete curing by distributing moisture throughout the mix. The most common aggregate is sand or grit, although crushed brick, seashells, lumps of old mortar, kiln slag, and ash are also found in historic lime mortars. Crushed limestone, a by-product of quarrying, can improve workability and carbonation of lime mortars.

A clean, well-graded "sharp" sand is heavy and course and adds strength to the mix while "soft" rounded sand can increase the flexibility. Pit or quarry sand is usually "sharp" or angular, which helps it key into the binder, improving performance. However, pit sands vary in quality and are often contaminated with clay. River sands tend to be cleaner, but also more rounded, and larger, with fewer fines. Sea sands vary in size and shape but tend to be rounded and contain bits of rock and shells. Unless washed well, sea sand is likely to be contaminated with salts that will draw moisture into the mortar.

Pozzolanic materials are named for the volcanic rock of Italy but include a variety of natural and man-made materials that have high quantities of silica and alumina. Naturally pozzolanic materials include volcanic rock, ash, or tuff. Manufactured pozzolans include clays or minerals that have been fired or calcinated then ground, including burned brick or clay, fired tile or pottery, minion (calcinated iron ore), kiln residue and fired dolomitic limestone.

When heated naturally in a volcano or artificially, as in a kiln, pozzolanic materials form reactive silicates and aluminates. The hydrated silicates and aluminates combine with calcium hydroxides ($Ca(OH)_2$) to form a strong hydraulic binder with good adhesion, flexibility, and porosity.

Metakaolin, a highly reactive aluminosilicate pozzolan, is produced by firing china clay (kaolin) at temperatures in excess of 750°C. Metakaolin reacts with any free lime in cement or mortars to form hydrated silicates and aluminates. It has been added to non-hydraulic limes used in conservation work to prevent lime leaching and blooming, especially in wet locations.

Figure 2.5.5 Restored lime render at San Juan, San Antonio, Texas. (Photo by Daniel Thomas, 2021.)

Natural fibers were traditionally added to lime and renders to increase flexibility and strength by controlling shrinkage and bridging gaps or cracks. Animal hair, especially those with a rough texture, was most common. Vegetable fibers such as hemp, straw, and grass are more common in clay mortars but also found in early lime renders. Synthetic fibers of minerals, such as fiberglass and basalt, or polymers have been used since the 20th century to reduce shrinkage and increase strength in lime products.

During the 20th century, fibers of mineral asbestos were added to thousands of building products, including lime and cement renders, to increase strength, fire protection and insulation and acoustic properties. It is most often found in textured render. If asbestos is suspected, materials should be appropriately tested and handled with care to avoid serious health concerns.

There is a long history of using organic additives to add plasticity, workability, or strength to lime mortars and renders. Blood was used for 2,000 years as a thickening agent. Egg and dairy products, animal glues, urine, dung, beeswax, resin, and beer were all commonly used additives. Rosin, casein, and animal glues were believed to improve bond; citric acid was believed to slow setting. Fig juice accelerated setting, and molasses increased hardness. Oils, mastics, linseed oil, tar oil, turpentine were used as additives to make both lime and cement renders more waterproof. While most of these are no longer used, beeswax is being used in Canada to improve water resistance, and cactus juice is used today in the American southwest to improve workability and reduce shrinkage.

Synthetic additives were introduced during the 19th and 20th century in the quest to improve strength, durability, water resistance, or workability. The success of these additives varied depending on the type of mortar, construction techniques, and the local weather. Contemporary additives, including accelerators, retarders, air-entrainment, polymer modifies, and bonding agents, should be used with care because they can inhibit the bond, increase density, trap moisture, and alter workability. Air entrainers improve workability and frost resistance but can reduce strength.

The color of mortars and coatings can be changed by adding colored sand, fly ash or minerals to the mix, or by applying pigments to either wet or dry surfaces, and generally do not affect the durability. However, pigments applied alfresco to the wet render will bond to the curing lime and are more durable than those applied to a cured surface. Alum was added to give a brilliant character to the lime, but also improved water repellency.

Workmanship

The durability of good lime mortar or render relies on good craftsmanship and careful timing at every step of the process and was heavily influenced by an understanding of the impact of temperature and humidity. Skilled practitioners know how to adjust the amount and temperature of water, mixing, and aeration based on humidity and temperature. Mixes must be blended well to break up any clumps, distribute water, sand, and aggregates evenly, and properly aerate the mix. Beating the mix improves cohesion, reduces water (which reduces shrinkage), and improves plasticity. Aeration is critical to introduce pores that increase permeability required for complete chemical reactions and curing, and to allow excess water to evaporate quickly. Small batches were mixed to ensure that they could be applied before they set.

Proper detailing can enhance the durability of lime materials. Lime mortars and renders rely on a good bond with a supporting substrate with compatible qualities of porosity and permeability, flexibility, and thermal expansion and contraction. Substrates should be damp before applying lime mortars or renders so that the substrate does not draw moisture out of the render, causing the render to shrink and pull away from the substrate. The bond is improved by keying into gaps between laths, or around protrusions, such as stones embedded in mortar joints. Fibers, including the traditional animal hairs and modern fiberglass and basalt can also improve keying.

Masonry walls of adobe, brick, or stone were laid in a bedding of clay or lime that bonded units to each other, filled any voids in the irregular units, and blocked infiltration of air or water. The more irregular the masonry, the thicker the mortar bed was likely to be. Chinking or pinning the gaps between large stones with small stones provided stability to thicker joints of rubble or roughly coursed stone walls and helped reduce shrinkage across large joints. The exposed surface of mortar joints and render were sometimes protected with small stones (gallets) embedded in the exposed surface of the mortar. Although decorative, gallet also reduced the exposed surface of the lime mortar or render.

Thick masonry walls were often built with regular courses on the exterior and interior faces, and then filled with a more flowable mixture of mortar and rubble. These lime mortars, protected from exposure to water and carbon dioxide, can survive hundreds of years without ever fully curing.

Mortar Joints

Mortar joints were designed to resist weather. Although the terms "pointing," "tuckpointing," and "repointing" are often used interchangeably in modern restoration work, historically they had slightly different meanings.

Regular blocks were laid with thin, uniform mortar joints that extended to the face of the masonry, where they were "struck" with a tool to compact the surface and shed water. Irregular surfaces required different approaches.

"Pointing" originally referred to the practice of raking the bedding mortar a few inches from the surface and filling the exposed joint with a more durable hydraulic lime mortar. This process was more labor intensive but protected softer and less expensive clay or non-hydraulic lime bedding mortar. It relied on a good bond between the bedding mortar and the pointing mortar to avoid creating a gap that trapped water.

"Flush pointing" was a quick way to build a wall when using rubble or roughly coursed masonry. The bedding mortar was struck flush with the surface of the wall, often covering as much as half of the wall surface. This mortar was left exposed, or later covered with layers of lime render or whitewash to protect the feathered edges of the bedding mortar. The smooth surface eliminated edges that would collect water and allowed water to run quickly down the wall, limiting erosion and infiltration. The large surface area of exposed mortar also allowed rapid evaporation of moisture in the wall.

"Tuckpointing" originally referred to the practice of filling irregular joints with mortar that matches the color of the masonry, raking the mortar to create a uniform groove, and then "tucking" a contrasting mortar into the groove to create the illusion of a regular joint. It was labor intensive, but less expensive than using gauged brick or dressed stone.

"Repointing" refers to the practice of replacing deteriorated surface mortar with new mortar, regardless of whether the original mortar at the exposed surface was originally bedding mortar or pointing mortar.

Lime Renders and Washes

Exterior stuccos and renders are used over many substrates, including rubble or ashlar stone, adobe or fired brick, and wood lath or wattle-work. The term "render" is used here to apply to any exterior lime plaster but is sometimes used to distinguish a smooth finish from rough finishes like harling. Lime plasters were also used to create interior and exterior sculptures, columns, or balusters, as well as architraves, quoins, and bands.

Lime plasters and renders are a mix of lime, sand, or other aggregates, and may include fibers or other additives. Lime washes are generally a thinner coating of lime and water or other viscous liquid such as cactus juice. Limewater and nanolime are thinner solutions designed to penetrate and strengthen existing lime plasters, renders or limestone by depositing the calcium carbonate ($CaCO_3$) particles that remain after the solvents evaporate.

Lime coatings are generally, but not always, applied in two or three layers. A single rough coat may be sufficient, or multiple layers might be required to infill uneven conditions or build up projecting shapes. Multiple layers might be added over time to maintain a protective coating. A lime wash is often applied to the finish coat, or to the entire face of a pointed masonry or timber framed wall.

The base course is generally thicker and coarser so that it can key into and fill any irregularities in the substrate. Fibers are often added to provide strength and flexibility and absorb any subsequent movement of the substrate. This coat is applied forcefully, sometimes literally hurled at the wall, to create a good bond. The surface is then scratched to create a key for the second coat.

The second coat, also called the "floating" or "straightening" cost is generally thinner and less course, filling small voids to provide a smoother surface. Unless this is the finish coat, it is also scratched to provide a key for the finish coat.

Depending on the aesthetic requirements, a third finish coat is applied to provide a smooth uniform finish. A cross-grained wood trowel created a slightly textured surface, while a metal trowel produced a smoother surface. On roof surfaces, projections, drainage canals, or other surfaces subject to heavy water runoff, the finish coat was often polished with smooth stones or thick animal hides or similar materials to consolidate the surface, making it denser and more water repellant.

Harling, also known as rough cast or wet dash, is a wet slurry of aggregate and hot lime that was thrown or hurled onto a rough substrate in a single coat. The resulting compaction released moisture and produced fewer shrinkage cracks. The rough texture resisted the impact of driving rain and had a larger surface area that increased evaporation. However, in urban areas, the rough surface tends to collect dirt and pollution.

Drivers of Deterioration

Most deterioration of lime can be attributed to conditions that cause failure to carbonate or cure, or cause subsequent breakdown. Many of these factors are caused by or aggravated by excess humidity and temperature during fabrication or after completion.

- Imperfections inherent in natural materials or in its fabrication
- Inappropriate use
- Poor workmanship or design
- Environmental conditions

Imperfections in the Materials

A complex series of chemical reactions is required to produce consistent lime-based mortars, plasters, and renders. However, the materials and processes are rarely as perfect or complete as the formulas might indicate. Even in modern construction, very few processes can achieve perfect proportions and complete reactions between all materials. Uncombined materials and incomplete processes can produce material that is inherently weak. Friability is often an indication that there was too little binder in the mix, that the binder was not fully hydrated or carbonated, or that the binder has been washed away or broken.

Excess clay, salts, acid, and other minerals can introduce inherent weaknesses that make the final product more vulnerable to the weather. Aggregates can introduce impurities such as salts, non-reactive clay, reactive silica, or ash that are detrimental to both lime and cement mixes. Ferrous minerals and pyrites can cause corrosion that produces stains or mechanical stress.

Iron in the sand can cause staining. Unwashed sea sand frequently contains sodium chloride which is hygroscopic and attracts moisture to the surface of the mortar. Wet-dry cycles can cause salt crystals to accumulate, clogging pores and breaking bonds. Sodium chloride can also corrode ferrous reinforcement and anchors. Non-reactive clays can adhere to aggregates and interfere with the bond between binders and larger aggregates. Hygroscopic clays can swell when saturated.

Alkali-reactive silica is rare but found in some volcanic aggregates. This type of silica reacts with potassium and sodium hydroxides and other strong alkalis commonly found in cement, cement-lime mixes and concrete. The reaction produces an alkali-silica gel that expands in volume and disrupts the mix. Ash from coal burners, a common aggregate in the 19th century, contains sulfur that can form reactive solutions that corrode ferrous inserts and attack binders. Rust from pyrites in aggregates, ferrous inserts, or metal lath can expand and damage mortars and renders.

Soluble or partly soluble salts are found in cements or are formed during hydration of cement mixes. These cement hydrates include calcium hydroxide, sodium hydroxide, sodium silicate, sodium sulfate, and calcium sulfate. These salts can attack adjacent materials or can react with silicates in aggregates to reduce the integrity of cement mortars.

The size, grading, and type of aggregates, and the proportion of binder to aggregate are key to developing a strong and adequately porous mix. There must be a well-graded mix of aggregate sizes and enough binder to coat all the aggregate particles and fill the spaces between aggregates. A well-graded mix of aggregates with insufficient binders will have empty voids that absorb water and can suffer frost damage. On the other hand, a poorly graded mix of aggregates can perform well if there is an excess of binder sufficient to fill remaining voids. A mix with too many fine aggregates will require more water to become workable because the fine voids will rapidly absorb the water. The resulting mix is more likely to shrink and crack as the excess water evaporates. The mix will also dry more slowly, increasing the risk of frost damage.

Incomplete Calcination, Hydration, Carbonation, or Chemical Reactions

Calcination (firing) requires both specific temperatures and adequate duration to completely convert limestone and the clay to calcium oxide and reactive silicates. Incomplete burning of limestone can result in lumps of unfired limestone and non-reactive clay. Unfired limestone particles may simply act as aggregates in the final mix. However, large pieces or a high proportion could reduce the strength and bond of the mix. Unfired clay reduces the percentage of reactive silicates, which reduces the bonding strength and hydraulic qualities of the future mortar or render. Unfired clay will also increase the percentage of fines which reduce porosity and interfere with the bond. Over-burned dolomite will hydrate more slowly and produce a friable mortar.

Burned lime is very unstable and must be carefully hydrated and stored to prevent air slaking, incomplete slaking, or slaking in place. Each of these events can reduce the percentage of the binders, and produce a weak, friable mortar or render. Air slaking, a premature hydration of the lime, is caused by improper storage that exposes dry lime to moisture in the air and can penetrate a stockpile over time. Dry slaking can produce inert binders that will not hydrate or carbonate. Incomplete slaking is caused by insufficient or poorly distributed water. Clumps of un-hydrated quicklime in slaked lime or hydrated lime powder may hydrate later, reducing the water available for carbonation and chemical curing. Slaking in place can occur when burned lime or dolomite hydrates and expands after mortar or renders are placed. The impact of delayed hydration and swelling may be minimal in a mortar bed but can disrupt the surface of a plaster or render.

Carbonation requires enough water and carbon dioxide to create the acidic solution that converts calcium hydroxide to calcium carbonate (CaO_2 to $CaCO_3$). Too little water limits the formation of acid; too much water blocks the distribution of carbon dioxide throughout the pores. Excess heat or low environmental humidity can cause rapid drying, increasing shrinkage, and reducing water available for carbonation. Incomplete carbonation produces a friable mix that does not cure or harden. Mortars and renders can crack or separate from adjacent materials as water evaporates and the mix shrinks.

The conversion of reactive silicates to hydrated calcium silicates and hydrates also requires sufficient water. Insufficient water results in a less hydraulic and weaker mortar or render. Hydration or carbonation often continues long after application and can strengthen the mortar.

Inappropriate Use, Detailing, and Workmanship

Different binders and processes are appropriate to different environmental conditions. Early builders and artisans understood this and used materials, accordingly, with appropriate levels of wetting, protection, regular maintenance, and periodic restoration. Although non-hydraulic lime mortars remain relatively soft, they have functioned well for centuries when used in protected locations or protected by sacrificial layers of lime render or lime wash. Use of non-hydraulic lime without appropriate protection or maintenance will result in deterioration.

The hydrated silicates in hydraulic lime and cements make them harder than non-hydraulic limes that rely solely on carbonation of the lime, and are appropriate for locations with more exposure. However, hydraulic lime and cement mortars can transfer movement or thermal stresses to softer adjacent materials, causing those materials to crack or to separate from the mortar. Impervious mortars can divert moisture and salts into and through adjacent masonry. Hence, the strength of the mortar or render must be appropriate to its substrate. As stone and masonry becomes weaker with age and weathering, it may be necessary to use mortar mixes that are less hydraulic than originally used.

Cement mortars and renders are very durable and strong in the right applications. However, cement products can be incompatible with softer clay, brick, and stone substrates. Impervious coatings can trap water and lead to moisture damage. Efforts to improve flexibility and porosity by mixing lime and cement have had mixed results. Adding small amounts of cement to lime mixes can reduce the strength of the mix. Adding too much cement to lime mixes can lead to excessive strength and reduced permeability or porosity.

Environmental Factors

Moisture, chemicals, biological agents, and climate conditions all work together to create and destroy limestone. Even lime mortars and renders that are well produced and appropriately used are subject to gradual deterioration from environmental conditions ranging from everyday conditions to seasonal or exceptional weather events. Fluctuations in humidity and temperature cause expansion, contraction, and chemical changes that can break bonds in both non-hydraulic and more hydraulic lime materials.

Unfortunately, changes in the climate are moving toward extremes – greater daily temperature swings, longer periods of rain or drought, and more wet-dry and freeze-thaw cycles – that are more damaging than previous conditions. Deterioration from environmental conditions is usually iterative. Stresses cause cracks or voids that admit water and salts. Water intrusion creates larger cracks or voids that allow more water, and so on. Salt attracts more water and more salt, creating its own cycle of escalating damage.

Water

Moisture is not necessarily detrimental to lime materials that have the capacity to "breathe" – to absorb and release moisture. However, over time, water can dissolve and leach the lime out of mortars and renders, leaving behind unstable unbound sand. This damage is accelerated when moisture is trapped in the pores by impervious consolidants, coatings, or when the lime binder was unable to carbonate due to excess water in the environment. Alternating cycles of wetting and drying carry and deposit soluble salts through a wall; frequent cycles accumulate salt crystals in the pores until they exert pressure, disintegrate the binder and spall the surface. Lime plasters have historically been used to draw salt out of more porous substrates, sacrificing the plaster to protect less easily replaced substrates.

Heavy and prolonged rains will increase the depth and duration of saturation. The longer a wall is saturated, the more likely it is to suffer deterioration. Water expands in volume by 9% when it freezes, producing ice crystals that fill pores and break internal bonds, presenting a formidable risk to damp mortar that has not set up, or cured mortar with small pores. Repeated freeze-thaw cycles have a compound effect, as each cycle creates enlarged cracks and provides access for more water.

Sunlight, Heat, and Fire

Temperature affects lime mixtures in two ways. First, by directly causing expansion and contraction as temperature changes, and second, by driving the movement and expansion of water.

A certain amount of expansion and contraction is easily absorbed by lime. However, when materials with different coefficient of expansion are adjacent, the difference in movement can cause the materials to separate or can exert compression forces that crack the weaker material. Metal anchors and reinforcement will expand more rapidly than renders, which can break the bond between the materials, or crack the weaker lime materials. Sunlight can exacerbate this condition by selectively and rapidly heating exposed surfaces, causing differential movement between the surface and substrate, or between one exposure and adjacent walls.

Rising temperatures evaporate water from the pores of lime materials; the drying effect can be beneficial if there is too much water in the material, but detrimental if it causes the lime to shrink and crack. In extreme cases, the loss of water can cause chemical changes in the lime binders. Rapid temperature changes can rupture bonds in lime materials by expanding water in small pores more rapidly than it can evaporate.

Lime renders are often used as a fireproofing because they are noncombustible and can absorb enough heat to prevent combustion of underlying materials such as wood. However, fire can damage lime by causing calcination. At temperatures over 850°F, carbon dioxide and moisture are driven out of calcium carbonate, leaving friable calcium oxide. Water added while extinguishing a fire can quickly slake the lime, converting it to a dry hydrated lime powder. Depending on the duration of the fire, this conversion might occur only at the surface, and not affect bedding mortar.

Wind

Wind can directly damage lime renders by eroding the surface, separating binder from aggregates. It can also inhibit curing of the surface render and increase long-term weathering of lime renders by drawing moisture from the surface. Wind can create positive or negative pressure through a wall that drives water vapor or rain more deeply into the pores of a wall. Combined with more frequent or longer storms, wind driven rain can increase the depth of damage from salt crystals, frost, and wet-dry expansion.

Chemicals and Salts

Water, CO_2, destructive salts and other chemicals are introduced by movement of liquid or vapor water or air vapor through and into wall assemblies. Rainwater, air, and water vapor can introduce pollutants from the air, including acids and salts. Fog and other persistent humidity can carry pollutants gradually but deeply into porous materials like lime. Rising damp is particularly likely to introduce salts from the soil, ground, or surface water.

These chemicals can deteriorate lime products by dissolving calcium carbonate binders, corroding ferrous inserts or forming salt crystals in the pores. Corroded minerals can also cause staining.

Limestone and lime products are relatively pure, with low levels of inherent salts. Cement products, however, contain significant alkaline salts, soluble calcium sulfates and sodium salts that can leach out over time. Aggregates can include impurities of salt and minerals. Even natural stones can contain minerals and salts.

Acidic and alkaline salts and corrosives can also migrate into the renders and mortars from adjacent materials. Sulfates are introduced from contaminates in soil, groundwater, seawater, or pollution or may migrate from adjacent brick, gypsum, or Portland cement. Chemicals can also be introduced by organic or biological agents. The hair added to binders can break down into alkaline materials if the mortar is not cured shortly after mixing. Certain bacteria, molds, and fungi can excrete acids that break down lime products. Pigeon droppings, all too common on urban facades, can be very acidic.

Water contains dissolved carbon dioxide, forming a mild carbonic acid. These carbonic acids react with calcium and magnesium carbonates in lime to create calcium and magnesium bicarbonates that are easily dissolved and leached out of mortars and renders, leaving behind loose friable aggregate. Weak acids in ground water can dissolve calcium carbonate in foundations. Sulfates in acid rain and air pollution react with calcium carbonate in pores in the lime to create calcium sulfate dihydrate, which is gypsum. Gypsum is highly soluble in water, and easily leaches from lime mortar and is deposited on the surface where it forms a brittle crust of sulfates that traps pollutants. Hence acid rain produces a black crust on both limestone and lime mortars and renders.

Acid rain has a similar reaction with the magnesium in dolomitic lime, creating magnesium sulfate. Magnesium sulfate is even more soluble in water than gypsum and tends to migrate deeper into the mortar where it crystallizes. Sulfates also react with hydrated calcium aluminates in natural and artificial cements to create the mineral ettringite, calcium aluminum sulfate. The transition to gypsum or ettringite involves swelling, a change in volume that is harmful to hardened mortar or render. It also decalcifies calcium silicate hydrates, weakening those binders. This type of sulfate damage is rare in cement or in hydraulic limes, but it is more common in lime-cement mortars because sulfates inherent to the cement can attack the lime.

Soluble salts can recrystallize as either efflorescence (deposits on the surface) or crypto florescence (crystals deposited within the pores or just under the surface). Efflorescence is caused when water carries soluble salts to the surface and evaporates, leaving the salts on the surface. Crusts of calcite are unsightly but usually harmless. However, deposits of gypsum or c calcium carbonate can form a crust that attracts or traps moisture, leading to deterioration or delamination of the surface behind the crust. Crypto florescence occurs when the surface dries rapidly due to low humidity, solar radiation, or high winds. As the water rapidly evaporates from the pores, soluble salts crystallize below the surface. Crystal build-up, especially in small pores, can cause a loss of cohesion and surface delamination. Many residual salts are hygroscopic and will attract water from the atmosphere even after the initial source is removed.

Biological Agents

Materials with a high lime content have an inherent level of resistance to some biological agents, one of the benefits of using limewash to protect interior and exterior materials. However, rough, porous, and damp lime surfaces can be an inviting environment for a variety of organisms.

The minerals in mortars provide nutrients for microorganisms. Rough surfaces can hold enough moisture to provide a suitable habitat for a variety of algae and cyanobacteria. Both bacteria and fungi can produce acid that further etch the surface by dissolving carbonates. The hyphae of fungi can penetrate deep into porous materials while the surface vegetation (thallus) can expand and contract, enlarging cracks on the surface. Lichens can penetrate a few millimeters, excreting acid, while mosses can penetrate a few centimeters. Much of this damage is cosmetic, but over time, moisture and cracks increase, allowing colonization by larger mosses and lichens, and eventually ivy and the roots of other opportunistic vegetation.

Ivy can cause deterioration by holding moisture against the wall, and blocking sunlight that would otherwise cause drying, although it can also protect the render surface by protecting it from weather extremes. Ivy that attaches with a disk-like adhesive pad is less damaging than ivy that is secured by aerial roots that penetrate the lime surface.

Sound lime mortars and renders are generally inhospitable to insects; however, masonry bees have been known to bore into soft mortars to deposit their pupal cells. A few holes can begin a cycle of water penetration and breakdown of the mortar, leading to opportunistic tunneling by ants and other insects. Insects that penetrate the protective lime render can also attack organic fibers or wood substrates.

Mechanical Stress

Mechanical stresses can be introduced by expansion and contraction discussed previously, or by larger events such as lightning, earthquakes, or settlement of the geological substrate. Stresses usually pull the wall apart where the materials are weakest. The weakest point, by design, is the lime mortar or render.

Strategies for Responding to Climate-Based Threats

Lime mortars and renders are a durable traditional material when used and maintained appropriately. More importantly, lime renders can successfully protect a variety of substrates from external weathering, biological damage, and mechanical stress. As a sacrificial coating, it can draw salts out of masonry structures protect substrates that are more easily replaced. Efforts to replace lime products with stronger, impermeable cement-based products have largely failed to protect underlying materials. Sooner or later, rigid cement materials develop at least small cracks, that allow water to enter but prevent its escape.

The largest dilemma facing the caretaker of an historic building with lime is whether or not to replace missing render coatings. The missions at Tumacácori and San Antonio are being maintained in a semi-ruined state, stabilizing and patching plaster remnants rather than restoring a complete protective layer. This is a philosophical approach that will delay but not prevent continued deterioration of the remaining materials. Another dilemma is the desire to retain and display original decorative features. In the past, new coats of lime wash or render and decorative paint would have been applied over previous coatings. Today, caretakers may prefer to retain original decorative fabric through consolidation or reattachment rather than by reapplying lime washes, as was traditionally done at Mission Concepcion in San Antonio.

Conservation standards require that both products and procedures more closely mimic the performance of the original materials in appearance, strength, porosity, and permeability. While "reversibility" has long been a criterion for changes and treatments, "retreatabilty" is now an important criterion for consolidants and other penetrating treatments that cannot be reversed. Retreatabilty does not require that prior treatments be removed but allows the surface can be treated again (retreated) with the same or a different treatment in the future. This philosophy acknowledges that few treatments are truly reversible, but that new treatments are always being developed.

Investigation

Efforts to restore original material must begin with an understanding of the historic conditions that led to its current state. After obvious sources of water damage are eliminated or controlled, the conservator must understand how prior conditions may have changed the chemical composition and condition of remaining substrates and lime mortars and renders. Salts and clays that remain can be hygroscopic, attracting water that can cause expansion and contraction.

The dynamics of thermal and moisture movement through the wall under various climate conditions must be considered. New heating, air conditioning, and ventilation systems can dramatically impact the movement of water and water-soluble minerals and salts.

Direct observation cannot identify the composition of mortars and renders, and even experts may not be able to distinguish between a hydraulic lime and a cement render. Simple field tests can identify aggregates and the presence of lime, but lab analysis is generally required for more detailed information.

Maintenance and Repair

Ongoing restoration of lime coatings should be expected. Limewash is a protective coating, a dilute solution of hydrated lime (calcium hydroxide) that is applied in one or two coats, and usually reapplied annually to restore and protect lime coatings.

Patching materials should be compatible in color, texture, porosity, and strength. Ideally the patching material should be a lime mix that is chemically identical to the original. Synthetic patching materials should be used as a last resort where lime-based materials are not adequate, and only after appropriate testing and consultation by experts. Where existing material is sound but is detached from the substrate, conservators prefer the use of adhesives or nano-limes to secure delaminated materials.

Consolidants

The success of coatings, consolidants, and other chemical treatments to treat limestone and lime coatings is very much dependent on the composition and condition of the original materials and substrate, and the environmental conditions of temperature and humidity of the environment. The impact of treatments on existing porosity, and the potential for interaction with existing salt and gypsum deposits must be considered.

Limewater is a traditional consolidant produced by allowing the suspended particles in limewash to settle out, producing a clear solution of pure calcium hydroxide. Although used since Roman times, the process can require 30 to 40 coats which can introduce substantial amounts of water that dissolve existing binders, swell clay, and mobilize salts.

During the 20th century, a variety of synthetic adhesives and consolidants were developed, including vinyl, acrylate, silicone, and epoxy resins and other synthetic materials in alcohol solvents, copolymers, and alkoxysilanes. These materials were very adhesive and worked quickly but did not penetrate deeply. Many formed a hard impervious surface layer or coating that changed the dynamics of water migration and quickly led to deterioration of the underlying historic material.

Nano-lime technology is a popular approach based on the limewater approach. Fine nano-particles of alkaline-earth metal hydroxides (100 times smaller than slaked non-hydraulic lime) in an alcohol solvent penetrate more deeply, dry more quickly and require fewer applications. The nano-particles carbonate, reinforcing the existing binder without changing the chemistry of the original lime, but can decrease porosity by increasing the percentage of binder. Nano-lime has been used with ethyl silicate to achieve greater strength and penetration, but it is unclear if the incremental increase warrants the introduction of potentially incompatible silicate materials. Concentrated dispersions of nanoparticles can also be injected to fill voids or cracks or to reattach loose fragments. Successful carbonation will rely on appropriate temperature, humidity, and access to carbon dioxide and moisture in the pores. Both lab and on-site testing prior to specific applications is critical to understanding the effect on the porosity of existing pores especially if reduced porosity could increase damage from salt crystallization. The use of lime nanoparticles in dolomitic stones can cause an alkali-domolite reaction, reducing carbonation in the short term.

Other potential treatments include the use of calcium oxalates, calcium phosphates, and calcium alkoxides. Calcium carbonate, barium hydroxide, artificial calcium hydroxide, and tartrate have also been tested to consolidate porous calcium-based stone and render. Each method was successful in treating specific problems. Barium hydroxide is effective in counteracting the effect of gypsum deposits. Ethyl silicate and methyl silicate have been used to stabilize salt in wall paintings. However, their interaction with natural stone and other chemicals can be unpredictable.

More novel approaches are testing bacteria and enzymes to produce calcium carbonate. While it shows promise, it currently requires an impractical number of applications.

Repellants

A variety of waterproofing agents were developed in the 20th century to reduce water infiltration, including silicone-based and polyvinyl acetate copolymers dispersed in water. Waterproofing is a surface application with a limited life. It can wear unevenly and leave a residue that can affect the appearance and interfere with later treatments. Synthetic coatings should be used only where lime-based materials are not adequate, and compatibility and permeability have been confirmed by appropriate testing and consultation by experts.

Potassium Silicates

Potassium silicate coatings use an inorganic mineral-based binder that chemically bonds to mineral-based substrates like lime, masonry, and concrete. They form a hard but breathable, UV-resistant coating that resists moisture and biological growth. They are also useful as a penetrating coating that chemically reacts with mineral-based substrates, including brick, terracotta, and metals, calcium hydroxide in cements, and silica in mortars and sandstones. Proper surface preparation and testing of the substrate are critical to avoid damage from moisture and other materials remaining in or on the substrate.

Conclusions

Lime-based materials have a long history, and with care and thought, a place in our future. The ease of manipulating calcium, magnesium, and silicates provides many opportunities for creative ways to protect or maintain historic lime in the future, and to use lime coatings to protect other historic masonry. To do so responsibly, we must understand how our traditional lime artifacts are created and destroyed, and how future manipulations can shorten or extend their life.

Our recent history suggests we should rely on traditions and techniques with proven history. However, traditional techniques have always evolved and changed. Changes to our climate are occurring rapidly, and demand that we learn to respond more quickly and cautiously at the same time.

Bibliography

Allen, Edward, and Joseph Iano. 2013. *Fundamentals of Building Construction: Materials and Methods*. Hoboken: Wiley.

Andrady, Anthony L, Hamid S. Hamid, and Ayako Torika. 2003. "Effects of Climate Change and UV-B on Materials." *Photochemical and Photobiological Sciences*. 2(1):68–72.

Ashurst, John. 1997. *The Technology and Use of Hydraulic Lime*. https://www.buildingconservation.com/articles/hydraulic/hydraulic.htm.

Basu, Sudeshna, Scott Allan Orr, and Yasemin D. Aktas. 2020. "A Geological Perspective on Climate Change and Building Stone Deterioration in London: Implications for Urban Stone-Built Heritage Research and Management." *Atmosphere*. https://www.mdpi.com/2073-4433/11/8/788/htm.

Bowen, R. L. 1956. "Use of Epoxy Resins in Restorative Materials." *Journal of Dental Research* 35 (3):360–369. http://doi.org/10.1177/00220345560350030501.

Desilvey, Caitlin. 2017. *Curated Decay, Heritage Beyond Saving*. University of Minnesota Press.

Edison. 2015. *Everkote 300 Mineral Coating Systems*. December 17. https://www.edisoncoatings.com/Potassium_Silicate_Masonry_Pai/potassium_silicate_masonry_pai.html.

Elert, Kerstub, Carlos Rodriguez-Navarro, Eduardo Sebastian Pardo, Erid Hansen, and Olga Cazalla. 2002. "Lime Mortars for the Conservation of Historic Buildings." *Studies in Conservation* 47 (1): 62–75. https://www.jstor.org/stable/1506835.

English Heritage. 2011. *Mortars, Renders & Plasters*. Edited by Alison Henry and John Stewart. New York: Routledge.

Friedman, Donald. 2010. *Historical Building Construction*. New York: W. W. Norton & Company, Inc.

Granziani, Gabriela, Enrico Sassoni, and Elisa Franzoni. 2015. "Consolidation of porous carbonate stones by an innovative phosphase treatment: mechanical strengthening and physical-microstructural compatibility in comparison with TEOS based treatments." *Chemistry Central*. https://heritagesciencejournal.springeropen.com/track/pdf/10.1186/s40494-014-0031-0.pdf.

Harvey, Donald W. Jr, and Michael F Schuller. 2010. "Nondestructive Evaluation: Structural Performance of Masonry." *APT Bulletin, The Journal of Preservation Technology* Practice Points 09 https://www.apti.org/assets/docs/PracticePoints9.pdf

Jang, Jean. 2016. *Performance Evaluation of Commercial Nanolime as a Consolidant for Friable Lime Based Plaster*. Master's Thesis, Philadelphia, PA: University of Pennsylvania. https://repository.upenn.edu/hp_theses/597/.

John Canning & Co. n.d. *Types of Plastering Finishes and How to Identify Them*. https://johncanningco.com/blog/types-plastering-finishes/?utm_source=Newsletter&utm_medium=email&utm_campaign=December.

Kondam, Namratha. 2021. *Infrared Thermography as a non-destructive method to measure moisture patterns in lime-plaster finishes of historic free-standing adobe walls*. Thesis (Historic Preservation), Philadelphia, PA: University of Pennsylvania. https://repository.upenn.edu/hp_theses/715/.

Licht, Sue, interview by susan Pranger. 2023. *Historic Preservation Architect* Corvalis, Oregon (January 12).

Lim, Alex, interview by Susan Pranger. 2021. *Conservator, NPS* (November 29).

Lovett, Gary M., Marissa Weiss, Andrew M. Liebhold, Thomas P. Holmes, Brian Leung, Kathy Fallon Lambert, and David A. Orwig, et al. 2016. "Nonnative Forest Insects and Pathogens in the United States: Impacts and Policy Options." *Ecological Applications* (Wiley on Behalf of Ecological Society of America) 26 (5): 1437–1455. https://www.jstor.org/stable/24818127.

Lstiburek, Joseph. 2006. "BSD-1006: Understanding Vapor Barriers." *Building Science Corporation Digest*. https://www.buildingscience.com/documents/digests/bsd-106-understanding-vapor-barriers.

———. 2005. *Builder's Guides*. Westford: Building Science Press.

Means, Eric, interview by Susan Pranger. 2021. *Eric Means Constructions* (November 30).

Meek, Tim, and Tom Addyman. 2019. *Historic Environment Scotland Technical Paper 31: Historic External Lime Finishes in Scotland*. Historic Environment Scotland.

Myjer, Ivan, interview by Susan Pranger. 2021. *San Antonio Missions* San Antonio, Texas, (November 23).

Rodriguez-Navarro, Carlos, and Encarnacion Ruiz-Audo. 2018. "Nanolimes: from Synthesis to Application." *Pure and Applied Chemistry* 523–550.

Sickels-Traves, Lauren B., and Philip D Allsopp. 2005. "Lime and its Place in the 21st Century: Combining Tradition, Innovation and Science in Building Preservation." *International Building Lime Symposium 2005*. Orlando, Florida.

Simpson, J. W., and P. J. Horrobin. 1970. *The Weathering and Performance of Building Materials*. Beverly Farms: Aylesbury Press.

Woods, Amy Lamb. 2021. "Historic Mortar Replication and Repointing." *Webinar*. International Masonry Institute.

Young, Robert A. 2008. *Historic Preservation Technology*. Hoboken: John Wiley & Sons, Inc.

CHAPTER 2.6

Concrete and Cement

Introduction to Concrete and Cement

The appeal of concrete has always been its strength, ease of installation and the ability to retain complex shapes. It has been used to construct everything from buildings and infrastructure to furniture and art. The process of manufacturing cement binder, especially the proportion and type of ingredients and the firing temperature, is tightly controlled so that the strength and porosity of the finished concrete mix is predictable. However, subsequent processes to mix, transport, and place wet concrete are less controlled, and can adversely affect the durability and quality of the final concrete product.

Properly prepared concrete is very hard, but not impervious. This rigidity makes concrete vulnerable to mechanical stresses that create cracks. Even the highest-quality concrete can develop cracks from shrinkage, thermal expansion, and contraction. The surface of uncured concrete is particularly vulnerable to damage from premature drying, salt, and ice crystals. While surface damage is initially cosmetic, it can begin a cycle of continuing deterioration and larger cracks.

All concrete allows moisture and carbon dioxide to slowly penetrate its pores. Water and carbon dioxide create a carbonic acid that converts calcium hydroxide to calcium carbonate. Carbonation does not make the concrete weaker, but it reduces the alkalinity that protects ferrous metals from corrosion. Expansion of corroded metal causes most damage to concrete structures. Even microcracks in the concrete can accelerate the rate of carbonation and other surface deterioration.

Preservation efforts should first seek to manage water infiltration. Concrete restoration falls into one of three approaches:

- Treatments to restore or strengthen the matrix of the cement.
- Coatings to block penetration by water, carbon dioxide, and salts.

DOI: 10.4324/9781003160489-11

- Electrochemical treatments to restore the alkalinity of the concrete or reverse the flow of chloride ions to halt deterioration of the ferrous metals.

Impacts – Past and Future

History

Cement and concrete were developed by the early Romans. The word concrete comes from the Latin *concretus*, which means "mixed together." Roman concrete was of a very high quality, and much survives in good condition today, in large part because ferrous inserts were rare and because the concrete was often protected from the weather by a coat of lime plaster or a veneer of stone or brick. Vitruvius described the production of cement in *The Ten Books on Architecture*.

The use of concrete and ferrous reinforcement grew in the United States in the 19th century, gradually replacing the use of natural lime and natural cements as engineering projects demanded improved strength and consistent performance of cements and mortars. By 1920, Portland cement or composition mixes almost completely replaced the use of natural. However, artificial cements were less resistant to frost damage until the introduction of air entrainment improved porosity.

Concrete was traditionally covered with a veneer of stone, masonry, plaster, or paint that improved aesthetics and slowed the carbonation of the concrete. The use of exposed concrete in modern architecture increased in the 1960s, before an understanding of carbonation lead to the use of coated and non-ferrous anchors and reinforcing.

Figure 2.6.1 Boston Government Service Center (BGSC) built 1962, Paul Rudolph Architect. (Photo by Holland, 2023.)

Current Social, Cultural, and Climatic Impacts

The value of modern concrete architecture, originally rooted in social movements rather than aesthetics, is controversial. Exposed concrete is generally appreciated more by architectural professionals than by the general public. The difficulty of repairing or modifying exposed exterior concrete also makes preservation difficult but is offset by the high cost of demolition and the difficulty of recycling.

Concrete is used across the globe due to the ease of access and simplicity of fabrication and long life. It is strong, durable, weather resistant and very fireproof. Unfortunately, the extraction, transportation, grinding, and firing of raw materials for cement is energy and resource intensive and concrete is difficult to recycle because it cannot be returned to its original components. Regardless of the fact that concrete is not very sustainable, one fact remains: maintenance of existing concrete structures is the best way to reduce their life cycle cost.

The lifespan of reinforced concrete is limited by the gradual carbonation of concrete that allows reinforcement bar to corrode in wet climates. Many concrete structures have reached a state of full carbonation, are at risk of metal corrosion, and will soon require significant repairs or protective action.

The risk of concrete deterioration is higher in wet areas because water is necessary for both carbonation and corrosion. Carbonation and chlorides contribute to the corrosion of steel reinforcing in concrete. Studies in Australia suggest that the increases in atmospheric carbon dioxide by 20% to 40% by 2100 would increase corrosion by more than 16%. Atmospheric changes in China are expected to increase the depth of carbonation by a mean of 45% to 65% depending on the temperature.

Predicted increases in intense rainfall are likely to saturate exposed concrete structures more deeply and for longer periods, leading to damage from salt and ice crystals, increased carbonation, and corrosion of ferrous reinforcement. The combination of higher temperature and humidity allows carbon dioxide, chlorides, and other chemicals to penetrate more rapidly into the pores of concrete.

Understanding Concrete

In geology, "cement" refers to naturally occurring materials that bind other particles together to form sedimentary or conglomerate stone. In construction, "cement" refers to natural or artificial materials that bind sand and other aggregates to form mortar, plasters, or concrete. Concrete is a mix of cement, water, aggregates, and other additives.

The strength, porosity, and flexibility of a cement depend primarily on its ingredients and the firing temperature. The performance of concrete is influenced by the type and proportion of cement, water, and aggregates, and the processes used to mix, place, and cure the concrete. It is important, but difficult, to match these qualities when repairing or patching historic concrete materials.

Ordinary Portland Cement (OPC) is a modern cement named after the Portland stone of similar color. The strength of OPC evolved during the 20th century with improvements in technology, so early OPCs are not as dense or strong as modern OPC. Specialty cements are manufactured with different additives to enhance specific qualities.

The Nature of Cement

The production of natural and artificial cement binders follows a process similar to the lime cycle, but with key differences. The proportion of clay, other minerals, and pure limestone (calcium carbonate) and the temperature are carefully controlled so that all of the lime is chemically combined with clay to create reactive silicates and aluminates. These binders are much harder than calcium carbonate binders and cannot be converted back to their calcium components.

Modern cement mixes use a carefully controlled and proportioned mix of limestone, clay, and other additives. The proportion of clay or slag is carefully controlled at around 35% to ensure that firing will combine all calcium, silica, and alumina into complex silicates and aluminates, while leaving little or no excess calcium oxide ("free" lime). Pozzolans are added to cement prior to firing or during hydration to form reactive silicates and control the hydraulic quality. The reactive silicates combine with water to create hydrated silicates and aluminates that combine with calcium hydroxides to form a strong hydraulic binder with good adhesion and flexibility as well as a good pore structure. Naturally pozzolanic materials include volcanic rock or ash such as Pozzolana from Italy, and Trass, a volcanic tuff from Germany. Manufactured pozzolans include clays or minerals that have been fired or calcinated then ground, including burned brick or clay, minion (calcinated iron ore), kiln residue, and fired dolomitic limestone.

At around 1,200°C, the silica and alumina in clay will melt and recombine with some of the calcium oxide to form dicalcium silicate (belite), reactive silicates and aluminates that, when hydrated, will become cement binders. At slightly higher temperatures, around 1,260°C, the clay and calcium oxides form tricalcium silicate (alite) a different form of silicate and aluminate, which produces harder, denser cements when hydrated. Modern cements are fired at temperatures in excess of 1,400°C to consistently produce alite.

The fired clinkers are ground and combined with other materials to create a variety of different cements. When water is added to the dry cement, it triggers a rapid but complex chemical reaction that quickly sets the cement. Because artificial cement does not rely on carbonation of free lime, it can set up completely under water.

Ordinary Portland Cement

Ordinary Portland Cement (OPC) clinkers are bricks of artificial cement created by burning relatively "pure" crushed limestones) with a combination of clay or shale (silicates), and blast-furnace slag (pozzolans). This combination produces clinkers of alite, belite, aluminate, which give cement its strength, and ferrite, which reduces the temperature of the clinker during firing. Varying the proportion of these compounds produces cements with slightly different qualities. The OPC clinkers are crushed and mixed with gypsum or other calcium sulfates, or blast-furnace slag to produce dry OPC. The gypsum or other calcium sulfates are added to slow the subsequent hydration process by reacting with aluminates to create ettringite, a calcium-aluminum sulfate that forms a thin skin of around the globules of cement.

Materials can be added after firing limestone and clay to enhance certain weather-resistant qualities of OPC.

- Oleic acid or stearic acid creates a water repellant film around the unhydrated cement particles that prevents premature hydration during storage or transport.
- Sodium, aluminum, or other stearates make the cement more waterproof by increasing the binding action of the OPC.
- Sulfo-aluminate is used to create expansive cement that offsets the natural shrinkage as water evaporates from the concrete mix, allowing the mix to maintain a stable volume as it cures.
- Pozzolana cement is made by adding more pozzolana (up to 30%) to create a very hydraulic cement used for dams and weirs.
- Air entrainment additives form tiny discrete air bubbles to resistant to freeze-thaw damage. (wood resin, calcium, vegetable oils, or animal fats were originally used.)
- Bauxite (aluminum ore) is fired with limestone to create a hard high alumina cement, or calcium aluminate cement that sets in 24 hours, even in cold weather. Alumina cement was banned from structural use in the 1970s because high water-cement ratios led to a loss of strength over time.

Fly ash, slag, or volcanic ash are used to replace some of the OPC in concrete to reduce the economic and environmental cost of producing OPC. However, blended cements may hydrate slowly and gain strength more slowly even though they can reach a long-term strength comparable to OPC. When fully hydrated, blended cement may be more resistant to attack from chlorides or alkali-aggregate reactions.

Concrete Materials

Concrete is formed when cement reacts with water to create a binder that cements or binds the sand and aggregates to each other. The components should be well proportioned, hydrated, mixed and cured to complete all the desired chemical reactions. Aggregates should be well graded so that the space between large aggregates is filled by smaller aggregates, and all aggregates are coated with cement. Air is captured in the mix to create pores within the hardened concrete. Admixtures can be added to the mix during or after hydration and curing to control various physical properties. Concrete mixes are also adjusted to improve flowability or stiffness, setting time and curing time, to accommodate different working or climatic conditions.

The size of the cement particles affects strength, setting time, and porosity. Early OPC particles were larger, hydrated slowly and acquired strength slowly. Modern OPC has finer, more consistently sized particles that hydrate and acquire strength more quickly. As modern cement particles were decreased in size, the total surface area of the cement particles increased and more water is bound up by the cement, reducing flow and increasing stiffness of the wet concrete. Additives such as super-plasticizers are used to increase the flow of wet cement.

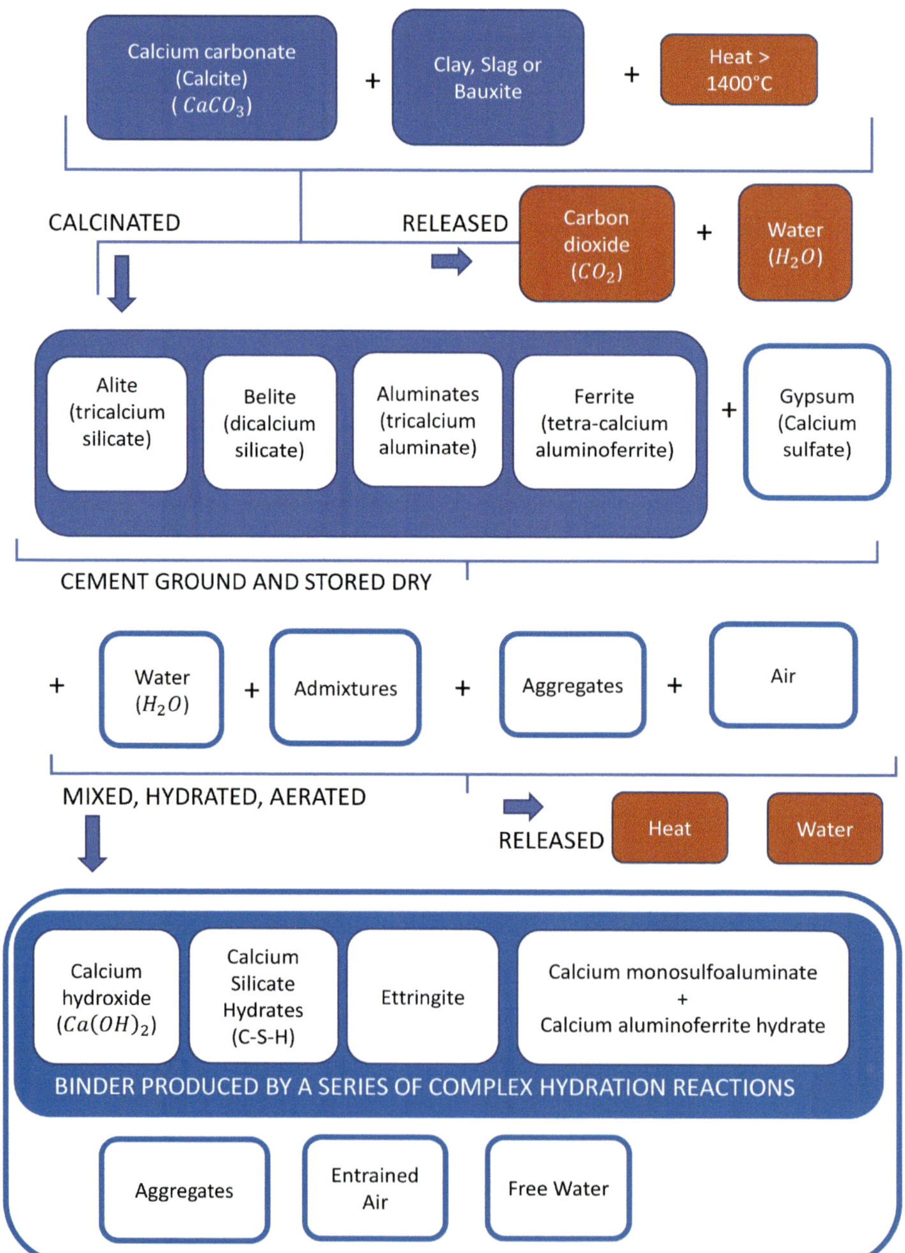

Figure 2.6.2 The concrete process. (Diagram by Susan Pranger, 2023.)

About 50% of the hydrated cement binder are hydrated calcium silicates and calcium hydroxides that are formed quickly by the chemical reaction of alite, belite, and water, and give cement its strength. About 15% to 25% of the binder are calcium hydroxides that reacts more slowly with water and carbon dioxide over time. Calcium hydroxides can also react with sulfates and water after the concrete has set, creating damaging particles of ettringite that swell in the pores of the concrete. Some but not all calcium hydroxide and ettringite produced by the primary reaction combine with other compounds in the cement to produce tetracalcium aluminate hydrate and calcium monosulfoaluminate. However, if sulfates are present, the calcium monosulfoaluminate crystals revert to ettringite, and expand by 2½ times which can damage the concrete.

Aggregates add compressive strength, reduce the amount of binder needed, add porosity to the mix and reduce shrinkage. Reducing the percentage of aggregate relative to binder can improve workability and flow but decrease strength and increase shrinkage. Sand comes in a variety of colors, sizes, and shapes. Unwashed sea sand is likely to be contaminated with salts that will draw moisture into the mortar. Both sand and aggregates can be contaminated with clay, chlorides, and other impurities.

Although there is a long history of using organic fibers in natural cements, artificial cements typically used asbestos, fiberglass, and other synthetic fibers to increase strength, reduce shrinkage or add fire protection, insulation, and acoustic properties. More recently, polymers and basalt fibers have been used.

Ferrous and non-ferrous anchors have always been used in natural cement but steel reinforcement became more extensive and widespread with the development of concrete structures. In sound, alkaline concrete, these ferrous materials are protected by a passive layer of calcium oxide that inhibits corrosion.

A wide range of synthetic additives were introduced during the 19th and 20th centuries in the quest to improve strength, durability, water resistance, or workability of cements and concrete. Contemporary additives include accelerators, retarders, air-entrainment, polymer modifiers, and bonding agents. These should be used with care, as they can inhibit the bond, increase density, trap moisture, and alter workability.

Calcium chloride ($CaCl_2$) was a popular additive found in mid-century concrete because it accelerated the concrete set and reduced the risk of frost damage. Its use ended in the 1970s, when it became apparent that $CaCl_2$ increased metal corrosion.

Plasticizers were introduced in the United States in the 1930s to improve strength and permeability by reducing water, or to improve workability. Superplasticizers introduced in the 1980s and 1990s led to the development of flowable, self-compacting concretes.

Concrete Workmanship

Although concrete can be mixed and placed with a minimum of skill, there are still opportunities for missteps that affect the strength and durability of the finished concrete. Strong, durable concrete requires a proper water to cement ratio, proper mixing, and adequate compaction. The appropriate concrete mix for the application must also be carefully considered.

The concrete mix must be blended to break up any clumps, distribute water, cement, sand, and aggregates evenly, and properly aerate the mix. Aeration is critical to introducing pores that increase permeability required for complete chemical reactions and curing, and to allow excess water to evaporate quickly.

The water must be clean, and free of excess salts and minerals, and sufficient to complete the chemical reactions and to facilitate mixing and placing the concrete without adding air bubbles that will weaken the concrete. Dry mixes are generally stronger and more durable if well compacted while wet mixes can be weaker but are more flowable and easier to place and compact. Excess water will evaporate, but excessively wet mixes will trap bubbles and be more porous.

After mixing, concrete is usually poured into a form that determines its shape and exterior texture. The finish and durability of the surface is affected by the texture and porosity of the formwork because joints and gaps can cause uneven evaporation and curing.

Even well-blended mixes begin to segregate as they are moved, placed, and compacted. Concrete was originally placed in small batches, and spread by shoveling or vibrating, which caused heavy aggregates to settle and moved sand, water, and binders to the edges. Pumped concrete, introduced in the 1970s, allowed more even distribution of the concrete mix, but required a wetter mix.

After placing, the mix must be tamped to compact the concrete and to fill any voids in the formwork. Originally, concrete was tamped in layers; later it was vibrated. Vibration can cause cement and water to migrate toward formwork, producing a fine surface without bubbles. However, if the formwork is removed before the cement is fully hydrated, water can create capillary voids as it moves to the surface, and the surface can dry out prematurely, creating a friable surface. As concrete cures, aggregates continue to settle and sand, cement and water rise to the surface. A visible change in color and texture can often be seen between one lift of concrete and the next, as was seen in the original finish of Unity Temple.

Concrete sets up in a matter of hours but can take a month to reach its final strength. Care must be taken to prevent premature drying of the surface from the evaporative effects of the heat of hydration, the sun, and wind. Removing formwork before the concrete has sufficiently cured can cause cracking. Concrete must also be protected from freezing during hydration.

Although cured cement may appear to be inert, chemical changes continue to occur. Carbonic acids continue to react with calcium hydroxide and calcium silicate hydrates to produce calcium carbonates ($CaCO_3$) that increase the strength and chemical stability of the concrete but reduce alkalinity. The loss of alkalinity increases the potential for corrosion of ferrous reinforcement.

Drivers of Deterioration

Concrete made with OPC is resistant but not impervious to cracking and shrinkage; it is considered waterproof but is still porous, and will allow water, carbon dioxide, and salts to move gradually through pores, capillaries, and microcracks. Most deterioration is related to chemical changes in the concrete that are accelerated by weaknesses inherent to the concrete, external stress, or environmental conditions.

The primary causes of deterioration are similar to those in lime products:

- Imperfections in the cement or concrete mix or fabrication process
- Inappropriate use
- Poor workmanship or design
- Corrosion of ferrous metal reinforcement or inserts
- Environmental conditions

Durability – the ability of concrete to resist deterioration – is strongly correlated to porosity and density. Porosity, capillary systems, and cracks control the depth to which water, chemicals, and biological agents can penetrate concrete aggregates and binders. Concrete that is dense and less permeable will be more watertight, and more resistant to penetration by water, sulfates, carbon dioxide, and other chemicals.

Figure 2.6.3 Efflorescence and staining on concrete at BGSC. (Photo by Holland, 2023.)

Types of Deterioration

Concrete deteriorates when the cement binder is dissolved, weakened, or broken and no longer binds the aggregate materials. Failure of the cement binder leads to friability, surface scaling, and cracking which can increase the rate of carbonation and corrosion.

Friability in new concrete occurs when the sand and aggregates are not well bonded, and often indicates that there was too little cement in the mix, that the binder was not fully hydrated, or that the binder has been washed away or broken. Premature drying of the surface cement can prevent complete hydration of the cement and produce a weak surface. Too much binder and fine aggregate on the surface can cause the surface to craze.

Microcracks often begin before the concrete is fully cured, as the binder dries and shrinks, and the heat of curing evaporates water more quickly from the surface than from the interior. Microcracks do not affect strength or durability, but even minor cracking can increase the depth and rate of carbonation, corrosion, rust expansion, and further cracking. Surface scaling and flaking are caused by freeze-thaw or expansion of salts, acids, biological growth, or erosion, and is primarily cosmetic. However, scaling and flaking can increase water infiltration and increase carbonation, and the rough surface is more conducive to biological growth.

Erosion, expansion from ice or salt crystals, biological growth, settlement and undermining of foundations, thermal expansion and contraction, and differential movement between concrete and adjacent materials can all cause cracks to form and expand in cured, usually by breaking the cement bond between aggregates, but sometimes by breaking the aggregates themselves. Freeze-thaw cycles can saturate and deteriorate porous aggregates.

Corrosion is an electrochemical reaction between ferrous materials, chloride ions, oxygen, and water. This reaction causes an expansion in the iron of up to seven times, exerting tremendous force on the adjacent concrete. Ferrous material corrodes when its passive coating fails, either because the concrete is no longer alkaline, or due to the presence of chlorides that penetrate the coating. Corrosion is also triggered by electrochemical reactions between ferrous metals in the concrete and dissimilar materials such as copper, aluminum or zinc-coated steel.

Carbonation makes concrete stronger and more chemically stable but can lead to shrinkage and can reduce the alkalinity both of which contribute to corrosion and the growth of microbiological organisms. When alkalinity adjacent to the steel falls below a pH of 13, the steel loses the passive alkaline coating that protects ferrous materials from corrosion. Many exposed concrete structures built in the early to mid-20th century have now reached a level of carbonation and de-alkalization that is allowing reinforcement to corrode. Increases in temperature, humidity, rain, and freeze-thaw are all contributing to more rapid carbonization and corrosion.

Chlorides in the original concrete mix, or those introduced by the environment through salt water, marine air, or deicing salts, do not affect alkalinity but can penetrate the protective passive layer and cause corrosion if the ratio of chloride ions to hydroxyl ions exceeds 0.6.

Staining is caused by soot, biological growth or corroded metals that collect in the pores of the concrete. Ferrous materials can also leach to and stain the surface.

Imperfections in Materials

A strong and adequately porous mix is best achieved with adequate binder and a well-graded mix of aggregate sizes. Since cement binder is generally more expensive than the aggregates, most mixes seek to minimize the proportion of binder. A well-graded mix with insufficient binders will have empty pores that will absorb water and be vulnerable to frost damage. A poorly graded mix of aggregates can perform well if there is enough binder to coat the aggregates, but a mix with too many fine aggregates will require more water to become workable and is more likely to shrink as the excess water evaporates. A wet mix will also dry more slowly, increasing the risk of frost damage.

Incomplete calcination (firing) of clay will increase the percentage of unfired clay fines, reduce bonding strength and hydraulic qualities, and increase porosity. Dolomite that is overburned will hydrate more slowly, and continue to hydrate after installation, producing a friable mortar.

Delayed ettringite formation (DEF) can be caused by reaction of sulfides and hydrated calcium hydroxides that causes damaging swelling after the concrete has set. This type of sulfate damage is rare in cement or in hydraulic limes, but it is more common in blended lime-cement mortars where sulfates inherent to the cement can attack the lime.

Aggregates can introduce impurities such as salts, non-reactive clay, reactive silica, or ash that can be detrimental to both lime and cement mixes. Non-reactive clays can adhere to aggregates and interfere with the bond between binders and larger aggregates. Clays are also hygroscopic and will swell when saturated. Unwashed sea sand frequently contains sodium chloride which is hygroscopic and attracts moisture to the surface of the mortar. Wet-dry cycles can cause salt crystals to expand, clog pores, and break bonds. Sodium chloride can also corrode ferrous reinforcement and anchors. Iron in sand can cause staining.

Minerals in the aggregates can react with the alkali hydroxides in the concrete to cause an alkali-silica (ASR) or an alkali carbonate reaction (ACR), a crystalline gel that expands when wet, cracking the concrete. Alkali-reactive silica is rare but found in some volcanic aggregates. This type of silica reacts with potassium and sodium hydroxides and other strong alkalis commonly found in cement, cement-lime mixes, and concrete Cement with high alumina (HM) can deteriorate over time.

Concrete in the early 19th century often used often contains ash from coal burners, coal cinders, or crushed brick as aggregates. These materials are weak, porous, and tend to absorb water, leading to freeze-thaw damage. Coal cinders also contain sulfur that can form reactive solutions that trigger corrosion of ferrous inserts and sulfate attacks of the binders.

Imperfections in Fabrication

A precise mix of materials and processes are required to achieve the desired strength and durability in concrete. The complex chemistry of modern concrete is balanced with rigorous measurement and testing during construction. However, these processes were not as well understood when exposed exterior concrete was first introduced. Even in modern construction, very few processes are capable of achieving perfect proportions and complete reactions between all of the materials in the field.

Hydration is a slow chemical process that usually allows time for the cement to be mixed with aggregates, transported, and placed before it hardens. However, if heat from the initial exothermic reaction, or low humidity in the air evaporates water from the surface before the cement is fully hydrated, incomplete hydration will leave an excess of calcium carbonate and a soft surface that is more vulnerable to salt and to frost damage.

Concrete that was mixed, placed, and compacted by hand is more likely to be unevenly mixed and poorly consolidated, reducing strength and increasing the likelihood of voids and honeycombs that absorb water. Insufficient or inconsistent thickness of the cement cover over reinforcement can lead to early corrosion. Joints between lifts of concrete can dry prematurely, or have high cement content if aggregates settle, causing a weak spot that can absorb water.

The rate of carbonation is higher in poorly compacted or cracked concrete that more rapidly admits carbon dioxide and water. Even tiny shrinkage cracks and surface crazing can expedite the rate of carbonation. Carbonation could take 50 years to penetrate 10 mm, or 200 years to penetrate 20 mm in a well hydrated, dense concrete. However, it could reach 20 mm in less than 10 years in poor-quality concrete.

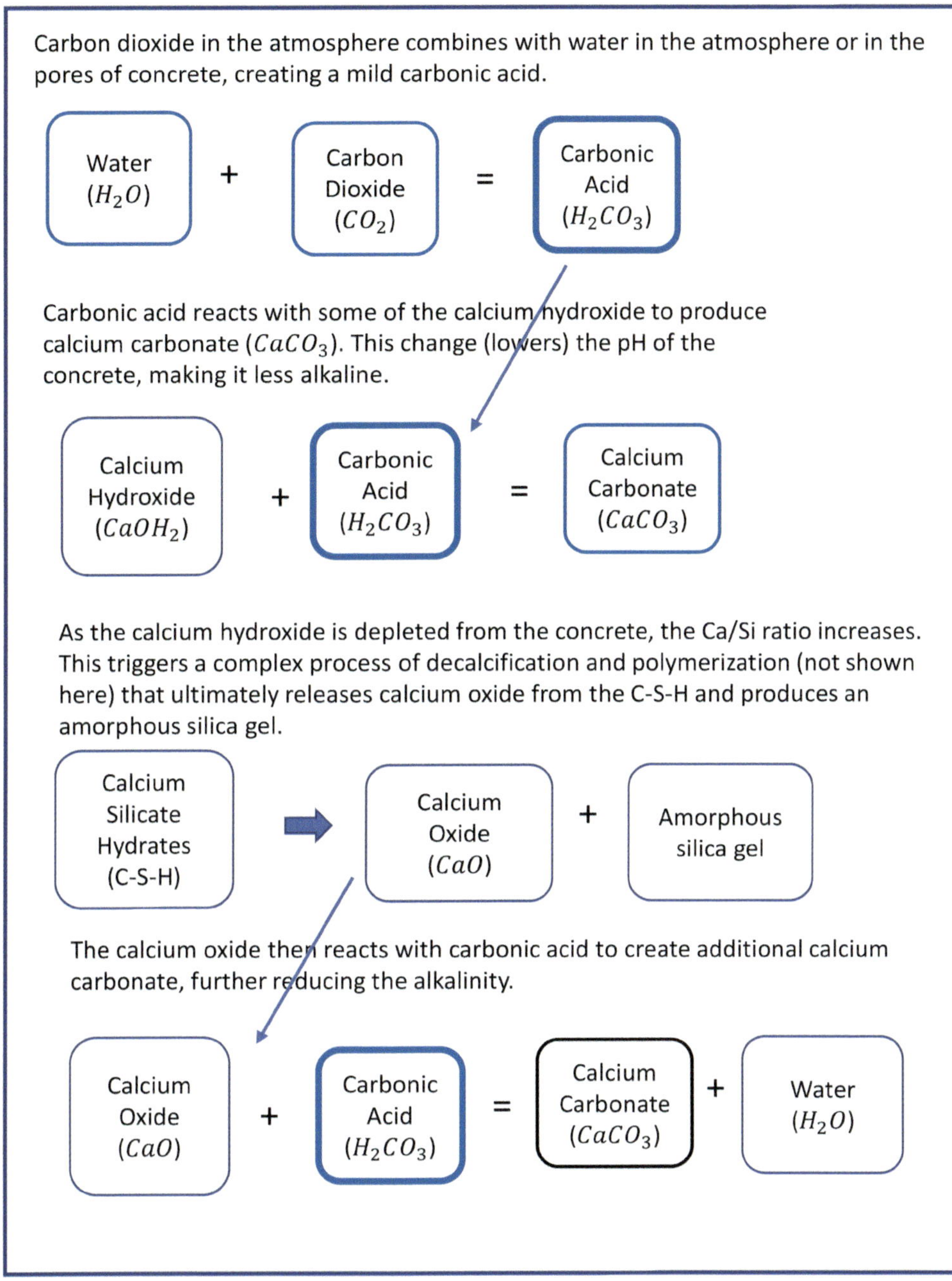

Figure 2.6.4 Carbonation of Portland cement concrete. (Diagram by Susan Pranger, 2023.)

Environmental Factors

Concrete that is well produced and appropriately used is still subject to gradual deterioration from seasonal weather and exceptional events. The best quality concrete products, thoughtfully detailed and maintained, will still absorb water, carbon dioxide and chemicals that can cause stresses from the expansion of ice, salt crystals or corrosion. Greater daily temperature swings, longer periods of rain or drought, and more wet-dry and freeze-thaw cycles are likely to accelerate this damage.

The hard, inflexible nature of concrete makes it vulnerable to cycles of iterative damage. Fluctuations in humidity and temperature cause expansion and contraction cracks. Cracks admit water and salts that cause larger cracks, and so on. Salts attract more water and more salts, creating its own cycle of increasing damage.

Water

Water typically strengthens the chemical bonds of cement binders, but it can also facilitate carbonation, the transport of salts, and the formation of ice crystals. Water also combines with ferrous minerals to cause corrosion. This damage is accelerated when moisture is trapped in the pores by impervious consolidants or coatings.

Rain falling on a concrete surface is absorbed until it creates a surface layer that causes additional rain to sheet down the wall. However, the longer and farther water travels down a wall, the more likely it is to penetrate. Heavy or prolonged rains will increase the depth of water saturation.

Concrete materials with tight, dense pores will absorb less moisture than more porous materials but are also slow to dry. Sunlight, low humidity, and high temperatures all encourage evaporation of water from the surface, but water that has penetrated more deeply will dry more slowly. The longer the wall is saturated, the more likely it is to suffer deterioration.

Alternating cycles of wetting and drying can carry and deposit soluble salts through concrete. More frequent cycles promote the buildup of larger salt crystals that eventually fill pores and exert pressure that causes disintegration of the binder or spalling on the surface.

Heavy rains also increase the buildup of ground and surface water that can saturate foundations and rise up into walls, often carrying soluble salts. Excess water in concrete can support corrosion and biological damage.

Longer rains allow concrete materials to slowly absorb moisture that is slow to evaporate. Rapid changes in temperature can cause freeze-thaw cycles that produce ice crystals that fill small pores and break the bonds, causing accelerating deterioration over time. Water expands in volume by 9% when it freezes, presenting a formidable risk to concrete that has not set up, or cured concrete with microcracks or larger cracks.

Sunlight and Heat

Temperature affects concrete in two ways. First, by directly causing expansion and contraction of the concrete, and second, by driving the movement and expansion of water within the concrete.

A certain amount of expansion and contraction is easily absorbed by most materials except where the material is constrained, or when there is a difference in the coefficient of expansion. Extreme temperatures can break the chemical bonds within cured concrete. Metal anchors and reinforcement expand more rapidly than concrete, which can break the bond between the two materials, or cause cracking of the relatively weaker concrete. Expansion of mass concrete can cause cracking at restrained corners, as seen at both Baha'i and Unity Temples in Chicago.

Fluctuations in temperature and humidity can cause condensation on the surface or within pores of concrete. A drop in temperature can cause ice crystals to form within pores, extremely hot temperatures can also expand water in small pores more rapidly than it can evaporate, or deposit salt as it evaporates, all of which can rupture the bonds within the material.

Wind

Wind can directly damage concrete by eroding the surface, separating binder from aggregates or by causing rapid drying that interferes with final curing. Wind can create positive or negative pressure through a wall that drives water vapor or rain more deeply into the pores.

Extreme weather events that combine more intense wind and heavy rain will likely cause deeper penetration of moisture into concrete walls.

Chemicals

Chemical damage to concrete is primarily caused by migration and deposits of salts, carbonation caused by carbon dioxide, and other and electrochemical reactions that cause corrosion. The cement binders are alkaline, with a pH between 12.5 and 13.5, which makes it attractive to deterioration by aggressive acids that naturally occur in the environment. Soluble salts can recrystallize as either efflorescence (deposits on the surface) or crypto-florescence (crystals deposited within the pores or just under the surface). Many residual salts are hygroscopic and will attract water from the atmosphere even after the initial source is removed. Ferrous minerals and pyrites can cause corrosion that produces stains or mechanical stress.

Soluble or partly soluble salts are found in cements or are formed during hydration of cement mixes. These cement hydrates include calcium hydroxide, sodium hydroxide, sodium silicate, sodium sulfate and calcium sulfate. These salts can attack adjacent materials or react with silicates in aggregates to attack cement.

Destructive chemicals can be intrinsic or extrinsic. They can exist in raw materials added during manufacture, aging, or weathering, or migrate from one material to another. Cement can contain alkaline salts, soluble calcium sulfates and sodium salts that can leach out over time. Aggregates can include impurities of salt and minerals. Natural stone can contain minerals and salts that can migrate into adjacent cement materials. Chemicals can also be introduced by organic or biological agents. Pigeon droppings, all too common on urban facades, can be very acidic. Certain bacteria, molds and fungi can excrete acids that break down cement binders.

Water, carbon dioxide, and destructive salts and other chemicals are carried through concrete by the water, water vapor or air. Rain, fog, and persistent humidity can carry atmospheric pollutants, including acids and salts, gradually but deeply into porous concrete. Rising damp is particularly likely to introduce salts from the soil, ground, or surface water. Water and water vapor in marine environments can contain salts. Ammonium nitrate fertilizers can attack concrete. Deicing chemicals for pavements include sodium chloride, calcium chloride, magnesium chloride, and potassium chloride.

Sulfates found in acid rain, air pollution, and ground water react with hydrated calcium aluminates in natural and artificial cements to create the mineral ettringite (calcium aluminum sulfate).

Chlorides introduced into the concrete mix by calcium chloride additives (an anti-freeze additive used in the 1920s), contaminated aggregates, or salty mixing water, are released within the concrete as the concrete carbonates. Chlorides cause a general deterioration of the concrete, but the primary impact is corrosion of rebar in the concrete. Excess chloride in the pores on concrete will break down the protective alkaline layer around reinforcing steel and remain on the surface of the steel, allowing corrosion to continue indefinitely. Leaching and staining is an early indication that serious corrosion is occurring.

Calcium chloride and magnesium chloride deicing salts combine with water and portlandite (calcium hydroxide) in the concrete to form calcium hydroxy chlorides (CAOXY). The formation of CAOXY causes damaging expansion in the concrete.

The ammonium nitrate found in fertilizers can trigger a series of damaging reactions that quickly damage cement binders and reinforcing steel. As little as 0.5% ammonium nitrate in the concrete can cause brittleness and direct tensile failure of steel reinforcement. Ammonium nitrate combines with calcium hydroxide in cement binders to produce calcium nitrate and ammonia, which are very soluble and leach out of the concrete, enlarging pores and weakening the concrete. The reduction in calcium hydroxide also makes the binders less alkaline, which encourages corrosion. The calcium nitrate combines with hydrated aluminates in the cement to produce calcium nitro aluminate, which is even more soluble, and can attract water from wet surfaces to dry areas, spreading salt to new areas of the concrete. The accumulation of salt can crack or blister the concrete.

Biological Actors

Microorganisms, including bacteria, cyanobacteria, fungi, and algae, easily colonize rough, porous concrete surfaces. The damage is initially cosmetic, but subsequent cracks and moisture allow colonization by larger mosses and lichens, ivy and the roots of other opportunistic vegetation. Microorganisms also create biofilms that affect the surface porosity, texture, and mineral and chemical composition of the concrete.

Microorganisms consume minerals in the concrete and produce metabolites including acids, CO_2, and sulfuric compounds. These compounds react with the concrete to create materials that are soluble (calcium nitrate), expandable (ettringite), or lack cohesion (gypsum).

Both bacteria and fungi produce acids that etch the surface by dissolving carbonates. Autotropic bacteria convert sulfur and nitrogen compounds in the air into sulfuric or nitric acids. The hyphae of fungi can expand and contract, enlarging cracks on the surface. Lichens, a symbiosis of algae and fungi, can penetrate a few millimeters by excreting acid, or burrow more deeply with hyphae or roots.

Fire

Concrete is considered fireproof and will not combust. However, concrete can become friable from prolonged exposure to high temperatures that rapidly expand water in the pores or cause chemical changes by forcing water from the cement binders.

Mechanical Stress

Concrete is very inflexible and will crack in response to the mechanical stresses caused by expansion and contraction, salt and ice crystals, and biological growth. Stress can also be caused by larger environmental events such as flooding, mudslides, and earthquakes that cause vibration or differential movement of concrete foundations and structures.

Strategies for Responding to Climate-Based Threats to Concrete

As with all other materials, it is important to identify the cause of weather-related damage rather than just addressing the result. Responding to weather-related damage in concrete is complicated by several factors:

- Hard, inflexible concrete is vulnerable to damage that might be absorbed by flexible materials.
- Many historic concrete structures have reached a level of carbonation that allows corrosion of ferrous reinforcements.
- Internal damage is difficult to identify and rectify.
- Patches and repairs are not easily concealed.

- It is difficult to replicate historic concrete materials.
- It is not easy to modify a concrete structure without changing its appearance or properties.

There are many options for slowing the process of carbonation and corrosion in new concrete, but the options for treating existing carbonation and corrosion are more limited:

- Patching cracks to reduce infiltration.
- Replacement or treatment of corroded reinforcement and damaged concrete.
- Consolidation and other surface treatments.
- Electrochemical extraction (ECE) of chloride, and cathodic protection.

Ultimately the development of a maintenance or restoration plan should include experts who are able to provide a holistic interpretation of visual assessments and other testing, and who understand how to effectively intervene in the complex interactions of chemical and structural changes that occur as concrete weathers.

Investigation and Testing

Past history, current observations, and test results must be evaluated by both structural and corrosion experts who understand the relationships between chemistry, material properties, structural issues, and environmental conditions. The dynamics of thermal and moisture movement through the concrete, and their effect on water-soluble minerals and salts, must be considered when evaluating existing conditions and considering future treatments.

It is critical to understand the location and condition of ferrous inserts and the extent of concrete carbonation to identify existing issues and predict future problems. The amount of concrete cover, the depth of carbonation and the level of chloride-ion concentration should be determined. Preventive responses will be much less invasive and expensive than extensive repairs after corrosion has occurred.

The degree of carbonation of the concrete is a good indicator of the risk of corrosion. Copper-sulfate half-cell tests and linear polarization techniques can provide information on the levels of carbonation or chloride and help predict the rate of corrosion. Rainbow indicator and phenolphthalein are two fast and accurate test results to measure carbonation but can be inaccurate if the sample is not tested promptly, if the matrix is complex, or the aggregates are colored.

Linear-polarization resistance testing can identify the extent of actual corrosion, but results must be interpreted by someone qualified to interpret the impacts of environmental conditions. Epoxy or galvanized steel, a variety of historic admixtures, patches, and parging can also distort the results. Corrosion modeling can provide additional information.

Laboratory tests will usually require sample cores. Chemical analysis can determine chloride, sulfate, and alkali levels, identify deleterious aggregates, and determine the depth and degree of carbonation. Petrographic evaluation of samples under a microscope should be conducted by a geologist that specializes in construction materials, and can provide detailed information on the air content, water-cement ratio, cement content, and type of aggregates. Compression studies can determine the strength of concrete. Laboratory tests can identify damage from freeze-thaw cycles, alkali-aggregate reactions, or chemical attacks.

Maintenance

Maintenance of concrete surfaces should begin with monitoring and correcting any potential sources of water infiltration, including roofs, gutters and downspouts, sealant joints and coatings. Structural or settlement cracks should be repaired or sealed to avoid water infiltration, after correcting any underlying causes.

A regular inspection and testing program should be implemented to monitor for corrosion or deterioration before treatment becomes prohibitive or hidden failures cause more catastrophic failures. Inspections should involve both structural and corrosion experts.

Cleaning may be appropriate to remove biological growth and chemicals, but chemical cleaners, especially strong acids, and even low-pressure particulates can change the appearance of the concrete or remove or etch the cement surface.

Repairing Concrete

Any attempt to patch concrete must be preceded by an understanding of the cause of the original damage. This often will require addressing the cause and result of any steel corrosion. Patching materials must then be selected that will be compatible with the properties of the adjacent materials in strength and porosity as well as texture and color. Trial repairs should be conducted and evaluated for technical performance and aesthetic appearance prior to proceeding with widespread repair work.

Patching materials should be selected to prevent water infiltration but not restrict seasonal thermal movement. Ideally, the patching material should be chemically identical to the original cement and concrete mix. Synthetic patching materials should be used as a last resort where cement-based materials are not adequate, and only after appropriate testing and consultation by experts. Sealants may be more appropriate where cracks may continue to move.

The goal is to retain as much original material as possible and to avoid enlarging existing cracks however damaged or friable concrete should be removed to provide a sound substrate for new patches. Exposed reinforcement must be carefully cleaned to remove existing corrosion and treated to inhibit future corrosion. New non-corrosive materials, such as mild steel, epoxy-coated steel or stainless steel may be required to reinforce patches. All patches must be carefully cured to avoid premature drying, shrinkage, and cracking.

Modifications

The durability of concrete also depends on how shrinkage, water, and other stresses are managed by details of the concrete and the overall construction. Control joint and expansion joints determine whether and where cracks will occur. Stable foundations and substrates can minimize stress from settlement, earthquake vibration and thermal expansion. Concrete surfaces should be designed to drain and not trap water. Water that sits on the surface will migrate into the pores where it can cause ice or salt damage or facilitate biological growth. These are conditions that are not easily modified without changing the appearance of exposed concrete but may be required to prevent further deterioration.

Coatings and Surface Treatments

Surface treatments and coatings have been used to reduce carbonation, conceal repairs, and consolidate friable surfaces. However, even clear coating will change the appearance of the concrete, and thicker coatings will conceal the original texture. Impervious coatings eventually trap moisture in the concrete unless they are carefully specified and maintained. Advances continue to be made in the use of natural and synthetic treatments and coatings, but success is very dependent on the specific concrete, and environmental conditions.

Surface treatments today include the following:

- Paints and other film forming surface coating that span rather than fill surface pores.
- Penetrating sealers that cover the surface and fill surface pores.
- Treatments that chemically react with the concrete.

Paint and Surface Coatings

Coatings that form a film on the surface of concrete significantly change the appearance and performance of concrete and are generally not recommended where they previously did not exist. However, they are frequently installed to provide a more uniform appearance or to protect or conceal patches.

Unlike penetrating sealers, film forming coatings sit on the surface of the concrete and do not fill pores. If not maintained, coatings can allow water to penetrate and build up behind the coating, causing the concrete to deteriorate or the coating to delaminate. Preliminary studies suggest that acrylic-based surface coatings, specifically vinyl-acrylic and silane coatings, have the potential to significantly reduce the rate of carbonation by reducing permeability if they are uniformly applied and consistently maintained over the service life of the concrete. However, polymer-modified cementitious coating shows a negligible reduction in carbonation.

The naturally alkaline surface of new concrete is not compatible with most paints, especially those produced before 1939. Around 1930, paint manufacturers recommended that concrete be allowed to cure for 12 to 18 months before applying paint.

Repellants and Sealers

Sealers, stabilizers, and neutralizers penetrate more deeply than paint and are used to stabilize friable or dusty surfaces, or to provide a primer for paint. Many are clear but will still leave a surface sheen that affects the appearance of the concrete. Clear coatings also have the potential to attract and hold soot or dirt, which also changes the appearance. If the film wears unevenly, the surface can become streaky or splotchy in appearance.

A variety of penetrating sealers were developed in the 20th century to reduce water infiltration. These sealers provide a clear, hydrophobic coating that protects fine cracks and pores but does not bridge cracks. Most sealers are silicone based or polyvinyl acetate copolymers dispersed in water. In general, waterproofing is a surface application with a limited life, and requires reapplication to remain effective. If not maintained, sealers will wear unevenly, leaving remnants that can produce a splotchy appearance in the rain, and will interfere with later treatments of dissimilar materials. As with any modification, they should be used as a last resort and only after appropriate testing and consultation by experts.

Chemically Bonded Treatments

Acid stains and mineral silicate paints stains both penetrate and react chemically with the concrete.

Acid stains rely on a reaction with the initially alkaline surface of fresh concrete and may not perform well when applied to aged concrete where the alkalinity is both reduced and uneven. Acid stains generally do not improve the durability of the concrete although reactive polymer stains use acrylic and inorganic silicate polymers that bind to the cement to provide a durable, UV-resistant coating.

Potassium silicate coatings are being used successfully for mineral based substrates like limestone, lime renders and concrete. The soluble silicates dry to form a hard, breathable, UV resistant coating that resists moisture and biological growth and also combines with unreacted calcium hydroxide to form hard calcium silicate hydrates.

Nanotechnology, already used to consolidate lime products, is being explored as a method to use nano-sized particles to penetrate more deeply into concrete to reverse carbonization. Similar research is exploring the use of enzymes. Nanotechnology is already being used to develop more effective polymer-modified concrete coatings and patching materials.

Electrochemical Extraction

Electrochemical rehabilitation is used to reverse the carbonation and corrosion in existing concrete. An electrical current is run through existing concrete to trigger an electrochemical reaction that removes chloride ions and restores the alkalinity of existing concrete. A variety of methods have been developed that are applied short or long term, including cathodic protection (CP), electrochemical realkalization (ER), electro deposition, electrochemical chloride extraction (ECR) and inhibitor electro migration anticorrosion technology.

Electro-chemical extraction (ECE) is widely used to reduce chloride concentrations in existing concrete, but the results vary with the method used, the arrangement of rebars, the depth of cover and the duration of treatment. Preliminary studies indicate that ECE can reduce the concentration of chlorides in existing concrete by 20% to 90%, with a corresponding reduction in corrosion of 60% to 90%.

In ECE, a high-voltage direct current (DC) is connected to a temporary electrolyte mesh in solution on the surface of the concrete (the anode) and to the existing reinforcement mesh (the cathode). The chloride ions on the rebar become negatively charged and migrate toward the positively charged anode mesh where they are trapped in the electrolyte or released as chlorine gas. Hydroxyl ions are formed on the rebars, increasing the alkalinity of the surrounding area, and decompose water at the anode. Calcium hydroxide solutions are sometimes used in lieu of potable water as an electrolyte because it is less likely to become acidic and produce chlorine gas.

Realkalization restores the alkalinity of concrete by soaking it in an alkaline solution, using an electrical current (similar to ECE) to force it toward the reinforcement steel.

Cathodic protection has proven to be a successful method of preventing corrosion of ferrous reinforcement. It is similar to ECE, except that the anode is a permanent mesh and coating, and the current is run continuously. The installation must be monitored and maintained indefinitely.

Cathodic activity is very moisture sensitive. Monitoring by the Oregon Department of Transportation has confirmed significant variation in the flow of water and ions in the wet winter versus dry summer.

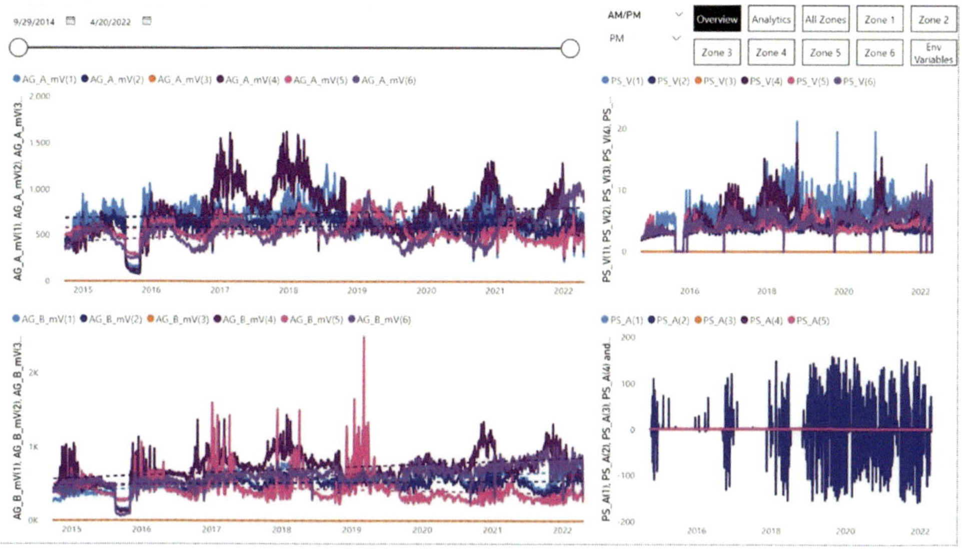

Figure 2.6.5 Screen capture from ODOT monitoring of cathodic protection at Big Creek Bridge, Oregon. (Screen capture courtesy of Oregon Department of Transportation, 2022.)

Conclusions

Although a variety of environmental conditions cause erosion, spalling and other damage to concrete, the processes that accelerate corrosion of ferrous metals in the concrete cause the most damage. The natural process of carbonation reduces the passive protection of ferrous materials. Concrete that is very old is less likely to have extensive ferrous reinforcement; however, early reinforced concrete is now in the 50- to 100-year age range and is already at or is approaching full carbonation. Increases in temperature and humidity in many regions will accelerate the process of carbonation and corrosion. Increasing cycles of freeze-thaw can damage concrete by opening cracks and admitting water, further increasing the rate of carbonation. Chemicals and water driven into the concrete by driving rain or pressure differences will also increase corrosion.

Good maintenance practices that manage water, avoid salt contamination, and repair surface cracks in a timely manner can reduce the rate and impact of carbonation, corrosion, and freeze-thaw damage. Coatings, consolidants, and electrochemical treatments can reduce or reverse damage in some circumstances. In any case, maintenance, repair and new treatments must be preceded by proactive investigations of the character and condition of the concrete, and appropriate testing to determine the impacts of proposed treatments.

Bibliography

Armbruster, Robert, interview by Susan Pranger. 2022. (November 21 and December 15).

Armbruster, Robert F. 2014. "Secrets of John J. Earley's Mosaic Concrete on the Barh'i Temple."

Basu, sudenshna' Orr, Allan Scott, and Yasemin D. Aktas. 2020. "A Geological Perspective on Climate Change and Building Stone Deterioration in Londan: Implications for Urban Stone-built Heritage Research and Management." *Atmosphere.* https://www.mdpi.com/2073-4433/11/8/788/htm.

Bibi, Tayyaba, Amir Ali, Jianhua Zhang, Amjid Naseer, and Shams Ul Islam. 2020. *Microscopic Analysis of the Deleterious Effects of Ammonium Nitrate Fertilizer on Concrete.* Amsterdam: Elsevier Ltd.

Birgisson, Bjorn, Anal K Mukhopadhyay, Georgene Geary, Mohammad Khan, and Konstantin Sobolev. 2012. "Nanotechnology in Concrete Materials." *Transportation Research Circular*, December.

Crevello, Gina. 2015. "Data Interpretation and Early Corrosion Test Program for Historic Concrete Buildings." *APT Bulletin Special Issue on Modern Metal Finishes* (Association for Preservation Technology International) 46 (1): 1–8. http://www.jstor.org/stable/43390259.

Edison. 2015. *Everkote 300 Mineral Coating Systems.* December 17. https://www.edisoncoatings.com/Potassium_Silicate_Masonry_Pai/potassium_silicate_masonry_pai.html.

Elert, Kerstub, Carlos Rodriguez-Navarro, Eduardo Sebastian Pardo, Erid Hansen, and Olga Cazalla. 2002. "Lime Mortars for the Conservation of Historic Buildings." *Studies in Conservation* (Taylor & Francis, Ltd on Behalf of the International Institute for Conservation of Historic and Artistic Works) 47 (1): 62–75. https://www.jstor.org/stable/1506835.

English Heritage. 2012. *Concrete.* Edited by David Odgers. London: Routledge.

———. 2011. *Mortars, Renders & Plasters.* Edited by Henry, Alison, and John Stewart. New York: Routledge.

Gaudette, Paul, and Deborah Slaton. 2007. *15 Preservation of Historic Concrete.* Preservation Briefs, National Park Service.

Gilmore, Alice Louise, and Frank Matero. 2018. "Performance Testing of Acid Stains on Historic Concrete." *APT Bulletin: The Journal of Preservation Technology* 49 (4): 5–15. https://www.jstor.org/stable/10.2307/26632383.

Hes, Dominique, and Margaret Bates. 2003. *Concrete and Sustainability - Supporting Environmentally Responsible Decision Making.* BPD Environment Design Guide, Royal Australian Institute of Architects, 1–12. https://www.jstor.org/stable/10.2307/26148410.

Jang, Jean. 2016. *Performance Evaluation of Commercial Nanolime as a Consolidant for Friable Lime Based Plaster.* Master's Thesis, Philadelphia, PA: University of Pennsylvania. https://repository.upenn.edu/hp_theses/597/.

Kosmatka, Steven H., Beatrix Kerkoff, and William C. Panarese. 2002. *Design and Control of Concrete Mixtures, EB001.* 14th ed. Skokie: Portland Cement Association.

Krishna. 2018. *14 different types of cement you must know.* February 27. http://civilread.com/14-different-types=cement-must-know.

Lee, Hosin "David", Jungyong "Joe" Kim, Steve Jacobsen, and Rosanne Edwards. 2005. "Effectiveness of Electrochemical Chloride Extraction for the Iowa Avenue Pedestrian Bridge." Public Policy Center, Civil and Environmental Engineering, University of Iowa.

Macdonald, Susan. 2003. *The Investigation and Repair of Historic Concrete.* Parramatta: NSW Heritage Office.

Mao, Jianghong, Kaiqi Yu, Yidong Xu, Xixi Wu, Weilinag Jin, Chen Xu, and Chonggen Pan. 2016. "Experimental Research on the Distribution of Chloride Ion Migration in Concrete Cover during Electrochemical Chloride Extraction Treatment." *International Journal of Electrochemical Science* 11: 4079–4083. www.electrochemsci.org.

Means, Eric, interview by Susan Pranger. 2021. *Eric Means Constructions* (November 30).

Ministry of Works. 1944. *The Painting of Buildings, by a committee convened by the Paint Research Association 1944.*

Miron, Luis Emilio Rendon: Diax. 2017. "The Effect of Microorganisms on Concrete Weathering."

Monkman, Sean. n.d. *Types of Concrete Carbonation.* Technical Note, Carbon Cure Technologies, Inc.

Oregon Department of Transportation, 2022. *Big Creek Graphite Based Cathodic Analysis Program.* ODOT Bridge Preservation Program

PCA. 2019. *Freeze Thaw Resistance.* Portland Cement Association.

Pearson, Candace. 2020. "Can Concrete Become Carbon Neutral?" *Building Green.* November 9. https://www.buildinggreen.com/newsbrief/can-concrete-become-carbon-neutral.

Peng, Lizhengli, and Mark G. Stewart. 2014. "Climate Change and Corrosion Damage Risks for Reinforced Concrete Infrastructure in China." *Structure and Infrastructure Engineering.* https://doi.org/10.1080/15732479.2013.858270.

Pinehiro, Sayonara Maria de Moraes, and Moema Ribas Silva. 2003. "Alteration of Concrete Microstructure by Biodeterioration Mechanisms." *Conference on Microbial Impact on Building Materials.*

Rendon Diaz Miron, L.E., and M.R. Lara. 2017. "The Effects of Microorganisms on Concrete Weathering." In *Concrete Durability*, by Koleva D. Rendon Diaz Miron L.E. Springer, Cham.

Rodriguez-Navarro, Carlos, and Encarnacion Ruiz-Audo. 2018. "Nanolimes: from Synthesis to Application." *Pure and Applied Chemistry* 523–550.

Sevelsted, Time F., and Jargen Skibsted. 2015. "Carbonatio of C-S-H and C-A-S-H Samples Studied by .MAS NMR Spectroscopy." *Cement and Concrete Research* 71: 56–65. https://doi.org/10.1016/j.cemconres.2015.01.019.

Sharp, Stephen R., Gerry G. Clemena, Y. Paul Virmani, Glenn E. Stoner, and Robert G. Kelly. 2002. *Electrochemical Chloride Extraction: Influence of Concrete Surface on Treatment.* U.S. Department of Transportation, U. S. Government, 49.

Sickels-Traves, Lauren B., and Philip D Allsopp. 2005. "Lime and its Place in the 21st Century: Combining Tradition, Innovation and Science in Building Preservation." *International Building Lime Symposium 2005.* Orlando, Florida.

Siry, Joseph M. 1996. *Unity Temple.* New York: Cambridge University Press.

Stewart, MArk G., Xiaoming Wang, and Minh N. Nguyen. 2012. "Climate Change Adaptation for Corrosion Control of Concrete Infrastructure." *Structural Safety.* https://doi.org/10.1016/j.strusafe.2011.10.002.

Swift, D. G. 1981. "The Use of Organic Fibers in Cement: Some Structural Considerations." *Composite Structures* (Springer, Dordrecht) 1–41. https://doi.org/10.1007/978-94-009-8120-1_41.

Talati, Shuchi, Na'im Merchant, and Chris Neidl. 2020. *Paving the Way for Low Carbon Concrete.* White Paper, Carbon180.

Talukdar, S., N. Banthia, J.R. Grace, and S. Cohen. 2012. "Carbonation in Concrete Infrastructure in the Context of Global Climate Change: Part 2 - Canadian Urban Simulations." *Cement & Concrete Composites*. Elsevier Ltd. www.elsevier.com/locate/cemconcomp.

Vitruvius. n.d. *The Ten Books on Architecture*. Translated by Morris Hicky in 1914 Morgan. New York: Dover Publications, Inc.

Weerdt, Klaartje De. 2021. "Chloride Binding in Concrete: Recent Investigations and Recognised Knowledge Gaps: RILEM Robert L'Hermite Medal Paper 2021." *Materials and Structures*, 54–214. https://doi.org/10.1617/s11527-021-01793-9.

Woods, Amy Lamb. 2021. "Historic Mortar Replication and Repointing." *Webinar.* International Masonry Institute.

CHAPTER 2.7

Metals

Introduction to Metals

Very few metals exist in nature in their pure elemental form. Most are bound in complex ores and must be extracted, refined, and shaped before they can be used as building components. Many metals are mixed with other metals to create alloys with properties more appropriate to their use.

Both pure and alloyed metals tend to revert to their natural state as oxide-ores when exposed to the atmosphere and weather. This electrochemical process of corrosion can happen very slowly or very quickly, depending on the nature of the specific metal, the character of surrounding materials, and the atmospheric conditions of temperature, humidity, and pollution.

Shiny, malleable metals are used for many functional and decorative products. Metals used in traditional building exteriors include the following:

- Exterior finishes of copper, aluminum, tin and zinc, and alloys of bronze, brass.
- Coated exterior ferrous iron, wrought iron, and cast iron, and alloys of stainless steel.
- Concealed ferrous and non-ferrous metal pins or reinforcement.
- Metallic coatings of cobalt, zinc, titanium that are chemically or electrically bound to other metals.

Impacts – Past and Future

History

Gold, silver, copper, and lead are found pure in nature, so it is not surprising that they were the first to be used by early man. Bronze alloys of copper and tin were

DOI: 10.4324/9781003160489-12

made as early as 3500 BCE, and were commonplace by 1400 BCE, so much so that the period from 3300 to 1200 BCE is known as the Bronze Age. However, extraction and refining were time and energy intensive, so early metals were used strategically and selectively for decorative elements and small fastenings or pressed into thin sheets for cladding or roofing.

Ferrous metals were not used until 1200 BCE because of the rigorous and complex processes required to extract iron from iron ore. Cast iron was being produced in China in the 6th century BCE, in Europe by the 14th century, and the Americas by the 17th century. Iron was used more extensively for structural framing after more efficient furnaces were developed for smelting in the 16th century, culminating with the Bessemer converter in 1851 and electric arc furnaces in 1900. The 20th century brought standardized production of wrought iron, the development of stainless-steel alloys, and thermal sprays to prevent corrosion. Ductile iron, a more flexible form of cast iron, was introduced in 1943. Today there are thousands of different alloys, each developed with specific qualities for unique applications.

Social, Cultural, and Climatic Impacts

Alloys and coating have been developed specifically to resist the weather that returns most refined metals to their original oxidized state. However, increased acidity during the first and second industrial ages and increased precipitation in the 20th century have increased corrosive conditions in the atmosphere that can destabilize protective passive coatings and increase corrosion of metals.

Extreme rain events, wind, and UV are all expected to increase in most areas and have the potential to increase visible and hidden corrosion of metal materials. In some areas, drought and fewer days of rainfall will decrease the natural cleansing effect of rain.

Understanding Metals

There are over four dozen "pure" metallic minerals, including alkali metals, alkaline earth metals, transition metals, and metalloids. They produce metals that are lustrous, ductile, fusible, malleable, and easily conduct heat and electricity. Each of these metals and metal alloys exhibit unique characteristics and mechanical properties. They may be weak or strong, brittle, or flexible, and soft or hard, depending on both the type of metal or alloy and on how it was refined and shaped. These characteristics determine whether the metal can be successfully cast or forged, and whether it is appropriate for specific uses. These factors also determine how the material will respond to weather conditions.

Emissivity (the ability to absorb and emit infrared energy) and reflectivity influence the temperature of a metal, and will vary depending on the type of metal and the texture of the surface. A polished surface will be more reflective and have a lower emissivity than a rough or oxidized metal.

Electrochemical Reactions

Most metals exist in nature as stable oxides of elemental minerals combined with oxygen. Metal oxides are converted to "pure" metals by refining, usually with heat, to drive off the oxygen ions. However, refined metals are unstable and seek to recombine with oxygen to return to their natural state as a metal oxide. The type of reaction depends on three factors: the type of metal, the character of adjacent materials, and atmospheric conditions. Understanding the relationship between these factors is key to avoiding or controlling damaging corrosion.

Chemically, metals are a specific type of mineral that tends to form positive ions (cations) while losing electrons. The positive ions tend to move from less reactive areas, called anodes, to more reactive areas, called cathodes. Depending on the type of metal and the electrochemical environment, the surface will react in one of three ways:

- The balanced flow of positive ions to and from the metal produces a stable passive coating.
- More positive ions returning than are lost, leading to cathodic protection.
- More positive ions are lost than return, leading to corrosion.

The flow of ions to and from the anodic metal depends on the character of the metal and the character of the electrochemical environment (electrolyte). Small amounts of salts, acids or alkalis increase the number of ions in water, increasing corrosion.

Corrosion occurs when positive ions flowing from the metal combine with oxygen to form metal oxide (rust) and do not return to form a passive layer. Because there is no passive layer, the metal will continue to lose positive ions, leading to a continuous loss of metal at the anode, and a buildup of metal oxides at the cathode. This loss can be slow or rapid, depending on the metal and the surrounding cathodic conditions. Corrosion increases in the presence of water and oxygen and is accelerated by the presence of salts. Porous metals such as iron and steel will trap water and salt and corrode quickly. If left unprotected from the atmosphere, corrosion accelerates over time until the materials lose structural integrity.

Ferrous materials will typically corrode if left exposed, and must be protected by painting, galvanizing, anodizing, or electroplating. Other materials, such as copper, bronze, and brass, will naturally "passivate" by forming a protective coating that inhibits additional corrosion. Some materials are inert and slow to react with the atmosphere, such as stainless steel and aluminum. Electroplating is similar to electrolysis in that it uses an electrical current to cause migration of metallic salts from the bath to the surface of the metal.

Types of Corrosion

Ions migrate from anodic to cathodic areas that exist across the surface of a single piece of metal. As the anodic areas erode, they become more cathodic relative to adjacent areas, reversing the flow of ions. Eventually a coating of corrosion is uniform over the entire surface. Intergranular corrosion can occur when anodic and cathodic zones meet at the boundaries of the metal crystals or at variations in the orientation of metal grains. Severe deterioration by intergranular corrosion is most common in stainless steel where it can cause whole grains to separate.

This corrosion can take several forms:

- Uniform corrosion
- Concealed corrosion
- Pit corrosion
- Selective leaching
- Galvanic corrosion of dissimilar materials

Concealed corrosion occurs when metal that is concealed in cracks or voids or covered by joints, fasteners, or dirt becomes anodic. Positive metal ions travel away from the concealed areas to adjacent areas where they combine with oxygen and are deposited as metal oxides. A similar corrosion can occur when oxygen is absent, but chloride is present, as in marine climate. The creation of hydrogen gas on exposed surfaces can also cause crack corrosion by removing positive ions from exposed and unprotected metal surfaces, preventing them from migrating back to concealed negatively charged areas, which then continue to corrode.

Pit corrosion forms when the passive coating has broken down in spots. Pits are usually deeper than they are wide and can become deeper as the metal becomes increasingly anodic from the lack of oxygen. Pitting is common on lead, zinc, bronze, and some aluminum alloys and stainless-steel alloys.

Selective leaching occurs through micro-galvanic activity, when a metal separates from the rest of the alloy through a corrosion process. Common examples are when iron is lost from the carbon matrix of cast iron and "dezincification" when zinc is leached from brass.

Galvanic corrosion occurs when two dissimilar metals are connected by an electrolyte solution, and ions flow from the more anodic metal to the more cathodic. The anodic material will lose electrons and corrode, while the cathodic material will gain more electrons than it loses and be protected. Saline environments can accelerate galvanic action of very cathodic metals such as copper, causing rapid corrosion of concealed fasteners.

The electrochemical series identifies the tendency of ions to flow from one material to another, based on their relative properties. Hence a specific material may be anodic relative to one metal, but cathodic relative to another. The larger the differential, the more potential for corrosion of the anodic material. Gold and silver are "noble metals" that do not easily combine with other elements. Noble metals are low on the electrochemical scale, indicating that they will not corrode but can cause other materials to corrode. Iron, being in the middle of the scale, will corrode when exposed to more anodic metals, such as copper, and will cause the corrosion of less anodic metals, such as zinc.

ANODIC	Magnesium	Aluminum	Zinc	Chromium	Iron (FE2+)	Nickel	Tin	Lead	Iron (FE3+)	Hydrogen Reference	Copper	Silver	Gold	CATHODIC

Figure 2.7.1 Electrochemical series of pure metals. (Table by Susan Pranger, 2023.)

Copper

Copper is relatively noble and will not corrode from contact with most other metals or react to clean cement, mortars, or wood. However, corrosive agents can cause pitting or more extensive damage. Contact with copper can cause more anodic metal materials to corrode, and runoff can stain stone or cementitious materials. Fasteners and clips should be made of copper or compatible bronze or brass alloys.

Copper corrosion creates an impermeable patina layer of copper oxide (cuprite) that retards additional oxidation. Depending on atmospheric conditions, copper will oxidize within 6 months, forming a dark brown patina of copper hydroxide with salts of sulfate and chloride, and small amounts of nitrate and carbonate salts. After 5 to 10 years, the patina becomes a mottled or even green. After 70 years, the patina converts to a mineral similar to atacamite or brochantite that prevents further corrosion. Chlorides in coastal environments contribute to a green patina, while sulfates in rural and urban areas form a bluish patina.

Patinas forms more rapidly in polluted areas. The natural process of acquiring a patina can be artificially accelerated with chemicals or avoided by periodically applying a lacquer finish with corrosion inhibitor. Without the inhibitor, corrosion can occur below the lacquer finish and at any gaps in the coating.

Figure 2.7.2 Copper patina, Boston, MA. (Photo by Daniel Thomas, 2023.)

Lead

Lead is soft and malleable, suitable for filling gaps between other materials or coating other metals including copper, iron, and steel. Lead sheets are used for roofing, and alloys of lead and tin are used for soft solders. Lead is initially shiny but rapidly oxidizes, developing a light gray coating of lead oxide and lead carbonates in clean atmospheres. In polluted environments the lead oxidizes into a dark gray or black patina that includes lead sulfates and carbon particulates. Both patinas are highly resistant to corrosion.

A lead carbonate coating provides excellent resistance to the corrosion of both marine and heavily industrial environments. Unlike copper, runoff from lead will rarely stain other materials. However, lead is highly cathodic and can cause corrosion of more anodic materials, especially aluminum.

Lead will corrode quickly if condensation or erosion prevents a patina from forming or through contact with organic and man-made acids or highly alkaline materials, such as damp, uncured mortar, and concrete. Corrosion from mortar usually stops after the mortar cures, however lime mortar can continue to corrode and deteriorate lead flashings as the lime slowly carbonates.

Lead sheets must be carefully detailed and supported to provide ventilation and prevent trapped condensation. Lead is easily shaped but is not self-supporting except over very short spans and will settle onto any supporting material. Care must be taken to avoid any projections or sharp edges in the substrate that would penetrate the material. Lead also has a high thermal coefficient of expansion compared to most other metals, so loose-lock seams must allow for expansion and contraction. Fittings are usually copper or brass that is coated or covered with lead patches.

Zinc

Zinc is anodic relative to many metals. In clean atmospheres, zinc forms a zinc carbonate coating that protects it and less anodic metals from corrosion. Zinc is compatible with lead or aluminum materials but should not be used with copper or copper fittings. Runoff from copper roofing can corrode zinc.

The rate of corrosion can be 10 times faster in damp industrial environments than in dry rural ones. In wet, polluted environments, sulfurous and sulfuric acids combine with the zinc carbonate to form a soluble zinc sulfate that is easily washed away by rainfall and replaced by a new layer of zinc carbonate. This can cause a cycle of corrosion and erosion in wet, polluted environments.

Flat, horizontal areas that would allow acidic water to sit should also be avoided. Even more rapid deterioration can also occur in areas where the zinc is kept wet, but not exposed to the atmosphere. Concealed areas become anodic but produce a white powder known as "white rust" that does not protect the underlying materials and can destroy zinc sheets in mere weeks. Condensation can accelerate the deterioration of white rust in concealed areas, and corrosion in other areas.

Zinc can be applied as a protective coating to less anodic materials like iron using a variety of processes, including hot dipping, electrolytic, metal spraying, sherardizing (dipping in melted zinc dust), and painting with zinc-rich coatings. Hot dipping is especially effective because a zinc/iron alloy is formed between the iron and the zinc coating.

Zinc will suffer a temporary surface corrosion when used with clean mortars and cements. However, sulfates and chlorides in mortar or concrete will continue to attack zinc, especially in wet conditions. Zinc does not react with non-acidic woods, but can be damaged by contact or runoff from acidic woods like cedar and oak or wood treated with zinc chloride preservations.

Aluminum

Aluminum is strong, lightweight, and very corrosion resistant. Aluminum cladding installed on the cupolas of the Church of San Gioacchino in Rome in 1897 revealed only minor corrosion when inspected in 1966 and in 2013. A thin layer of oxidation naturally forms on the surface of aluminum to retard further corrosion and will form continuously over any scratches. The coating is very thin (4 to 10 mm) but occurs within seconds of any contact with air or other oxidizing agent. The thickness of this protective surface film can be increased through the anodizing process of electrolytic oxidation.

Temperatures from 70° to 150°C will increase the rate of general corrosion of aluminum immersed in water but may decrease the depth of pitting. Intercrystalline corrosion causes swelling at temperatures above 150°C and is destructive at temperatures above 250°C. Aluminum will experience galvanic corrosion when exposed to dissimilar metals that are more cathodic, such as iron and especially copper. This reaction is increased in saline or marine environments.

The oxide film is stable in most environments with a neutral pH of 6 to 6.5, but can pit when the temperatures are higher, and the pH level is more acidic. Pitting decreases after the first few weeks of exposure. Aluminum is resistant to common industrial pollutants like ammonia and sulfur gas, but chlorides effectively penetrate the protective oxide layer, and calcium, magnesium and sodium sulfates increase the depth of pitting. Spots of white crystalline corrosion can form in aggressive industrial climates and turn dark gray in areas of heavy soot. Aluminum can also react with direct contact or runoff from unseasoned cedar and oak, and wood preservatives that contain water soluble copper or mercury salts. It can react with concrete and masonry cement, especially damp materials. Regular washing by rainfall or maintenance will reduce corrosion.

Figure 2.7.3 Aluminum patina, Alcoa Building, Pittsburgh, PA. (Photo by Daniel Thomas, 2022.)

Tin

Tin is a ductile and very crystalline metal commonly added to other alloys or used to coat ferrous metals. It is malleable but become brittle when cooled. It resists corrosion, forming a silvery gray oxide. Tin, together with copper, is present in most solders. In a marine environment, tin will slowly oxidize to form tin oxychloride.

Ferrous Metals

Iron and steel are among the strongest of metals; the metal of choice where strength and durability are required. Due to the difficulty and high cost of fabrication, iron was traditionally used only for fittings, nails and other small components where strength was required. However, as iron and steel became more readily and economically available, they were used for a wide range of structural, decorative, and protective elements including exterior sheet metal facades and roofing, cast iron components, and eventually steel windows. Cast iron is strongest in compression, while wrought iron is stronger in tension, due to its fibrous nature. Alloys tend to be strong in both tension and compression.

Ferrous metals are the most susceptible to corrosion. Iron forms a protective layer in dry air, but hydrates into iron oxide in the presence of moisture. As iron or steel rusts and converts back to iron oxide, the corroded layer expands up to seven times its original thickness. In thick components such as structural members or reinforcing bars, the proportional loss of original material is minimal, and has little impact on structural capacity. However, the expanding rust can exert considerable pressure on adjacent materials (rust jacking), and leaching rust can disfigure adjacent materials. Thinner materials such as sheet metals can fail when rust penetrates the depth of the material.

Ferrous reinforcement in concrete can produce a protective passive layer in concrete, but this layer is lost as the concrete ages and becomes less alkaline. Steel alloys containing nickel, copper, or chromium are more corrosion resistant. All other iron and steel, historic and most modern steel requires an effective paint or metallic finish to resist corrosion. Galvanized zinc coatings can be applied by hot dipping or electrodeposition, or by using zinc-rich paints. Often a field or factory finish is applied over the zinc coating. As discussed previously, zinc is anodic relative to steel and will corrode first, protecting the underlying steel.

Metal Alloys

Two or more pure metals can be combined to form an alloy with unique properties. Until the 13th century, bronze and brass were the most common alloys. Since then, thousands of metal alloys have been developed to meet specific performance criteria.

Most alloys are created by melting the metals together during the smelting process. Single-phase alloys combine the separate metals into a single crystalline structure. Multi-phase alloys retain crystals of the constituent metals. Multi-phase alloys are more likely to develop anodic and cathodic zones that lead to corrosion than single-phase alloys with a single crystalline structure. In either case, the alloy may be more or less ductile, strong and corrosion resistant, and may melt at a different temperature than the original metals. There is usually a tradeoff between strength and corrosion resistance that depends on the proportion and type of materials.

Copper Alloys

Bronze and brass are alloys of copper and various other metals.

Bronze is a hard, brittle alloy, stronger than copper, but less so than iron. Bronze was originally made from copper and arsenic but now contains 86% to 89% copper, 9% to 11% tin, and 1% lead. Trace amount of iron, nickel, and other metals are often present. Bronze rapidly develops a corrosion resistant reddish-brown cuprite (copper oxide) patina. A light gray layer of copper sulfates or carbonation may eventually form over the earlier patina in industrial exposures. Bronze alloys that contain aluminum, silicone, or tin can suffer from stress corrosion cracking. Micro-galvanic activity can separate the aluminum from aluminum bronze alloys.

Brass is an alloy of copper and zinc, harder than copper but softer than bronze. It was originally made by combining copper with calamine (a zinc ore) before zinc itself was isolated. A typical brass contains 70% copper and 30% zinc, sometimes with tin, nickel, lead, or silicon. Manganese brass, known as "architectural bronze," has a higher tensile strength. Brass is a muted gold when new, but very slowly forms copper sulfates and zinc sulfate in the range of greenish yellow to coppery red, depending on the proportion of zinc. Brass alloys are resistant to corrosion, but less so than other copper alloys. Brass can experience stress cracking and can corrode locally if zinc is dissolved from the alloy matrix.

Nickel silver is an alloy of copper and nickel, zinc and occasionally antimony, tin, lead, or cadmium. It is corrosion resistant and polishes to a silver finish with gold tints. Gunmetal (red brass) is a harder alloy of copper, tin, and zinc. It is strong and corrosion resistant and was used for cannons until cast iron became available.

Steel-copper alloys, known as "weathering steel" provide a slowly rusting exterior which provides a protective coating. However, rust particles from the coating will stain adjacent materials.

Stainless Steel Alloys

Stainless steel is an alloy of steel with a high percentage – 17% to 20% – of chromium. It forms a thin passive oxide layer which is self-healing and generally prevents corrosion or reactions with other metals. As a result, stainless steel is not only prized for exterior cladding, but also for fittings, fasteners, and reinforcement, and to anchor repairs in plaster, masonry, and stone.

However, stainless steel is cathodic relative to aluminum or zinc, and can cause them to corrode in very aggressive marine or industrial environments. Unless low-carbon steel is used, stainless steel will corrode around welds where the heat of welding forms chromium carbide from chromium in the alloy.

Stainless steel used in construction is generally one of three types:

- 17% percent chromium
- 18% chromium and 8% nickel (18/8) or
- 18% chromium, 10% nickel, and 3% molybdenum (18/10/3)

The first, 17% chromium, produces a magnetic, ferritic stainless steel, appropriate for mild or protected locations. The other two are non-magnetic, "austenitic' stainless steel. The most durable alloy is 18/8/3, which is used in exterior and aggressive environments, and where appearance is important.

Aluminum Alloys

Aluminum alloys are used extensively for sheet cladding and cast components. Alloys of aluminum with low levels of magnesium are used for corrugated metal roofing; higher levels of magnesium produce an alloy that is stronger, but not heat-treatable. Adding silicon or silicone and copper can produce a high strength, heat-treatable alloy.

Alloys used for casting include up to 12% silicon, a mix of silicon and copper, or a mix of silicone and magnesium. Increasing the copper increases the strength but reduces corrosion resistance, so high-strength aluminum alloys are often coated with more corrosion-resistant pure aluminum. Corrosion can significantly reduce the strength of aluminum alloys in a corrosive environment.

Extraction, Refining, and Manufacturing

Metals and ores are collected from the surface, or mined from surface pits or tunnels. Rock is crushed to extract the metal oxides, and then processed to separate pure metal from undesirable impurities and combined to create alloys. Metals can be heated, hammered (forged), rolled, cast, extruded through forms, or electroformed to create the desired shapes and thicknesses. These manufacturing processes contribute to the physical properties of the final product including weather resistance.

The earliest method of production was casting molten metal into a sand or wax form. Cast metals are identified by their sandy or rough texture, or by the presence of blowholes. Cast metals may be porous or have inclusions of metal oxides, with a crystalline structure that can be identified by partly cutting and then bending the metal until it breaks. The porous nature of cast iron makes it vulnerable to interior corrosion. Sintering was used to form small shapes when the high temperatures required for casting were not feasible. Particles of ground metal powder are compacted into small or detailed shapes and then heated to fuse the particles into a single strong component.

Brittle metals were heated prior to hammering, pressing, extruding, or rolling to reduce the risk of fracturing. Wrought iron can be identified by hammer marks or incisions on bars and sheets, and by undulating surfaces or the inconsistent thickness of bars. The fibrous nature of wrought iron can be revealed by partially cutting and then bending the metal.

Bending, drawing, or cold forming can cause steel to harden but also accumulate stresses and become brittle. These stresses are relieved by annealing, a process of heating the metal so that the atoms in the crystal lattice migrate and form new crystals without residual stresses. The metal can be slowly air cooled, or quickly cooled by plunging in water. The duration of heating and cooling will determine the size and composition of the crystals, and the level of ductility and hardness. Annealing can also improve magnetic properties and the ability of metal to conduct electricity. The annealing process is most common in producing steel, but can also be used on aluminum, brass, and copper. Metal shapes were traditionally cut and shaped while hot or filed while cold. After the industrial revolution, equipment was developed to mill or shape cold steel shapes.

The joints in metal exteriors are often the weak point in an assembly and can be the source of deterioration and hidden corrosion. Metal parts are connected by mechanical fasteners, or bonded by soldering, brazing, or welding. The use of dissimilar materials can cause electrolytic action and corrosion.

Drivers of Deterioration

Metals naturally resist the infiltration of water that deteriorates many other materials; however, they are susceptible to chemical and electrochemical corrosion that can be enhanced by water, UV, and pollutants. The primary cause of deterioration in metals is corrosion, although heat and mechanical stress can cause deformation or metal fatigue.

Atmospheric conditions can accelerate or retard corrosion. Humidity and standing water act as electrolyte solutions. Salts in marine environments, polluted air, and in uncured or contaminated cement and mortar can make these solutions more reactive and conductive. Humid climates with industrial pollution will be the most corrosive, followed by marine climates high in humidity and salts. High temperatures can accelerate chemical reactions. Cold, dry, and rural areas tend to be less corrosive.

Atmospheric or weather conditions determine the amount of water and oxygen available for corrosion, the presence of salts and pollutants that act as a catalyst, and other factors such as temperature and wind that can facilitate corrosion. These conditions take advantage of weaknesses inherent to the type of metal or weaknesses introduced by processing or construction detailing.

Pollution

Atmospheric particles of soot, dirt, smoke, and salts can accumulate on the surface of metals, especially areas that are not washed by rain. Uneven deposits prevent patinas from forming uniformly, and lead to pitting. Particles in the air can combine with water and salts to form crusts over patinas, and then be converted by acidic fog or mist into corrosives that attack the patina below.

Ultraviolet Light

There is some evidence that zinc, aluminum, copper, and iron corrode or pit more quickly when subjected to a combination of UV light and water. Metals like zirconium, titanium, nickel, 304 stainless steel and silver that have insoluble oxides did not corrode more rapidly when exposed to UV light in laboratory tests.

Temperature

Metal rapidly expands and contracts with changes in the temperature. The rate of expansion is different for each material, with zinc expanding almost three times as much as iron, and all metals expanding more than most other building materials. Concrete is an exception, with an expansion coefficient comparable to steel.

Very high temperature and abrupt changes in temperature can cause flaws in the structure or rupture protective patinas, leading to an increase of corrosion. The average increase in global temperature is unlikely to affect the performance of metals directly, however changes in temperature affect the movement of water and water vapor in ways that can promote corrosion. Higher temperatures have been shown to increase corrosion of metals that typically develop a passive oxide layer, such as aluminum.

Thermal stress in metals is caused by fluctuations in the ambient temperature or solar exposure. Thermal fluctuations affect metals more than materials with a lower coefficient of expansion and a higher thermal mass. Abrupt changes in temperature can be caused by from a sudden rainstorm that suddenly cools a metal surface heated by the sun. Sun exposure can expand one section of metal differently than an adjacent shaded area. When expansion or contraction of the metal is restricted by the shape of the metal component, or by connections to other materials that move differently, the stress of thermal movement can cause deformation of the metal structure, failure of joints or buckling of metal surfaces. Frequent cycles can lead to metal fatigue and increase corrosion.

Even though metals are generally non-combustible, prolonged exposure to fire and very high temperatures can cause deformation or structural failure by melting the metal. Oxidation of the surface can occur when the surface is repeatedly heated over a long period of time (such as a boiler tank). Some metals can burn, including aluminum which will release aluminum oxide. Thermal shock can occur with the sudden change in temperature that occurs when water is applied to extinguish a fire, and cause failure of brittle materials like cast iron.

Moisture

The cleaning action of rainwater can remove surface pollutants that encourage corrosion, but continuous or concentrated rainwater can increase corrosion by eroding the protective layer of passive corrosion or by transferring cathodic metal particles to more anodic metals. Condensation and standing water also increase corrosion.

Marine and industrial environments generally increase the percentage of salts and pollutants found in water and water vapor. Rainwater can carry metal particles from one metal to another, facilitating galvanic action if anodic materials are downhill of cathodic metals. Rainwater runoff can carry harmful acids from cedar or oak shingles, or from organic matter on the roof onto metal gutters, downspouts, or wall surfaces.

Atmospheric changes in temperature and humidity are inextricably linked. Because metal absorbs and loses heat quickly, rapid temperature changes can easily lead to condensation on the surface of metal especially in humid environments. Because the metal is not porous, this condensation may remain on the surface, forming an electrolytic film that can lead to general corrosion. If the condensation drains and collects in joints and cracks, it can lead to crack corrosion. Thin roofing panels of lead, zinc, and aluminum are all susceptible to corrosion from condensation.

Areas that previously experienced long periods of freezing weather may now experience more cycles of freezing and thawing. Although metal is generally not porous, metal shapes, and joints can trap water and be damaged by expanding ice. Joints are particularly susceptible to corrosion and metal fatigue.

In general, an increase in the volume, intensity, or duration of rainfall will increase the potential for corrosion by increasing the likelihood that surfaces, cracks, and joints will hold water. Because metal is not porous, rainwater or condensation that flows down the surface can collect in joints or crevasses. Condensation and other water in a wall assembly can easily be trapped behind metal cladding. Heavy, continuous, or wind-driven rain can erode the protective passive coating, causing additional corrosion of exposed metal.

A decrease in the frequency and overall days of rainfall can increase corrosion if pollutants and salts are allowed to accumulate on metal surfaces.

Figure 2.7.4 Sheet metal corrosion bleeding through paint, Boston, MA. (Photo by Daniel Thomas, 2023.)

Wind

Wind events are becoming more severe and can damage metal surfaces through impact and erosion. In extreme events, wind can propel objects that can deform or perforate metal surfaces, or cause deformation or mechanical stress by lifting lightweight metal roof or cladding. Wind can drive rain more deeply into cracks, crevices, and joints. Corrosion can increase where impact or erosion removes the protective patina or coating, or triggers electrolytic action.

Biological Damage

Increased rainfall and humidity are likely to increase damage by making conditions for biological growth more favorable. Many metals can be damaged by organic acid from damp timbers, moss and lichen, formic acid from decaying insects or animals, nitric acid from bird droppings, acetic acid from damp hardwood like cedar and oak. Birds can indirectly damage metal by blocking drains with nests and feathers, causing water to stand on metal gutters and roofs.

Guano is a combination of urine and fecal matter from birds. The insoluble uricite crystals found suspended in urine are dissolved by rain and redeposited as white stains. The fecal matter is coated with urea and uric acids that also can decay of metal surfaces. Soluble salts that support corrosion can leach out of rain-soaked excrement.

Bird excreta provides nutrients to lichen and mosses that in turn produce acids. The excreta is also colonized by bacteria that oxidize ammonia, nitrate, and sulfur.

The bacteria produce nitrous, nitric, and sulfuric acids that corrode metal surfaces. There is some conjecture that the corrosion caused by guano might be caused by the fact that it excludes oxygen from the metal.

Chemical Damage

Acids in polluted environments can attack the patina on metals, especially where a crust of soot and smoke has formed. Chloride, sulfates, and other salts react with metals to form metal oxides. Depending on the metal, this either facilitates the formation of protective coatings, or causes damaging corrosion. Increased acidity from pollution or biological matter can convert insoluble cuprite, the patina on copper and copper alloys, into damaging copper oxides.

Chlorides are more common in coastal, marine areas. Concentrations of sulfates are common in rural, agricultural areas, but highest in polluted urban areas. The hydrochloric or hydrofluoric acids are found in both polluted rainwater and in some masonry cleaners can be corrosive to metal surfaces.

Figure 2.7.5 Corrosion of cast iron deck, North Head Lighthouse, Cape Disappointment, OR. (Photo by Washington State Parks and Recreation Commission, 2013.)

Strategies for Responding to Climate-Based Threats

Research and Investigation

Treatment of metal finishes requires a thorough understanding of the properties of the specific material, and how it interacts with surrounding materials and the atmosphere. Trade publications such as manuals, such as the "Copper and Common Sense," and "Standard Practice in Sheet Metal Work" provide useful insights.

A preliminary visual identification can often be made based on the color and texture of the metal. The configuration of joints and welds, and the type of fittings can also provide clues. Cutting the metal to expose the uncorroded metal can provide additional information, as can a spark test that compares the sparks produced by a grinder to the known patterns produced by different ferrous metals.

In absence of definitive documentation, a variety of metallurgical tests can identify the structure, composition, impurities, grain size, stress fractures, and corrosion in metals. Most of these are minimally invasive or require a small sample of the metal. Field and laboratory tests can determine hardness, ductility, and tensile or compressive strength. A simple microscope can distinguish between the character of wrought or cast iron, while more sophisticated scanning electron microscopes can determine the elemental composition of alloys, layered metals, or coated metal. A magnet can be used to distinguish ferrous materials like iron and steel from non-ferrous metals like aluminum.

Microchemical tests can identify the alloy components, their physical characteristics, and performance. X-ray fluorescence (XRF) can be used in the field or the laboratory while more sophisticated tests examination of a sample in the laboratory. In all cases, expert analysis and interpretation is required.

The nature and condition of the paint and metallic coating should also be determined. Historic ferrous metals are likely to have multiple layers of coatings unless the metal was stripped and repainted. Multiple layers are most likely to be intact in areas that are protected or concealed. Identification of original colors, binders, and pigments generally will require laboratory tests and examination by a paint expert of a small paint sample. Coatings should be examined for adhesion to the substrate, and for the presence of trapped water or salts.

Maintenance

All metals should be cleaned and inspected regularly to identify corrosion. Joints and cracks that could conceal corrosion should be sealed to exclude moisture. Ferrous metals may require cleaning to remove rust and restore coatings. Lubrication with simple graphite powder is odorless and greaseless, and an effective way to maintain moving parts.

A thorough and methodical visual inspection of all exposed surfaces and joints is the first step to identify missing sections, corrosion, cracking, or staining, which can identify hidden corrosion or other problems. The general conditions of metal parts can often be assessed by testing the functionality of moving parts, including hinges, locks, screws, and fittings. Corrosion is most likely to occur at dissimilar materials and wherever water is present, trapped, or concealed.

Coatings should be examined for adhesion to the substrate, and for the presence of trapped water or salts.

Repairs and Modifications

Modifications should recognize the character of vernacular details and should focus on eliminating the potential for corrosion:

- Allow for thermal expansion and contraction.
- Allow for positive drainage and avoid capturing water.
- Avoid or separate dissimilar materials prone to galvanic action.

Cleaning

The decision to clean, and the method and aggressiveness of the cleaning method should be determined with the expertise of a metal conservator and should consider the value of the existing oxidation or finish.

In the case of non-ferrous materials, removal of metal oxides may not be required or advisable. Although water can contribute to corrosion, a simple regimen of washing metal surfaces to remove pollutants, especially with materials that have natural passive layers, can reduce corrosion in polluted environments.

Cleaning methods should use the mildest method suitable for the task, which may be to remove surface dirt and pollutants, rust, and corrosion, or to prepare for repainting. Water-based cleaning is generally preferred but chemical cleaning might be required to remove heavy corrosion.

Phosphoric or other organic acids are usually adequate for removing heavy rust. Hydrochloric acids and bleaches containing sodium hypochlorite can leave residual salts and should be avoided. Chemical dips should be used with caution because they are difficult to remove from cracks, imperfections, or pores even when the metal is completely disassembled. Low viscosity releasing oils, cream polishes, and other commercial gels have some success in removing stains or rust from stainless steel but may be difficult to remove.

Mechanical cleaning to remove corrosion, salts, and other contaminates runs the gamut from hand brushing, sanding, or grinding, to needle guns and abrasive blast cleaning. Testing is required to determine the most effective method of cleaning while preserving original metal.

An oxypropane or oxyacetylene flame that is run lightly over a rusted surface will expand the rust at a different rate that the underlying metal, causing it to separate. This method can be effective on wrought iron but should be avoided with cast iron which can crack if overheated. The process should only be used in a shop with fume extraction and experienced operators.

Finishes on Non-Ferrous Metals

While brass, bronze, and copper are usually not painted, varnishes, lacquers, and other clear coatings are often used to maintain their bright lustrous appearance.

Varnishes and lacquers were traditionally oil- or resin-based, but a variety of synthetic coatings are now available. Wax coating can successfully protect metal finishes.

Acid or chemical washes are sometimes used to accelerate the process of oxidation on material such as lead, brass, bronze, and copper that naturally form a protective patina. These finishes should be coated lacquer or wax to arrest ongoing oxidation, and to prevent run-off of salts.

Finishes on Ferrous Metals

Ferrous metals generally require a coating to prevent corrosion. If that coating is deteriorated, and rust has occurred under the coating, it may be necessary to remove the coating. However, cleaning that removes both original metal and historic paint coatings can be inherently destructive and is irreversible.

Surface preparation is key to success of a protective coating for ferrous metals. Cleaning must remove both corrosion and any other contaminates such as oil, dirt, or salts that can compromise bond of the new coating. Hygroscopic particles of chlorides or sulfates can attract moisture through the pores of a new paint coating. This may require that components be disassembled to access hidden surfaces.

There is no need to remove existing paint coatings, especially lead paint, if it is well adhered. However, mill scale – the thin layer of passive iron oxide – can flake and should be removed if exposed. If existing paint is sound and well adhered, it may be possible and preferable to apply new paint over the existing. The existing paint must be tested to confirm that it is well adhered, then cleaned and possibly roughened to bond to the new material. If ferrous metals are cleaned to bare metal, a primer should be applied immediately to prevent new oxides from forming.

Existing bituminous paint is not compatible with most new paint products, and must either be completely and thoroughly removed, or renewed with a similar bituminous product.

Metallic and Ceramic Coatings

Existing metal that has been thoroughly cleaned with acid can be hot dip galvanized or coated with metal or ceramic coatings. Hot dip galvanizing should never be used on historic cast metals because the acid used for cleaning can remain in cracks and voids and then attack the zinc coating, and the heat of the zinc bath can distort the cast metal. The molten zinc coating may not penetrate joints allowing for future crack corrosion. The zinc coating can also affect the appearance and performance of fine details. The zinc is irreversible and can interfere with the future use of traditional blacksmith repair techniques.

Flame spray processes can be used to apply melted metal or ceramic coatings to metal surfaces. This process can apply a uniform, well-adhered finish but can obscure fine details. It is generally not appropriate to conservation of historic metals but may be appropriate for protection of concealed metalwork in hostile environments.

Conclusions

Metals are durable materials that have been used successfully for centuries. Careful maintenance to protect existing materials from excess moisture, pollution, and biological damage should allow them to continue indefinitely. A thorough understanding of minerology, coatings, and surrounding conditions is required to understand the unique electrochemical processes, and how they are affected by changes in the future weather.

Intense rains and dramatic temperature swings have the potential to increase corrosion and deformation of metals, and may require an increase in maintenance and protection.

Bibliography

Britannica, T. Editors of Encyclopaedia. "copper." Encyclopedia Britannica, August 1, 2023. https://www.britannica.com/science/copper.

Britannica, T. Editors of Encyclopaedia. "cupellation." Encyclopedia Britannica, September 8, 2019. https://www.britannica.com/technology/cupellation.

Callcut, Vin. 2000. *Brief Early History of Brass.* https://www.copper.org/publications/newsletters/innovations/2000/01/history_brass.html.

Chiavari, C., Rahmouni, K., Takenouti, H., Joiret, S., Vermaut, P., and Robbiola, L. 2007 "Composition and electrochemical properties of natural patinas of outdoor bronze monuments." October 2007 *Electrochimica Acta 52*(27): 7760–7769 https://doi.org/10.1016/j.electacta.2006.12.053

English Heritage. 2012. *Metals.* Edited by Sophie Godfraind, Robyn Pender and Bill Martin. London: Routledge.

Forsyth, Jarred K. 2000. *The Effect Ultraviolet Light has on Aqueous Corrosion.* Master's Thesis, Materials Science and Engineering, University of Pittsburgh, Pittsburgh: University of Pittsburgh. https://www.ftmsa.org/sites/default/files/pages/effect_ultraviolet_light_has_on_metal_corrosion.pdf.

Harvey, Donald W. Jr, and Michael F Schuller. 2010. "Nondestructive Evaluation: Structural Performance of Masonry." *APT Bulletin, The Journal of Preservation Technology* (Association of Preservation Technology International) XLI, No. 2–3.

Hornbostel, Caleb. 1978. *Construction Materials, Types, Uses and Applications.* New York: John Wiley & Sons, Inc.

Licht, Sue. 2014. "North Head Lighthouse, Materials Investigation Report."

National Association of Sheet Metal Contractors. 1929. *Standard Practice in Sheet Metal Work.* Pittsburgh, PA: National Association of Sheet Metal Contractors.

OSHA. 2004. *Lead in Construction.* Occupational Safety and Health Administration, U.S. Department of Labor, Washington, D.C.: OSHA Publications Office. https://www.osha.gov/sites/default/files/publications/osha3142.pdf.

Simpson, J.W., and P. J. Horrobin. 1970. *The Weathering and Performance of Building Materials*. Beverly Farms: Aylesbury Press.

Spennemann, Dirk H. R., and Maggie J. Watson. 2018. "Experimental Studies on the Impact of Bird Excreta on Architectural Metals." *APT Bulletin: The Journal of Preservation Technology* (Association for Preservation Technology International) 49 (1): 19–28. https://www.jstor.org/stable/26452201.

Stacey, Michael. 2015. *Aluminum and Durability, Towards Sustainable Cities*. Cwningen Press. https://international-aluminium.org/wp-content/uploads/2021/03/Aluminium-and-Durability-TSC.pdf.

TWI. 2022. *What Is Annealing? A Complete Process Guide*.

Vargel, Christian. 2004. *Corrosion of Aluminum*. Elsevier. file:///D:/BOOK%202020-current/0%20RESEARCH/metals/vargel%20CORROSION-OF-ALUMINIUM.pdf.

World Heritage Organization. 2021. *Lead Poisoning*. October 11. https://www.who.int/news-room/fact-sheets/detail/lead-poisoning-and-health.

CHAPTER 2.8

Coatings

Introduction to Coatings

The earliest buildings were constructed of unpainted earth, wood, and stone. The first coatings were either clay, tar, or fish oils added to buildings and ships to delay deterioration from the weather or earth pigments that were purely decorative. The chemistry of coatings and pigment, and their use as decoration and protection, has constantly evolved over the centuries, not always for the better. In the 20th century, the desire to reduce maintenance and decrease water intrusion led to the use of impervious coatings that were not always good for the underlying materials. Today's interest in combining preservation and sustainability has increased interest in traditional coatings and in new technologies that provide flexibility, permeability, and renewability.

The Past, Present, and Future of Coatings

History

The earliest paints were purely decorative, pulverized earth mixed with water. By 6,000 BCE, earth pigments were being mixed with animal fat. By 4,000 BCE, the Egyptians were using egg, glue, milk, honey, and other binders.

The first protective coatings were resins or tars used from Greece to Scandinavia during the Iron Age. Tar mixed with beeswax, natural pitch, and pigments was used to waterproof wooden ships and later distilled to produce varnishes, lacquers, and turpentine. Ground pigments were added to create a type of gloss paint or lacquer. Roman manuscripts from 200 CE refer to paint comprised of walnut and poppy oil. Linseed oil appeared in 6th century literature and was used throughout

 DOI: 10.4324/9781003160489-13

Europe by the 11th century. By 1949, linseed oil and tung oil were the most common medium for both paint and varnishes.

For centuries, the primary sources of pigments were natural minerals, earths, and plant or insect extracts, obtained dry and then ground into oil. In 1856, Perkins developed a process to distill artificial dyes from coal tar. By the late 19th century almost all natural pigments had been replaced with chemical compounds that were commercially available as a stiff paste ground into oil.

Synthetic resins and binders were introduced in the early 20th century, rapidly followed by oil-based alkyds, water-based latex and acrylics, and synthetic resins. By the 21st century, a large variety of synthetic pigments and paints, many proprietary, had been developed. Many of these were impermeable, blocking water. The desire to remove volatile organic compounds (VOC) from solvents for the health of individuals and the planet led to the development of latex and other water-based coatings.

Current Social, Cultural, and Climatic Impact

Not all areas are receiving more rain, but many areas are receiving more periods of intense rain and high humidity. Thus the desire to protect porous building materials in an environmentally sensitive manner has led to the use of both traditional lime and linseed oil-based paints and modern permeable materials such as potassium silicates. Flexible, permeable versions of synthetic paint are also being developed.

Future Projections

Increases in temperature, UV, and rain events combine to: (1) increase the deterioration of many traditional building materials and (2) to decrease the effectiveness of protective coatings. Preservation and environmental concerns are driving research and development into several directions. There has been an increase in commercial scale production of traditional coatings including tar, mineral paint, linseed oil paint, and urushi lacquers. Existing commercial manufacturers have developed water-based paints, for example latex paints, that are more resistant to weather, and less dependent on fossil fuels. Others are researching and promoting the use of nanotechnology and new compounds to consolidate or protect permeable materials.

Understanding Coatings

Coatings combine different materials. Each component serves a specific purpose. They work together to establish long-term performance and durability:

- Pigments add color or opacity.
- Binders connect the pigments to form a continuous film.

- Driers facilitate drying or curing of the paint.
- Additives add specific qualities, such as mildew resistance.
- Extenders increase the percent of solids or add specific qualities.
- Solvents dissolve or thin other components for application, and then evaporate.

Coatings are distinguished not only by their components, but also by how they perform. Both penetrating and surface coatings rely on a good bond and sufficient elasticity to maintain a continuous film.

- Surface coatings remain on the surface
- Penetrating coatings that fill pores and voids
- Penetrating coatings that chemically react with the substrate material

Surface coatings adhere to the surface but may leave voids between the coating and the surface. A more viscous surface coating will fill voids and gaps on the surface, while a less viscous coating may span over voids. Shellac, wax, copal, varnish, modern paints, and water repellants tend to be surface coatings.

Penetrating coatings, including oil-based paints and varnishes, synthetic stains, and consolidants, will seep into the voids and pores of porous materials and either coat or chemically react with the substrate. The film left behind as the coating cures can affect the movement of moisture or vapor, and the resistance to weather and climate. Chemical reactions can strengthen or weaken the substrate depending on the impact on cohesion, density, and porosity.

Vehicles: Carriers and Binders

The vehicle is a combination carrier and binder and serves different functions over the lifespan of a coating.

The vehicle must:

- hold (carry) pigments in an even suspension without excessive settling or chemical reaction.
- be fluid enough to be spread by brush or spray.
- bind the pigments to the medium to create a cohesive film.
- dry or cure to a tough, but flexible film.

The primary purpose of the coating is to resist moisture while also being vapor permeable and flexible. Coating must also be elastic enough to avoid shrinkage cracks as materials dry, and movement cracks as the substrates expand and contract. Paints with more binder will be harder and glossier, and more resistant to abrasion and mildew.

Traditional oil and water-based vehicles served as both carrier and binder. Traditional water-based paints, such as milk paints, casein, and calcimine, were not durable enough for exterior use unless they were reapplied on a regular basis, as was the norm for lime washes.

Oil-based paints used natural oils, such as linseed. Varnishes were "oleo-resinous" (oil fused with resins), while varnish paints were a mix of oil and varnish.

Synthetic binders include oil-based alkyds, water-based latex and acrylics, synthetic resins such as water or petroleum-based urethane, epoxy and polyester resins, and other polyethylene and polythene synthetics. Water-soluble synthetic binders were developed that were durable enough to withstand exterior conditions.

Extenders

Extenders are solids that are added to make the coating more opaque. They are often added to replace some of the more expensive pigments, but can be beneficial. Calcium carbonate can make the paint more resistant to chalking. White silica and silicates can make the coating more resistant to abrasion and cleaning.

Traditional Oils

Traditional oil-based coatings are penetrating, and are considered drying, non-drying, or semi-drying, based on the rate of drying. Oils are best applied in multiple thin layers, to facilitate penetration and drying. Oil paints continue to dry as they age, and so must be replenished as often as yearly, or every few years, depending on the type of oil and exterior conditions.

Drying oils like linseed oil, tung oil, poppy seed oil, perilla oil, and walnut oil are liquid oils (triglycerides) that oxidize slowly when exposed to air, gradually changing from a liquid to a solid film. Oxidation produces heat that can cause spontaneous combustion in a closed environment. A thin coat may be dry to the touch in hours or days but will continue to cure over the course of weeks or months. Pure drying oils can take up to a month to dry. However, the process can be accelerated by various chemical or heat treatments. Oxidation continues over the life of the paint, until the film becomes dry and brittle and begins to disintegrate if not retreated.

Litho varnish and stand oil are "bodied oils" that have more viscosity and are produced by processing drying oils with heat.

Non-drying oils, such as almond oil, mineral oil, olive oil and other nut oils, do not oxidize and are less of a fire risk. However, they remain in a liquid form that is more vulnerable to weathering and must be reapplied over time. Because they do not harden, they are used for interior applications, and on materials such as leather, where flexibility is important.

Linseed oil is a drying oil that is extracted or expressed from flax seed. Hot pressing of crushed flax meal produces a better yield, but cold pressing produces a better-quality oil. Boiled linseed oil is heat processed to shorten the curing time to as little as 12 hours. It is darker, thicker, and produces a film that has more body and gloss and is more weather resistant than raw linseed. Stand oil is heat processed at a higher temperature, for a longer duration, and without driers to produce a very durable and elastic film that is very slow to dry. Linseed oil products should not be used on alkaline surfaces such as fresh Portland cement or lime plaster. The alkali in lime and cement will convert the fat in the linseed oil into soap (a fatty acid).

Tung oil is produced from the fruit pods of the Chinese tung tree and processed similar to linseed by grinding and then boiling the pods to extract the oil. Raw tung oil is heat treated and refined to produce an oil that cures more quickly than linseed and resists alkali and water. Tung oil penetrates wood grain and cures to a hard surface but can take 7 to 10 days to partially cure, and a month for a full cure. It is often combined with synthetic resins to produce quick drying enamels, boat varnishes, and alkali-resistant coatings.

Fish oils do not dry well, but their performance improves with heat treatment. Fish oils are often mixed with tung oil, synthetic oils or petroleum-based products. Sturgeon and cod oils have been used in the United States and in Norway; menhaden, sardine, and pilchard oil were used in Great Britain.

Untreated castor oil is non-drying vegetable oil that can be heat or chemically treated to produce dehydrated castor, a useful drying oil. Dehydrated castor oil mixed with phenol formaldehyde resins produces a durable, flexible, water- and chemical-resistant film.

Sulfurized oils are produced by treating linseed, perilla, fish oil, or similar oils with sulfur chloride. These oils set quickly and are used to apply multiple layers of paint in a "wet on wet" process to build up a thick, durable, composite finish.

Driers and Drying

Driers and solvents are added to the other components to improve application, curing, and penetration of the coating. Metal oxides or varnishes expedite drying of oil paints. The amount of drier is based on the type of oil, the strength of the drier, the character of pigments, temperature, and humidity. If too much drier is added, the surface of the paint may cure rapidly into a hard skin that prevents oxidation of the oil beneath the skin. The hardened skin will crack from the movement of the soft elastic paint beneath it. The soft paint can also bleed through these cracks creating a tacky surface. Trial mixtures and applications are generally required to determine the appropriate amount of drying agent.

Traditional drying oils were boiled with metal oxides or acetates of cobalt, manganese, lead, or zirconium to release the fatty acids in the oil and accelerate the chemical reactions that caused oils to dry. Metal oxides act differently and thus the paint dries at different speeds. Cobalt dries most quickly. Lead, which is no longer available, dries evenly, up from the bottom of the film to the surface, while manganese and cobalt dry from the surface down.

Today, paint manufacturers produce a variety of commercial driers, including paste driers, terebine and liquid driers. Paste driers are a combination of drying agents and extenders, ground into a paste and mixed with boiled linseed oil. Terebine is a liquid blend of drying agents (metal oxides) and linseed oil that dries quickly and produces a brittle hard film suitable for ironwork. Pale Liquid Oil Dryer and a more concentrated Strong Liquid Oil Drier are made from linoleates mixed with turpentine and oil to produce a more flexible and durable coating than terebines.

Several proprietary mixes use short oil varnish to accelerate drying without affecting the flow or gloss of the paint. Unlike metal oxide driers, varnish driers can be added to any type of oil paint or natural varnish without causing a chemical reaction or causing surface drying. Goldsize, also known as Japan drier, is a quick drying short oil used as a drying agent. Like terebine, it creates a hard finish that should only be used on hard stable surfaces to avoid cracking.

Oil Varnish

Varnish adds brilliance and shine. It also protects wood, paints and stains from abrasion, moisture, chemicals, and ultraviolet (UV) light. A wide range of oil varnishes with different qualities are produced by varying the type and proportion of resin and oil. Resin adds hardness and shine, while the oil provides durability and flexibility. Curing depends on both the evaporation of volatiles from the resin and the slower carbonization of the oils, so the proportion of oils will dictate drying time.

Exterior varnishes are "long oil" with three to four parts oil to one part resin. "Short oil" varnishes have less oil and dry more quickly to a harder and glossier finish but are not as durable. The high percentage of oil in long oils, such as spar oil, produces more flexible varnish that can accommodate the seasonal movement of wood.

Varnish is a combination of oils, driers, resins, and solvents that can penetrate or remain on the surface. Most failures of varnish are caused by poor surface preparation or application, incompatible materials, or a change in temperature. The lack of a consistent bond between the varnish and the substrate (material or base coating) leads to cloudiness, sagging, blistering, or crazing of the varnish. Varnish can age prematurely if exposed to coal gas, ammonia fumes or similar gases, or to extreme temperature changes.

Oil varnishes provide good protection against moisture and some protection from UV light, but require multiple thin coats during the initial application, and must be replenished every few years to prevent drying and cracking. The life of a varnish finish can be extended by using a base coat of epoxy paint to seal the wood and limit seasonal movement. The varnish, in turn, extends the life of the epoxy by providing protection from UV light.

Resins

The term "resin" is used loosely to refer to a range of materials:

- Resin, a tree sap that is soluble in oil or other solvents, but not in water
- Gum, a tree sap that is soluble in water, such as gum Arabic
- Shellac, a resin secreted by the female lac bug
- Synthetic resins

Exterior varnishes primarily use tree resins that are not soluble in water. Fossil resin is harvested from trees that have been buried for centuries, while virgin resin is collected from live trees. Resins that are mixed with drying oils are called "copals," and vary in hardness. The hardest resins, such as "animis" from East Africa, are used to produce the finest varnishes. Softer resins from central and west Africa are often mixed with animis to produce a good varnish. New Zealand kauri resin is popular because it mixes easily with oils. Copal is also produced in Guiana, Borneo, and the Philippines.

Traditional Japanese lacquer is a varnish produced from the resin of urushi trees (Japanese Sumac) native to Japan. It produces a varnish that is breathable, waterproof, and resistant to acid, alkali, and alcohol but is vulnerable to UV light. Varnishes imported from China, Thailand, and Myanmar are produced from other varieties of sumac and have less urushiol than Japanese urushi trees and produce a lesser quality varnish.

Rosin is a solid resin, a byproduct of distilling turpentine from pine olio resin. It produces a lustrous varnish that can crack with prolonged exposure. However, melted rosin mixed with slaked lime produces a hardened rosin that is mixed with tung oil to produce a durable varnish.

Synthetic resins, including phenolic, alkyd and urethane resins, were developed in the mid-20th century and are commonly used in modern varnishes. Varnishes made with 100% phenol-formaldehyde resins cure quickly by polymerization (a chemical reaction) rather than oxidation and produce a coating that is very resistant to alkalis and corrosion of marine climates, especially when made with tung oil. Alkyds are synthetic resin binders used in oil-based paints. Similar to a traditional oil paint, alkyds cure by oxidation.

Wax

Wax is a fatty, solid substance that never completely hardens, but can be polished to a varnish like finish. It can be sourced from beeswax, the leaves of carnauba palms, or paraffin, a petroleum product. Chinese wax is excreted by a tree insect. Beeswax is popular, but less durable than varnish, and more reflective than oils. In some regions, beeswax is added to lime wash to increase resistance to water.

Stains

Stains are penetrating coatings used to add color and protection to wood, stone, and concrete surfaces. Stains are often protected by more water and UV-resistant coatings. They are usually categorized by the type of solvent: water, spirit, oil, or varnish, and by the type of dye or pigment.

SOLVENT	DYE OR PIGMENT	PROS	CONS
Water	Aniline or Coal tar dye	Clear, transparent, and brilliant hue Resistant to UV fading	Raises wood grain
	Color pigment		
	Acid and alkaline chemical		
Spirit	Basic aniline or coal tar dye	Quick drying	Quick drying
		Brilliant and transparent	May fade in sunlight
		Does not raise wood grain	Do not penetrate deeply
Oil	Oil soluble aniline and coal tar dye	Slow drying Does not raise wood grain	Slow drying May be fugitive in sunlight
	Color pigments		

Figure 2.8.1 Dye or pigments in stains. (Table by Susan Pranger based on content from Vanderwalker, 1957.)

The speed of drying can be either an advantage or a disadvantage, depending on the circumstances. Quick drying spirit stains allow subsequent finishes to be added more quickly but are difficult to apply evenly. Oil stains dry more slowly but are easier to apply.

Water stains penetrate deeply to produce more permanent color, but the water raises the wood grain, which must then be sanded. Spirit stains tend to be fugitive: they can fade in sunlight in the presence of oxygen and moisture. However, the colors are permanent if the stain is immediately covered with varnish or shellac. Spirit stains do not penetrate as deeply as other stains because the solvent evaporates quickly.

Oil stains penetrate more deeply than spirits, but not as deeply as water stains, and do not raise the grain of wood. Traditional oil stains generally faded more than water stains, but today's manufactured oil stains include chemical additives that make the color more permanent. Oil stains can bleed into varnishes if not separated by a layer of shellac.

Thinners and Solvent

Thinners and solvents are both used to dilute paints, to make them more fluid and workable. Exterior oil and acrylic paints tend to use petroleum-based solvents, the most common being turpentine, white spirit thinners, naphtha solvent, and industrial alcohols. Thinners should evaporate completely as the paint cures, leaving no residue in the paint film. Petroleum-based thinners and solvent-based paints can dissolve existing paint coatings and should be used carefully on previously painted surfaces.

Thinners are not strictly speaking a drying agent, but they do allow paint to be spread more thinly so that it dries more quickly. Thinning the paint also allows the oil to penetrate more deeply and bond more fully to absorbent substrates.

Pigments and Dyes

The performance and durability of paint is largely determined by the vehicle, drier and thinners, but is also impacted by the type and percentage of pigments. UV light and sulfides and other chemicals in the atmosphere can fade or change the colors of pigments. Stable pigments can protect underlying materials by blocking UV light.

Pigments, especially artificially prepared pigments, vary in their resistance to acids and alkaline, their reaction to linseed oil, and their response to heat, and UV light. Some pigments assist drying of oil vehicles, while others retard drying, or are neutral. Pigments that absorb more oil will cause paints to dry more quickly. Other pigments cause paint to "feed" or thicken.

Natural "earth pigments" are minerals or mineral compounds of ochres and other iron oxides ranging in color from yellow and orange (raw and burnt sienna) to red and brown (raw and burnt umber). Most natural pigments are inert, very stable, and dry slowly in oil.

Carbon and graphite are black pigments derived from natural minerals, or from soot.

Chemically prepared pigments are produced by a range of processes, including corrosion and precipitation. Many chemically prepared pigments are nearly identical to the chemical composition of natural pigments. Certain metallic pigments, such as white and red lead and lead chromes accelerate drying. Antimony white, bone black and Vandyke pigment will retard the drying process.

Several historic pigments, including white lead, and zinc oxide, were produced artificially by submerging sheet metal in a mix of acid vapors and carbon dioxide to promote corrosion. The resulting carbonate is very stable and reacts chemically with the fatty acids in linseed oil to produce lead soap that is easy to apply, promotes drying, and produces a tough durable coating. White lead paint can react with sulfur found in industrial and coal burning environments. Lead paint is now largely banned due to serious health risks associated with ingesting lead paint or dust, especially in developing children.

Zinc oxide is a popular substitute for white lead, although it does not assist drying as much as white lead, and it produces a harder film that tends to become brittle. Mixing white lead and zinc oxide produced paint that had the best qualities

	PIGMENT	DRYING EFFICIENCY IN OIL	REACTIONS
MISC. CALCINING	Vegetable Black	Bad	Inert
	(Vegetable oil soot)		
	Bone Black	Bad	Inert
	(Ground and burned)		
	Ultramarine (calcined china clay, soda, sulphur and coal)	Poor	Destroyed by acids
EARTH PIGMENTS	Vandyke Brown	Bad	Fugitive (fades in sunlight)
	Yellow ochres	Poor	Inert
	Raw sienna	Poor	Inert
	Burnt sienna		
	Indian red	Neutral	Inert
	(Calcined copperas)		
	Vermilion	Neutral	Blackened by sulphur
	(Cinnabar)		
	Raw turkey umber	Good	Inert
	Burnt turkey umber		
PRECIPITATES	Zinc chrome	Neutral	Decomposed by acid
	White Lead	Good	Thickened by acid medium
			Blackened by sulphur
			Decomposed by acid
	Yellow chromes	Good	Blackened by sulphur
	Brunswick	Neutral	Destroyed by alkalis
	Prussian blue (oxidized sulfate)	Good	Destroyed by alkalis
	Chrome green	Good	Destroyed by alkalis
	Brunswick green (yellow chrome & Prussian blue)		
OXIDES	Zinc Oxide (white)	Good	Thickened by acid medium
	Red lead	Very Good	Blackened by sulphur
	(oxide of lead)		
	Red oxides	Neutral	Inert
	Venetian Red		
	(calcined oxides of iron)		
DYES	Crimson lake	Poor	Fades on reduction

Figure 2.8.2 Drying efficiency of pigments. (Table by Susan Pranger based on content from Table 2, "Pigment Data" (Albert Edwin Hurst, 1949.)

of each. Antimony white is produced by condensing the fumes produced by roasting the ore stibnite. Antimony is inert and less reactive to acid but it retards the drying of linseed oil and turns yellow when exposed to sulfides. However mixing antimony with zinc oxide produces a hard film that is resistant to sea air. Titanium oxide is produced from the black ore ilmenite; it is inert and does not chemically react with oils or acids or discolor in contact with sulfur but is flexible. Zinc oxide is added to titanium oxide to decrease its natural elasticity, harden the film, and improve durability.

Anilines and Other Coal Tar Derivatives

Aniline dyes are distilled from coal tar, and are soluble in water, oil, and spirits. Different distillations produce an almost complete range of colors and are used for stains, tinting colors, and prepared paints.

Coal tar colors are classified as acid, basic, spirit, oil, and direct colors, each with different characteristics. Acid and direct colors can be mixed with acid and with each other, but never with basic, spirit or oil colors. Some coal tar colors do not bleed and are very resistant to UV light, while others bleed or fade. Some colors that fade in bright light are protected by a varnish coating. Hot water, alcohol, and other spirit solvents will absorb more color from coal tar, but are more sensitive to temperature than coal tar colors in oil.

Modern Paints

Post World War 2 paints were developed to allow consistent mass production of paints and varnishes. The majority of paints available today are water-based latex paints which use a mix of 12 different liquid organic colorants. Solvent-based paints, which use a variety of proprietary pigments, have limited use because they do not meet EPA standards for VOC content. However, solvent-based paints were used extensively in the 20th century and may remain on historic materials. Modern paints and primers with synthetic resins may be incompatible with existing layers of traditional oil paint or 20th century solvent-based paints. Modern paints can perform well when applied over well-cured and sound paints but can accelerate delamination of weakly adhered paint. Solvents in synthetic paints can dissolve existing paints.

Twentieth-century solvent-based paints required petroleum-based thinners. They were more elastic, more water resistant, and more UV resistant than latex paints but are no longer in use in most building applications due to the high VOC content. Synthetic oils were mixed with natural oils or "oil reactive" synthetic resins to create coatings with specific qualities, such as a quicker drying time, more elasticity or hardness, or better water or chemical resistance. They do not develop the high initial gloss of oil paints but retain their gloss longer. Fresh synthetic paint is less likely to be damaged by rain and dust because added plasticizers allow them to dry quickly without becoming brittle.

Alkyd paints are solvent based, dry quickly, and offer good color and durability, and flexibility combined with a hard finish. The type of oil determines the drying characteristics and ultimate use of the alkyd paint. Alkyd resins perform like a drying oil but dry more quickly. They have limited use today but were used extensively in the second half of the 20th century.

Latex paints are water based and vapor permeable. They are more flexible and therefore able to accomodate movement of hard substrates without cracking. Latex paint was originally made from the sap of rubber trees, but modern latex is made from a blend of vinyl and acrylic polymers. The proportion of vinyl and acrylic binders will determine the durability, performance, and suitability for exterior use. Vinyl polymers are less expensive but less durable; acrylic polymers are more expensive but more weather resistant. Acrylic latex paint, which has 100% acrylic polymers (no vinyl polymers), is more durable than a latex paint with a blend of acrylic and vinyl polymers.

Polyurethane is a clear liquid plastic coating that was introduced at the beginning of World War II. Today it is ubiquitous as a durable, waterproof coating. However, it is also hard and brittle and may flake and turn amber as it ages. Polyurethane can be water or solvent based.

Urethane is a synthetic resin used as a binder with water- or petroleum-based carriers. Aromatic urethane is only used for interiors, but aliphatic urethane is UV resistant. Water-based urethanes combined with acrylic binders are abrasion resistant.

Drivers of Deterioration

Paint adds a layer of durability to most materials by blocking moisture, pollution, and UV rays, but it is also vulnerable to those forces and must be renewed periodically. The best coatings are those that do not peel or flake and can be renewed without extensive preparation or repair. Coating should also block moisture without trapping the water and salts that inevitably make their way into materials, and then try to escape.

The chemical composition, strength, flexibility, and thickness of a coating each contribute to a coating's ability to resist deterioration and protect its substrate. Many paints will continue to dry or lose solvents over time, which will change its appearance and durability. This durability is only as good as the bond between coating and substrate. Paints that penetrate the substrate material will wear differently than those that sit on the surface.

Various aspects of climate – heat, temperature, moisture, wind, and pollution – all affect coatings during fabrication, application, and its wearing life. Heat and UV exposure can accelerate the natural aging of coatings; heat and moisture can cause differential movement; acids, alkalis and particles in the atmosphere can also change the chemical composition of paints.

Figure 2.8.3 Wear of linseed oil paint, Rorros, Norway. (Photo by Daniel Thomas, 2022.)

Water

Water can erode coatings or break the bonds between coatings and substrate. Water can enter the substrate through cracks in the paint or through rising or falling damp, leaks in the roof, or interior humidity. Moisture that builds up between the substrate and the paint film can crack paint that is no longer elastic, or create bubbles in paint that is elastic. Moisture can also drive the movement of salts that damage paint coatings.

The Sun: Heat and Solar Radiation

Building exteriors experience both ambient heat from the environment and heat from solar radiation. The texture and color of coatings can affect reflectivity and emissivity.

UV light and other actinic rays can trigger photochemical reactions in paints and varnishes, breaking down oils and degrading pigments. Light can cause some pigments to fade when also exposed to air and moisture. Pigments that fade in oil can be protected by a varnish that is less porous. Other pigments, such as pure red (iron) oxide, absorb UV rays and reduce the negative impact on the oil or varnish medium.

Dry heat can be beneficial by aiding in the initial drying of paint films, and reducing the use of driers that make paints less elastic and less durable. However, in most cases, heat is detrimental to paint. Excessive heat can cause rapid drying and cracking if the surface dries more quickly than the underlying paint. Over time, excessive heat expedites the natural gradual drying of oils and solvents. Rapid changes in temperature can cause excess movement of substrates, especially metals with high coefficients of thermal expansion, and break the bond between the substrate and paint film.

Smoggy environments can also block some UV damage. Clear air and high ozone levels do not block the UV rays, which contributes to more rapid oxidation of paints in coastal environments. Coatings as diverse as the tar coating of Norway's stave churches and the lacquer finishes of Japan's shrines are showing accelerated deterioration from increased levels of UV radiation.

Freezing temperatures can damage paints that have not cured, but are generally not detrimental to cured films, unless moisture is trapped behind the paint film.

Figure 2.8.4 Paint deterioration, Boston, MA. (Photo by Daniel Thomas, 2023.)

Chemical Damage

Incompatible pigments, driers, or other additives can cause chemical reactions that compromise drying and flexibility. Pigments that are acidic or alkaline can react negatively with other materials in the paint, the substrate, or the atmosphere. Lead paints are discolored by sulfur fumes while ultramarine is attacked by acidic rain. Certain paints should not be used with alkaline substrates such as concrete.

Soot, smoke, sulfur, and other pollutants leave corrosive deposits on the surface of the paint, where they are activated by rain. These deposits can react both with the paint medium and with the pigments. Deposits can settle into the surface of the paint where they are not easily removed, and can also attack new coatings from below if a dirty substrate is not properly cleaned.

Mechanical Stresses

Differences in thermal or moisture expansion will create mechanical stresses in the weaker material, which is usually the coating. Paint that is less flexible than its substrate can crack or delaminate.

Biological Damage

Paint coatings can reduce the potential for mold, fungi, or bacterial growth by reducing the moisture level in porous materials. However, the paint itself can support biological growth, especially if the surface is porous or textured. Insects can penetrate paint coatings and burrow into the underlying wood, as was the case at the temples in Nikko. In fact, paint coatings can conceal ongoing damage.

Textured surfaces and organic matter in coatings, such as flour, glue size, casein, and oils can support algae, fungus in damp, humid conditions. Modern paints have chemical mildewcides or fungicides, although there is some doubt as to the effectiveness of these additives over time. The most effective solution, after thorough cleaning and removal of existing growth, is to reduce humidity and correct any conditions that collect or retain moisture.

Strategies for Responding to Climate-Based Threats to Coatings

Research and Investigation

Historical research can identify the types of materials typically used in the past, or specific information about past coatings or conditions. For example, historical research identified wind-driven beach pebbles as the probable source of chips in the relatively new mineral coating at Haceta Head Lighthouse.

Damage to coatings is usually very visible, but the cause is often less apparent. Damage may be due to poor materials, poor surface preparation or application errors; existing environmental conditions of moisture, temperature, or UV; or biological or mechanical stress. Checking, alligatoring, and cracking are usually due to loss of binder, or movement of the substrate. Peeling is likely due to an issue with the substrate such as incompatibility or moisture pressure. Chalking or wear might be due to exterior forces like rain or wind, or internal failure of the paint.

There are number of tests that can be used to identify the type of materials, layers of paint, and problems with incompatible coatings and substrates. Inspections should include any sources of moisture, including moisture in the substrate, roof leaks, and interior humidity.

Protect, Maintain, Repair, Replace, Adapt

Paints, stains, and other coatings are used to protect other materials. In that role, coatings may be considered sacrificial and should be maintained by regular replacement. The impact of heat and sun on traditional paints can be mitigated by regular repainting, which replenishes the oils, if the surface is properly cleaned and free of contamination.

Sometimes the coatings themselves have historic significance and warrant protection through consolidation or other methods of diverting water or stress. Traditional linseed oil paints can be restored by applying clear linseed oil, restoring the binder while maintaining the original pigments.

Details that divert water and reduce sun exposure can extend the life of most coatings.

Conclusions

Exterior coatings are both a historic material requiring protection, and a source of protection for other materials. They are the first line of defense against (1) weather conditions, (2) poor maintenance, and (3) weak substrates. Good maintenance of both coatings and their substrates are critical to their survival as weather conditions become more intensive. However, exterior coatings must first be appropriate to their substrate, and to the weather conditions.

Traditional coatings were designed to meet these criteria but may be challenged by more intense weather conditions. New coatings can be successful if they meet the criteria for flexibility, permeability, and UV resistance.

Bibliography

Bartus, Jan, William J Simonsick Jr, and Otto Vogl. 1995. "Oriental Lacquers IX. Polymerizable Ultraviolet Stabilizers for Oriental Lacquers." *Polymer Journal* 27 (7): 703–771.

Bednarik, Robert G. 1998. "The first stirrings of creation." *UNESCO Courier*, April. http://cogweb.ucla.edu/ep/Art/Bednarik_98.html.

Bock, Gordon. Editor *Old House Journal.* 1983. "Whitewash and Calcimine: Paints That Ain't." *The Interiors Handbook for Historic Buildings, 2:6-19-6-22 Historic Preservation Foundation. Washington, D.C.*

Canning. 2022. *5 Types of Wood Finish Used Throughout History.* https://www.merriam-webster.com/dictionary/resin.

Chase, Sara, interview by Susan Pranger. 2023. *Paint* (May 28).

Edison. 2015. *Everkote 300 Mineral Coating Systems.* December 17. https://www.edisoncoatings.com/Potassium_Silicate_Masonry_Pai/potassium_silicate_masonry_pai.html.

Hurst, Albert Edwin. 1949. *Painting and Decoration.* 8th ed. Stevenson Press, 2012.

Keim. 2023. "Keim Silicate Paints - the Mineral Principle." https://www.keim.com/silicate-paint/.

Lukey, C. A. 2001. "Drying Oils and Alkyds." In *Encyclopedia of Materials: Science and Technology.* Elsevier Ltd. https://www.sciencedirect.com/topics/chemical-engineering/drying-oil.

Minogue, Patrick. 2021. *Tung Oil Production in Florida.* https://edis.ifas.ufl.edu/publication/FR420.

Nakagoshi, Kosuke, and Kunio Yoshizumi. 2011. "Degradation of Japanese Lacquer Under Wavelength Sensitivity of Light Radiation." *Materials Science and Applications* (Science Research) 2: 1507–1515.

Okahisa, Yoko, Chieko Narita, and Tsuyoshi Yoshimura. 2019. "Resistance of Wood Coated with Oriental Lacquer (Urushi) Against Damage Caused by Subterranean Termite." *Journal of Wood Science* (Springer Nature) 65 (41). https://doi.org/10.1186/s10086-019-1820-6.

Popular Woodworking. 2020. *Outdoor Finishes.* https://www.popularwoodworking.com/finishing/outdoor-finishes/.

Real Milk Paint. 2019. *Drying Oil vs Nondrying Oil: Which Is Right For You?* https://www.realmilkpaint.com/blog/tips/drying-oil-vs-nondrying-oil-which-is-right-for-you/.

Urushi Nation Joboji. n.d. *Beyond the Time, Japanese Elegance.* http://en.urushi-joboji.com/urushi.

Vanderwalker, F. N. 1957. *Wood Finishing, Plain and Decorative.* Chicago: Frederick J. Drake & Co.

Wardell, Charles. 1999. "Understanding New Paint Products and Formulations." *Architectural Record*, November: 131–138.

Waterlox. 2022. *What Is Tung Oil?* which was also cultivated in Russia, America and in areas of the British empire during the 20th century.

Webster, Merriam. 2022. *Definition of Resin.* https://www.merriam-webster.com/dictionary/resin.

PART 3
United States Case Studies

CHAPTER 3.1

Covered Bridges in Oregon and New England

Introduction

Covered bridges are a fascinating combination of exposed and protected wood, built wherever timber was plentiful, the climate was wet, and rivers needed to be crossed. The trusses were almost always unpainted and often only partially protected by a wood enclosure very similar to siding and the roof on a house. The design of the enclosure was similar across the United States, with variations influenced by local weather conditions and construction traditions.

Wood can survive almost indefinitely if it is kept dry. While many bridges have been lost to natural fires, arson, flooding, or vehicular damage, most bridges deteriorated from a lack of maintenance of the enclosures. A comparison of covered bridges in Oregon to those in New England illustrates how the design reflected – or ignored – the climate.

History, Location and Climate

History

Covered wood truss bridges became popular in the United States in the early 19th century. As many as 10,000 covered bridges had been built in the United States by 1870. By the early 20th century, many aging wood bridges were being replaced by steel or concrete bridges that required less maintenance and could carry larger and heavier commercial vehicles. However, construction of wood bridges continued, especially when steel became scarce during World War I, and where old growth timber was abundant.. Interest in preservation and cultural

DOI: 10.4324/9781003160489-15

heritage led to advocacy and funding for the restoration of many remaining wood bridges, but those in rural areas with limited use and resources fell into disrepair.

In 2023, only about 750 covered bridges remain, with the largest concentrations in Pennsylvania, Ohio, Vermont, Indiana, New Hampshire, and Oregon. Many have been lost to highway expansion, intensive commercial development, and physical neglect as well as flooding and fire. Transfer from private ownership to local municipalities and counties has brought greater oversight and inspection, but funds for maintenance and capital repairs continue to lag, especially for those bridges that are not actively carrying vehicle traffic.

Location and Climate

Western Oregon, Vermont, and New Hampshire are at the same latitude, 43 north, and receive comparable amounts of annual precipitation. However, Oregon receives most of its precipitation as rain in the colder months of November through February, while the New England states get a mix of rain and snow, with precipitation falling more evenly and intermittently throughout the year. Thawing snowpack and coastal storms can cause intense flooding in both areas.

Oregon's long wet winters provide little opportunity for wood to dry once it gets wet. On the other hand, the dry spring and summer months do not encourage the growth of insect populations as much as the wet spring and summers of New Hampshire and Vermont. However, powder-post beetles and carpenter ants do swarm in high summer temperatures in Oregon.

Changing Patterns in Climate

While temperatures are generally expected to increase in both Oregon and New England, changes to precipitation are more difficult to predict. The consensus seems to be that the overall annual precipitation may be the same, but winters will be slightly wetter and summers slightly drier. Rain may be less frequent, but the intensity and duration of rain events will increase.

Longer rain events increase the likelihood that unprotected wood will have a higher moisture content for longer periods, leading to increased wood decay. Intense rain events, combined with an increase in land clearance for agriculture and forestry, have already increased runoff and flooding, which damages or destroys wood bridges each year. Fire damage is more likely to occur in areas that are becoming seasonally drier. Dry conditions and high winds increase the likelihood that fires will be more widespread, whether triggered by natural, accidental, or intentional events.

Construction Techniques and Materials

Covered bridges are constructed of heavy wood trusses designed to span rivers, usually without intermediate supports. Their initial construction required scaffolding across the river during low water seasons. Once constructed, if protected from the weather, the wood trusses could last almost indefinitely.

Several truss configurations have been used over the centuries. The four Oregon bridges included here use the Howe truss, which was strong and economical, using steel rods in lieu of wood for the vertical tension members. The top chord and diagonals are each in compression, while the bottom chord and vertical tie rods are in tension. Covered bridges in New England exhibit a wider variety of truss types, including the Town lattice used at Ashuelot covered bridge in New Hampshire and the Paddleford truss used at the Sanborn covered bridge in Vermont.

In the Pacific Northwest, trusses were constructed from large old-growth Douglas fir, which is very dense and not very porous. Douglas fir, unlike the species used on the East Coast, undergoes a permanent aspiration of inter-tracheid bordered pits as the wood dries, resulting in low permeability. Bridges in New England were constructed typically from dense, old-growth eastern white pine and occasionally from spruce or white oak.

The bridges were covered to protect the trusses, which were difficult to repair or replace, and the bridge decking, which became slippery when wet, from the elements. The roofs usually had wide overhangs in all four directions and were covered with unpainted wood shingles or shakes, usually western red cedar in Oregon. Atlantic white cedar, oak, or eastern white pine were prevalent in New England. A few bridges used sheet metal roofing. Wood shingles and shakes were installed over open strapping boards to maximize ventilation and drying from the underside.

Wood siding was typically installed only on the exterior side of the trusses to maximize ventilation and drying and provide access for visual inspections. The siding was sometimes continued on the first interior bays at each entrance, to protect from wind blown spray and vehicle splatter. The siding was made of local wood, either vertical board and batten or horizontal wood clapboard, both of which easily shed water. In Oregon, the siding was typically painted, likely to protect the wood from the constant rain. In New England, the siding was typically not painted except in village settings, where a more finished look was desired. The lack of paint has been attributed to Yankee frugality, but also ensures that the wood can dry between rain events. However, some siding details, such as horizontal strapping applied directly to horizontal beam, trapped water and promoted wood decay.

Openings in the sidewall enclosures provided visibility and light but also exposed the trusses to the weather. Designs that used louvered openings or a clerestory type opening tucked under the overhangs more effectively protected exposed trusses from weather. However, many bridges had long unprotected openings on the sides to maximize visibility, and relied on sills to drain water, and paint to protect the timbers.

Deterioration, Restoration, and Other Changes

The biggest threats to covered bridges are vandalism, wood deterioration (fungal decay and insect damage), flood damage, and fire. The presence, or the lack, of moisture increases the risk of all these threats with the possible exception of vandalism. High humidity in the atmosphere promotes the growth of mold and mildew, while high moisture content in the wood is necessary for fungal decay and insect damage. Heavy rains or sudden snow melts cause flooding that can literally knock a bridge off its abutment. Drought conditions can increase the likelihood of forest fires.

Salt-laden air or water spray can cause a breakdown of the surface fibers in wood, creating an appearance called "fuzzy wood". Salt in the wood can also attract more moisture to the wood, increasing the potential for fungal attack.

When roofs deteriorate, water accumulates on the horizontal surfaces of the top chords, or in upward facing "V" joints of cross bracing, creating damp environments that promote fungal and insect decay. Water can also run down the vertical and diagonal truss members to accumulate on the bottom chords. Windblown dirt, vegetation, bird nests, bat guano, and other debris can also accumulate in crevasses and trap moisture. Peeling paint can also trap moisture.

In Oregon, large old-growth Douglas fir timber is not readily available as it once was, and harvesting large diameter, old-growth Douglas fir is now heavily restricted. Less dense or small diameter wood is available and satisfactory to replacemost members but may not have the structural capacity required for large top and bottom chords with long spans or large highway loads. Structural capacity of existing members has been increased by adding post-tensioned rods. In Oregon, replacement with glue-laminated timbers (i.e., glulam) was approved by the State Historic Preservation Officer (SHPO) for use in locations not visible to the public.

Alternatives to replacement have included epoxy repairs on compression members, sistering or other external reinforcement.

Pressure treated Douglas fir used for replacement members can be identified by the pattern of puncture marks (incising) that allow wood preservatives to better penetrate the wood of Douglas fir and other refractory species. In Oregon, plugged holes are often seen in existing wood members where tubes of a solid fumigant are inserted on a regular cycle. The use of liquid fumigants is discouraged because they can leak out of checks and splits. Solid fumigants vaporize over time, releasing a gas that penetrates and protects the wood from fungi and insects, but does not leach out to contaminate the waterways below.

In New England, general service use, wood weathering from sun and rain, and biological degradation all cause deterioration, but deferred maintenance, flooding, vandalism, and arson cause the most damage. In addition to high, rapid water, spring floods can carry damaging ice. Unpainted siding is more vulnerable to UV damage, which is greatest on the south and west elevations. Although unpainted western softwood will only erode ¼″ per century, the damage is accelerated if the eroded material is washed away by rain. Western red cedar can erode as much or greater than 2/3″ to an inch of loss.

Because of the increased threat of arson and wildfires, fire retardants and intumescent coatings have been used on some painted surfaces, and a clear proprietary fire retardant is now available for use on unsealed and unpainted wood. Cedar shingles with a fire retardant are used in Oregon.

Oregon Bridges

Figure 3.1.1 Goodpasture Covered Bridge, Lane County, Oregon. (Photo by Daniel Thomas, 2023.)

Figure 3.1.2 Goodpasture Covered Bridge – louvered openings. (Photo by Daniel Thomas, 2023.)

Goodpasture Bridge, built in Lane County in 1938, is the second longest covered bridge in Oregon, with a span of 165 feet. The Howe truss is enclosed with a cedar shingle roof and painted vertical board and batten siding. There are eight shuttered openings on each side that protect the trusses with deep window returns. The bridge was repaired and repainted in 1973 and extensively restored in 1987. By 2010, heavy loads that exceeded the posted limits had damaged the bottom chords and caused the bridge to sag. Temporary closure was not an option so the chords were repaired in place and stengthened by adding post-tension steel strands on each side in 2013. In 2019, the cedar roofing and roof trusses were replaced after damage from an ice storm. There has been no new insect damage since Lane County began the use of solid fumigation plugs every five years. The Holiday Farm fire of 2020, driven by heavy winds and drought conditions, burned 173,000 acres surrounding the bridge but did not damage it. Today the bridge is in good condition, but frequent, high-load traffic from clearing 2020 fire damage has taken a toll on the bridge. There is significant advocacy for its continued operation and maintenance because it is the sole access to the communities beyond.

Crawfordsville Bridge, also known as the Calapooia River Bridge, is a 105-foot span built in Linn County in 1932. The Crawfordsville Bridge is now owned by the Linn County Parks Department and considered a "historical artifact". It is no longer rated for vehicle traffic, which limits the availability of funds for maintenance. Its Howe trusses are enclosed by a cedar roof and vertical board and batten siding, all of which are in poor condition. In this case, paint on the interior is no substitute for water management. Mold is visible on the cross members, the face of the upper chords, and even on the deck boards. Insect damage is not readily visible, but resistance drilling of the top chords below the four roof leaks confirmed internal decay: the sides of the top chord have 3 to 4 inches of solid wood on either side, concealing 6 inches of internal void caused by fungal decay. Temporary plastic panels were recently installed to divert the water under the holes away from the trusses. Repair is anticipated but not currently funded or scheduled.

Figure 3.1.3 Crawfordsville Covered Bridge, Linn County, Oregon. (Photo by Daniel Thomas, 2023.)

Figure 3.1.4 Crawfordsville Covered Bridge, temporary roof patch. (Photo by Daniel Thomas, 2023.)

Shimanek Bridge, also in Linn County, is much newer, constructed in 1966 with a Howe truss, cedar roof and painted board and batten siding on the sides and interior wrap arounds, and horizontal clapboard siding on the portals. It was renovated in 1990. The trusses are painted on all faces, and the siding openings are louvered, but the jambs are narrow, and do not protect the trusses as thoroughly as they do at Goodpasture. The wood roof shingles were old, not maintained, and had many holes. This led damp trusses through the long, wet Oregon winters and provided ideal conditions for wood decay. There were numerous insect damaged areas in the upper and lower chords. Replacement of the bottom chords with new glued-laminated timbers in 2022 was allowed by the Oregon SHPO because Douglas fir of the required size and strength was not readily available, and the members are not visible to the public

Figure 3.1.5 Shimanek Covered Bridge, Linn County, Oregon – louvered openings. (Photo by Daniel Thomas, 2023.)

Figure 3.1.6 Shimanek Covered Bridge – surface decay from water trapped by continuous strapping for siding. (Photo by Kevin Groom, 2022.)

Figure 3.1.7 Shimanek Covered Bridge – deck stringer damage due to insect decay. (Photo by Kevin Groom, 2022.)

Gilkey Bridge was constructed in 1939 and renovated in 1990, and again in 2017. By 2023, however, it was already showing wear and moss is growing on the exposed painted trusses. The wood sills, designed to divert water from the bottom chords, tend to collect water at joints with the timber diagonals. Attempts to divert water from the lower chords have been of limited success. Flashing, added to protect the joint between the sill and diagonals, was later removed because it trapped water. A silicone sealant is nominally more effective. Epoxy has been used to repair carpenter ant damage in compression members.

Figure 3.1.8 Gilkey Covered Bridge, Linn County, Oregon. (Photo by Daniel Thomas, 2023.)

Figure 3.1.9 Gilkey Covered Bridge – flashing damage, fumigation holes. (Photo by Daniel Thomas, 2023.)

New England Bridges

Ashuelot Covered Bridge was built in 1864 in Cheshire County, New Hampshire, using a Town Lattice truss design that spans 178 feet, supported mid span by a granite pier. The painted lattice trusses are relatively exposed to the elements as there is no siding, protected only by deep roof overhangs over cantilevered sidewalks, and by a guardrail height wall at the outer edge of the sidewalk. However, the individual lattice members are relatively small and easily replaced when damaged or decayed. The original framing used Eastern white pine with 2” diameter white oak “treenails” – wooden pegs commonly seen in New England barn construction. A comprehensive renovation in 1999 addressed years of damage including decayed and broken members, water damage from a leaking waterline added in 1947, and a shifting abutment. Some chords and lattice beams were replaced with Douglas fir members. Decayed ends on additional lattice members were repaired with epoxy or wood dutchman. A fire detection system and fire-retardant coatings were added in response to prior fire damage to Ashuelot and the loss of three covered bridges in New Hampshire to arson in 1993.

The Sanborn Covered Bridge, also known as Centre Covered Bridge, was built in 1869 in Lyndonville, Caledonia County, Vermont. It is one of only a few remaining Paddleford trusses, a type of multiple kingpost truss designed by Peter Paddleford of New Hampshire. The unpainted trusses are exposed on both sides, with a walkway on one side, but protected by 5-foot-wide eave extensions and projecting gable ends, and partial height sheathing of 5-foot-high vertical boards on one side. Painted horizontal clapboards provide some protection at the entrances. In 1960, the bridge was moved a mile upstream to allow for construction of a new bridge. Unfortunately, the bridge was set closer to the river level, and has suffered considerable damage from ice and flood debris to the lower chords, counterbrace extensions, and post extensions. The original bridge design had borderline capacity according to a 2021 analysis. Over a century and a half, gradual creep, neglect, and floor damage have combined to distort and fracture the overloaded columns and chords. A request for proposals was issued in 2023 for extensive restoration but to date no one has been willing to undertake the challenging logistics.

Figure 3.1.10 Sanborn Covered Bridge. (Photo by Nicole Gratton, 2023.)

Figure 3.1.11 Sanborn Covered Bridge: spring flood damage. (Photo by Nicole Gratton, 2023.)

Projected Impacts of Climate

Unfortunately, temperatures, heavy rain, and humidity are all expected to increase compared to past weather conditions, increasing the risk of fire, flooding, and biological damage. The population size and habitat of beetles and termites are expected to expand to areas not previously at risk. Prolonged warm, wet conditions will promote mold and wood decay.

Conclusions

The best protection against fungi and insect damage is water management. Water management requires a combination of protecting the trusses from precipitation and avoiding details that allow for accumulation of debris that traps moisture. Restoration of roofs and enclosures must be complemented by ongoing inspection and maintenance to remove debris and maintain paint, fumigation, and fire protection systems.

In some cases, more drastic interventions may be required to preserve and protect the bridge. For example, although it would impact the original appearance, low bridges may need to be raised to avoid rising flood levels.

Historic covered bridges require regular restoration and maintenance, which is unlikely to happen without regular use and advocacy. Funds are simply not available to maintain bridges as artifacts. A new threat to these bridges is a lack of knowledge and craftsmen who understand how the small details allow the wood to dry.

Bibliography

Andrews, Timothy, and Christopher Marston. 2023. Interview by Zoom with Timothy Andrews, Barns & Bridges of New England and Christopher Marson, US National Park Service regarding covered bridges.

Anthony, Ron. 2022. Telephone interview with Ron Anthony of Anthony Associates regarding wood and covered bridges.

Anthony, R. W. (2018). *Guidelines to Restoring Structural Integrity of Covered Bridge Members*. Forest Service, Forest Service Laboratory. USDA.

Ausland, Gregory, and Tony La Morticella. 2013. "Goodpasture Covered Bridge Rehabilitation." *International Conference on Timber Bridges*. https://intrans.iastate.edu/app/uploads/sites/12/2019/03/ID_80_Ausland_LaMorticella.pdf.

Ausland, LaMorticella, and Lange Sevits. 2019. "Case Study: Goodpasture Covered Bridge." In *Guidelines for Rehabilitating Historic Covered Bridges*, edited by Christopher Marston and Thomas Vitanza. Washington, D.C.: Historic American Engineering Record, National Park Service.

Beckman, Stephen Dow, Nick Cockrell, and Bill Cockrell. 1977. "NR Nomination – Goodpasture." https://catalog.archives.gov/id/77848091.

Beckman, Stephen Dow, Nick Cockrell, and Bill Cockrell. 1979. "NR Nomination Crawfordsville Bridge." *National Register Nomination*. https://catalog.archives.gov/id/77848166.

Botsai, Dost. 1990. *Wood Detailing for Performance*. Mill Valley: GRDA Publications.

Christianson, Justine, and Christopher H. Marston, editors. 2015. Covered Bridges and the Birth of American Engineering. Washington, DC: Historic American Engineering Record, 68. https://www.nps.gov/hdp/CoveredBridges2015med.pdf

Gratton, Nicole. 2023. Zoom interview with Nicole Gratton, Planning Director, Town of Lyndon, Vermont regarding covered bridges in Lyndon.

Groom, Kevin. 2023. Interview with Kevin Groom, PE Linn County Road Department regarding covered bridges in Linn County.

Henry, Hugh H. 1974. "NR Nomination: Centre Covered Bridge." National Register of Historic Places https://catalog.archives.gov/id/84285125

James, Sean T., Robert H. Durfee, and Timothy Andrews. 2020. "Case Study: Ashuelot Covered Bridge." In *Guidelines for Rehabilitating Historic Covered Bridges*, edited by Christopher Marston and Thomas Vitanza. Washington, DC: Historic American Engineering Record, National Park Service.

Lewandoski, Jan. 2021. "The Sanborn Covered Bridge, Lyndonville, VT Technical Assistance Survey." Technical Assistance Survey. https://www.lyndonvt.org/wp-content/uploads/2023/01/Sanborn_Report_condensed.pdf

Lewandoski, Jan. 2023. Telephone interview with Jan Lewandoski, owner of Restoration and Traditional Building, regarding covered bridges in New England.

Melis, Darleen. 1979. "NR Nomination Ashuelot Bridge." National Register Nomination Form.

Ohio State University (OSU). 2023. *Climate Impacts Research Consortium*. https://pnwcirc.org/science/precipitation.

Price, Zach, and Hailey Rein. 2021. "Oregon Wildfires: 20 of the state's biggest since 2002". Statesman Journal (July 20). https://www.statesmanjournal.com/story/news/2021/07/20/oregon-wildfires-20-biggest-since-2002/7985470002/.

Smart Fume. n.d. *Advanced Control of Wood Decay*. https://smartfume.com/.

Thomas, Todd. 2023. Telephone interview with Todd Thomas, Bridge Supervisor, Lane County, Oregon, regarding Lane County covered bridges.

Weather Service. n.d. *U.S. Climate Data*. https://www.usclimatedata.com/climate/united-states/us.

CHAPTER 3.2

Early New England Masonry

Introduction

The 1690 manor house at the Spencer-Peirce-Little farm and the 1713 Old State House are early examples of masonry construction in New England. Over three centuries, both buildings experienced the harsh New England weather and alternating periods of neglect and care, including the loss of potentially protective coatings. Modification to the Old State House reflected its civic status, changing architectural styles, and preservation priorities, while modifications to the Spencer-Peirce-Little House met the functional needs of its residents.

Today both buildings are on the National Register of Historic Places and are operated and maintained by non-profit organizations that sometimes struggle with the ongoing cost of maintenance and capital improvements. However, the Old State House is a prominent Boston Landmark owned by the City of Boston and has more access to funds. Its materials and systems are in good condition but are still vulnerable to the effects of weather, especially where original soft brick is exposed. The Spencer-Peirce-Little House has received substantial restoration and maintenance since it was acquired by Historic New England in 1986, but the exterior masonry is exposed and vulnerable to current and future weather impacts.

Location and Climate

The Spencer-Peirce-Little House and the Old State House are both located near the Atlantic Coast in the state of Massachusetts. The nearby ocean not only moderates the temperatures but also brings the occasional Nor'easter, a heavy storm that deposits moisture from the ocean inland. Higher ocean temperature can increase the amount of moisture carried inland by these storms. Although only 40 miles

DOI: 10.4324/9781003160489-16

apart, Boston experiences slightly higher temperatures than Newbury, in part due to the urban heat island effect.

The Spencer-Peirce-Little House is in Newbury, Essex County, a coastal town classified as "Dfb" (warm summer continental climate) on the Köppen Climate Classification scale. Temperatures range from an average high in August of 70.6°F and an average low in January of 26.1°F. Peak temperatures have been recorded as high as 100.0°F and as low as 0.0°F. The total average precipitation of 46 inches varies from a low of 3.2 inches in August to 4.4 inches in November. The annual snowfall is 54 inches, with 15.4 inches occurring in January.

The Old State House is in the center of Boston, Massachusetts, classified as "Dfa" (hot summer continental climate), with slightly higher average summer and winter temperatures (73.3°F in July and 29.3°F in January). Peak temperatures in Boston were 103°F in July and –12°F in January. Boston receives 42.2 inches of annual precipitation and 41.8 inches of snow, both slightly less than in Newbury. Rain occurs year-round, peaking in March.

New England has already seen higher temperatures, more precipitation, and more frequent storms. The intensity of severe rain events has increased more in the Northeast than in any other part of the United States. Precipitation during intense storms increased 71% between 1958 and 2012, including Tropical Storm Irene in 2011 and Winter Storm Nemo in 2013. Winter Storm Juno brought more than 30 inches of snow in three days in January 2015. Precipitation is expected to continue increasing as temperatures increase. Both Boston and the Newbury could experience a 50% increase in the frequency of tropical storms with heavy rainfall and high winds by 2090.

Average temperatures increased by 4 degrees between 1981 and 2010. By 2070, temperatures could increase by another 4 degrees with 39 more days above 90°F. The maximum temperature could increase by as much as 12.5°F by 2090. Rising temperatures cause a slight increase in evaporation of water over the land and a substantial increase over the ocean. Increased temperatures and rainfall will also affect nearby forest environments, increasing pest populations.

Both buildings are on low-lying land near the coast and will be subjected to flooding from hurricanes, storm surge, and sea level rise. Without intervention on a regional scale, a worst-case scenario of 7.5 feet rise by the end of the century will bring the shoreline to the edges of both the Old State House and the Spencer Peirce Little House.

Spencer-Peirce-Little House

History

The farm was established in 1635 by John Spencer, an early settler of the Newbury community. The National Register nomination suggests that the original stone house was built between 1635 and 1651 but dendrochronology confirms a date as

1690. In any case, the stone house is remarkably intact in spite of additions c. 1800, various periods of benign neglect, and several restorations after Historic New England took control in 1986.

Nathaniel Tracy found the stone house "amusingly quaint" when he bought it in 1778. In spite of the current fashion for Georgian styles, Tracy limited his exterior modifications to rebuilding the chimney and replacing the leaded casements with sash windows.

Captain Offin Boardman purchased the farm in 1797 and added a two-and-a-half-story Federal style frame wing with brick end chimney to the west of the original house for his wife, who suffered from poor health and found the original stone house "an aged stone pile harboring dampness and disease". Edward H. Little, a long-term tenant who purchased the farm in 1861, only modified the windows of the parlor. The last generation of his family, five unmarried women, lived frugally in the home, with little change and minimal maintenance, from the mid-20th century until 1986.

In 1971, the property was designated a National Historic Landmark and deeded to the Society for the Preservation of New England Antiquities (SPNEA), now Historic New England. SPNEA took control after Amelia Little's death in 1986 and began a long-term research and conservation project to restore and interpret the site, beginning with the stabilization and repair of the chimney and structural members throughout the house. Historic New England completed a five-year conservation and preservation project in 2012, which included comprehensive stormwater drainage for the property to address moisture issues in and around the buildings. Subsequent repairs have included window repairs and replacement of the roof on the main house. The manor house and the remaining 230-acre farm are visited by 26,000 people annually.

Figure 3.2.1 South Elevation of Spencer-Peirce-Little House, Newbury, MA. (Photo by Susan Pranger, 2021.)

Construction and Modifications

The original stone house is in a cruciform shape, with two large rooms either side of a large stone chimney, a small vestibule in the front, and a larger kitchen wing in the rear. The second floor is supported by heavy wood summer beams. On the north side of the main house, the kitchen wing has an exterior brick chimney on a ten-foot-wide stone foundation on its north wall. The first addition, c. 1800, was a two-and-a-half-story wing that preserved the original lime and masonry on west wall of the main house. The second c. 1800 addition was a wood frame addition on the rear of the original house and was eventually used as a separate tenant house.

The two-foot-thick exterior walls were built of local granite rubble and fieldstones, possibly from the lower end of Plum Island. Smaller gallet (or garret) stones were pressed into the lime mortar to stabilize the larger stones and fill wide and irregular joints. Brick was used for interior and exterior arches, at exposed chimney corbels, and for trim or patches. The exterior walls of the front vestibule and the second-floor gable ends are also brick. Both the stone walls and the brick vestibule walls are thicker at the first floor, creating a small shelf on the exterior. This ledge would have originally been protected by either a plaster wash or a stone cap, neither of which exist today. A 1986 archeological investigation revealed evidence of a stone drip apron in the front of the house, which would have protected the lime plaster from rain splash.

The original lime mortar and several layers of limewash remain where the gable end of original west elevation was covered by the c. 1800 addition. A remnant of beveled lime mortar extends from the ridge to the eaves and would have originally protected the bottom of the wood rake board. Lime mortar was also used to caulk the gap between the window casing and the surrounding stone. These traditional English techniques suggest that the lime coating was either original to the construction of the house or followed soon after. The lack of parging on this same elevation suggests that the exterior masonry was originally whitewashed, and that parging was not applied until some time after the c. 1800 addition.

Figure 3.2.2 West gable end of original Spencer-Peirce-Little House. (Photo by Susan Pranger, 2021.)

Traces of red oxide paint exist on some wood trim and on edges of brick where overpainting of adjacent wood may have occurred. Red oxide paints were often traditionally used as a wash over bricks and mortar to conceal variations in color and size.

The presence of at least seven different exterior plaster mixes in various locations and conditions indicate that parging was applied intermittently rather than all at once, and that there was likely a mix of limewash, parging, and bare stone on different areas of the house at any given time. A visual analysis in 1990 suggested that the first layers of tan lime parging likely contained natural cement and while later hard gray parging (now aged to a tan color) likely contained an early Portland cement. Experienced craftsmen at HNE believe that the hard cement is an early Portland cement (less strong than modern OPC), but this has not been confirmed by petrographic analysis. The 1990 analysis proposed that it would be historically appropriate to maintain the building in its present mottled state rather than apply a single finish that had never occurred (except as a limewash).

Most of the gallet stones were removed as the lime mortar was replaced with cement mortar, and the original plaster has worn away or been replaced. However, a few examples of the lime mortar, gallet stones, and remnants of the early lime plaster remain in protected locations. Original clay mortar still exists at the cellar foundation walls of the central chimney and might have been used at other locations protected from the weather.

Figure 3.2.3 Remnants of parging, whitewash and gallet stones at Spencer-Peirce Little House. (Photo by Daniel Thomas, 2021.)

Deterioration

The manor house is located less than two miles from the Atlantic Coast where it is subjected to storms with extreme severe wind and rain. This is especially true of the east wall, a regular source of leaks. Past damage to interior wood members by carpenter ants and powder-post beetles is evidence of that moisture penetrated the building interior. The combination of past improvements to site drainage and future improvements to building drainage systems is expected to relieve the moisture problems in the basements that facilitate insect and fungal decay.

Figure 3.2.4 Carpenter Ant Damage in the basement at Spencer-Peirce-Little House. (Photo by Susan Pranger, 2021.)

Remnants of cement coatings remain on the north and south elevations. In the late 1990s, early cement-based mortars and stucco was removed from the east wall and replaced with a harled finish of pure lime putty, sand, and hair, followed by a white limewash over the entire wall. The finish started to fail soon after but was not repaired until 2009, when HNE carpenters primed the entire wall with a limewash, replaced missing patches with a mix of lime putty, sand, and water, and then finished the entire wall with a limewash. The wall was primed, patched, and limewashed again in 2015 and in 2019. These finishes were again delaminating in October 2021. Repairs to the gutters and downspouts are scheduled for 2023 with modificationsto improve the water carrying capacity in response to increases in storm intensity. Another restoration of the wall coating is scheduled to occur in 2024 or 2025.

It has been suggested that the 1990 era lime parging did not fully carbonate before being subjected to freezing winter temperatures. This might have been due to the thickness of the application and premature drying of the original putty-sand mix. A combination of harsh weather conditions, eastern exposure to ocean winds, and the lack of regular whitewashing likely led to cracking and additional frost damage. This was further exacerbated by the failure of gutters and downspouts to adequately carry the increasing volume of rainwater.

Figure 3.2.5 Deterioration of harling at east elevation of Spencer-Peirce-Little House. (Photo by Daniel Thomas, 2021.)

The other stone and brick walls of the manor house are not fully parged but still have some residual patches of lime or cement parging. Efforts have been made to manage runoff with improved gutters, downspouts, and subsurface drainage; however, the rough, deteriorated surfaces retain water, and provide good anchorage for mosses and lichen. Regular maintenance is required to avoid the establishment of larger plants, especially as the weather continues to get warmer and wetter.

Figure 3.2.6 Biological growth at Spencer-Peirce-Little House. (Photo by Susan Pranger, 2021.)

The Old State House

History

The Old State House was used concurrently for civic and commercial activities from 1713 until 1830, when it became Boston's City Hall for 11 years. This era was followed by 40 years of commercial adaptive use and change. In 1880, the Bostonian Society secured a long-term lease from the City of Boston to operate and maintain the building. The 1881 renovation was intended to restore the building to its Revolution era appearance but mistakenly restored it to the 1830 plan. In 1903–1905, the basement and foundation of the Old State House were modified as a new subway line and station was constructed below. The 1909–1910 renovation attempted to restore its original provincial style. Another major renovation followed in 1930. A comprehensive preservation program by the National Park Service in the 1990s focused on preventing further deterioration of the exterior masonry and wood. A full-time caretaker was hired c. 2005 to coordinate repairs and ensure ongoing monitoring and maintenance.

In 2020, the Bostonian Society merged with the Old South Association to become Revolutionary Spaces, which is now responsible for operating and maintaining the Old State House and the Old South Meeting House. As a small non-profit, the Bostonian Society and its successor, Revolutionary Spaces, do not have extensive in-house assets or financial resources. Capital funds for major projects or renovations are solicited on a case-by-case basis from state preservation sources. Operating expenses are primarily funded by income from the Old State House store and programs.

A "rainy day fund" existed in 2014 to allow a quick response to small repairs but was depleted during the 2019 COVID pandemic because of a reduction in visitor traffic and related funds. The Director of Facilities and Preservation and his assistant are responsible for a wide range of tasks at both buildings, from completing small in-house repairs to overseeing outside contractors, and supporting programming, exhibits, and events.

Figure 3.2.7 View toward the harbor from the east end of the Old State House, Boston, MA. (Photo by Susan Pranger, 2023.)

Existing Construction and Modifications

After 300 years of use, the Old State House is a mix of original construction and several adaptations and reconstructions. In addition to periods of neglect and modifications, the Old State House experienced several catastrophic events including a fire in 1748 and the subway construction in 1904. Major restorations in 1773, 1830, 1881, 1936, the late 1950s, and the 1990s, while not always historically accurate, did serve to maintain the building.

The original gambrel roof was replaced with a gable roof with wood dormers and a tiered tower in 1748, after the fire of 1747 destroyed everything except the masonry shell. The roof, dormers, and cupola were rebuilt in 1881 and in 1936 and repaired in 1990 and in 2007. Repairs to the tower and cupola are anticipated to be part of the 2027 renovation.

The original wood windows had all been replaced by 1909. The dormer windows were replaced with replicas in 2007. Subsequent leaks were attributed to condensation from excess interior humidity, which was reduced by the addition of attic exhaust fans in 2010. The remaining windows were restored, including sill replacement, painting, and replacement of glazing putty between 2014 and 2016.

The original paints would have likely been linseed oil, but have been replaced by synthetic paints, most recently off-the-shelf latex paints. Preservation staff are transitioning to the use of more permeable potassium silicate paints on interior plaster and linseed oil on interior and exterior wood.

The original brick foundation walls were replaced with a granite water table and concrete foundation walls as part of the construction of the subway line and station in 1903–1905. The granite water table was replaced with a molded brick water table in 1909–1910.

Records indicate that paint was applied to the exterior brick in 1773 and removed in 1909. Most of the east and west walls were modified and rebuilt over the years, but about two-thirds of the brick on the north and south elevations is believed to be hand molded English bond masonry and lime mortar original to 1713. New brick has a hard finish. The original brick is porous, often spalled, and easily absorbs moisture, especially where the surface was damaged by fire or paint removal. The exterior masonry was completely repointed and repaired in a series of projects from 2014 through 2020. Appropriate lime mortars have been and will continue to be used.

Figure 3.2.8 Circa 18 century brick from North Elevation of Old State House. (Photo by Susan Pranger, 2023.)

The building was heated by fireplaces until steam boilers and radiators were added in 1881. At some point, steam heat was provided by a central distribution system. Roof insulation and mechanical ventilation, air conditioning, and humidity control were added in 1990. The heating, ventilation and air conditioning (HVAC) system was replaced in 2007 and modified in 2010. The chiller was replaced in 2022. The existing steam boiler system is scheduled to be replaced with a high efficiency gas boiler as part of the next major renovation.

A master plan is being prepared for a comprehensive seven-year restoration of both the Old State House and the Old South Meeting House. Phase one will include reopening the west elevation entrance of the Old State House so that both buildings open onto Washington Street. The main restoration is scheduled to begin in 2027, pending funding. A disaster plan is also being developed in phases. It currently only addresses evacuation of people and collections. The next step, in conjunction with the master plan, will be to identify procedures and renovations to address future flood and environmental impacts.

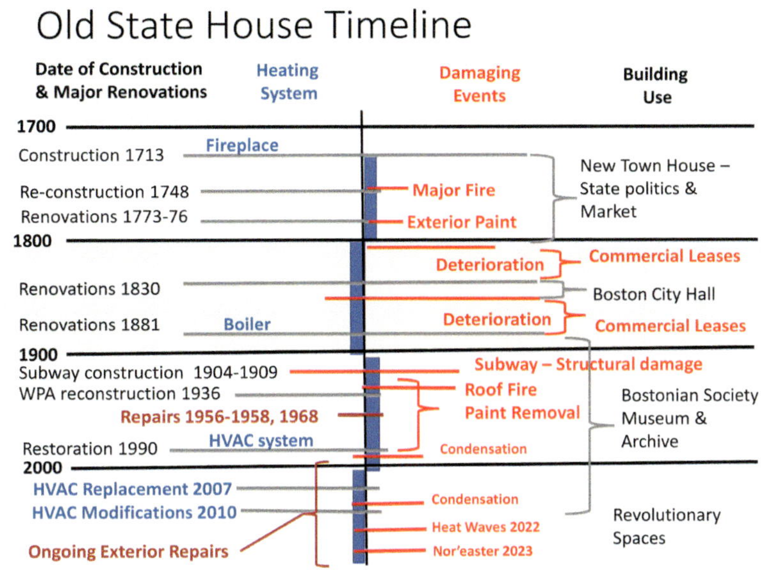

Figure 3.2.9 Old State House, timeline. (Diagram by Susan Pranger, 2015, revised 2023.)

Climate Impacts

Located near the harbor in the middle of a city, the Old State House is subjected to extremes of heat and ocean storms. Climate related damage is primarily due to high winds and driving rain off the harbor that has caused fungal decay in wood and deterioration of mortar joints in the masonry.

Damage to the windows and wood attic structure prior to 2010 was attributed to condensation from high interior humidity added by mechanical systems and by the lack of ventilation within the building. Careful monitoring of the humidity and temperature has successfully eliminated condensation.

Figure 3.2.10 Condensation leaking from interior of the Old State House. (Photo by Dr. Judith Selwyn, c. 2010.)

The northeast corner of the Old State House continues to be subjected to high winds and driving rain that funnels directly down State Street from the harbor and has been a continuous source of interior leaks for decades. The brick was repointed in 2007 and a "high-tech" vapor barrier was added in 2014 but the wall continues to leak. Water poured in through windows on the north and east elevation during a heavy storm in 2023. A temporary solution is being sought to reduce infiltration until the major work scheduled in 2027. Efflorescence is visible on all elevations but seems to be more pronounced near grade and at hard brick lintels on the south elevation, indicating the movement of moisture through the walls from both rising damp and rainwater.

Figure 3.2.11 Interior plaster damage at northeast corner, Old State House. (Photo by Susan Pranger, 2023.)

Increasing summer temperatures and humidity are taxing the mechanical systems. Boston experienced four heat waves in the summer of 2022, which exceeded the capacity of the aging chiller system. Temporary auxiliary commercial air conditioners and dehumidifiers ran concurrent with the existing chiller until it could be replaced in the winter of 2022.

Intermittent surface flooding of the impervious surrounding pavement has also caused damage to the existing masonry foundations in the past. The existing gutter system is inadequately sized and reportedly clogged so the rainwater overflows onto the pavement or makes its way to a basement drainage trench and sump pump. The extent of damage will not be uncovered until the future renovation proceeds; however, it is anticipated that new interior and exterior rainwater systems will be adequately resized to accommodate future increases in rainfall. Development of mitigation measures to address surface flood and sea level surge are also anticipated as part of the master plan process.

Figure 3.2.12 Efflorescence at hard modern brick, South Elevation, Old State House. (Photo by Susan Pranger, 2023.)

Conclusions

The increasing severity of wind and rain events will have the most immediate negative effect on the masonry at both Spencer-Peirce-Little House and the Old State House. Increased duration and depth of moisture penetration increase the risk of damage from cycles of wetting-drying and freeze-thaw, including condensation damage in wood, efflorescence and frost damage of masonry, and ice damming at roof lines. Masonry and wood surfaces are at risk of biological damage as increasing temperature and humidity increase the opportunities for biological growth. Sea levels at the end of the century will be close enough to both buildings to ensure that storm surge and surface flooding will cause moisture related damage to the masonry foundations. Flooding of drainage infrastructure is of particular concern in Boston.

The Old State House benefits from a high level of maintenance and oversight, which will hopefully continue as it faces increased threats from the weather. Their commitment to original materials and techniques combined with a proactive approach to mitigating future flooding should serve them well.

Spencer-Peirce-Little House clearly receives regular maintenance, but the site would benefit from a more comprehensive cleaning and restoration of the exterior stone and plaster. While the lichen and moss are picturesque, they have already established footholds for larger plants. Building and site drainage was improved in 2012 but should be monitored for new issues as rainfall quantities and intensity increase.

It is not known if the masonry at the Spencer-Peirce-Little House and the brick of the Old State House were originally protected but both were soon after their construction. As weather impacts become more severe, both buildings might benefit from a traditional limewash or similar breathable coating. Although both the Spencer-Peirce-Little House and the Old State House have gone through many changes in appearance over the past centuries, there would likely be debate and protest over any proposal to change the appearance. It may be one of many future discussions weighing survival of original material against historic appearance.

Bibliography

Archipedia NE. 2020. *Historic Masonry Finishes*. Edited by Brian Pheiffer. http://www.archipedianewengland.org/historic-masonry-finishes-sections/.

Beaudry, Mary C. 1998. "Farm Journal: First Person, Four Voices." *Historical Archaeology, Archaeologists as Storytellers (Springer)* 32 (1). http://www.jstor.org/stable/25616589.

Beaudry, Mary C. 1995. "Scratching the Surface; Seven Seasons at the Spencer-Peirce-Little Farm, Newbury, Massachusetts." *Northeast Historical Archaeology* 24 (Article 4). https://doi.org/10.22191/neha/vol24/iss1/4.

Boston, City of. "Climate Ready Boston Map Explorer". www.boston.gov/departments/environment/climate-ready-boston-map-explorer

BLC. 1994. *Report on the Designation of the Old State House as a Landmark*. Boston: Boston Landmarks Commission, City of Boston. http://www.cityofboston.gov/landmarks/publications/.

Blunden, Jessica and Derek S. Arndt. 2014. "BAMS The State of the Climate in 2013." *Bulletin of the American Meteorological Society* 95: S1–S279. http://dx.doi.org/10.1175/2014BAMSStateoftheClimate.1.

Chase, Sara B. 1978. *A Brief Survey of the Architectural History of the Old State House. Boston, Massachusetts.* Reprinted for the Bostonian Society from "Old Time New England", Vol 68, Number 251, pp. 31–48. (Winter–Spring 1978). Published by Society for the Preservation of New England Antiquities. https://www.historicnewengland.org/explore/collections-access/gusn/169861/

CMZ. 2017. *Massachusetts Sea Level Rise and Coastal Flooding Viewer.* Massachusetts Office of Coastal Zone Management (CMZ). https://mass-eoeea.maps.arcgis.com/.

EPA. 2016. *Adapting to Climate Change Northeast.* https://www.epa.gov/sites/default/files/2016-07/documents/northeast_fact_sheet.pdf.

Goody Clancy Architects. c. 1990. "Amendment to Old State House Historic Structures Report."

Heintzelman, Patricia. 1978. *Spencer-Peirce-Little House.* National Register of Historic Places Inventory – Nomination Form, United States Department of the Interior, National Park Service.

Horton, Radley. 2014. *Thinking ahead: New York communities consider risks of extreme rainfall events.* October 6. http://www.climate.gov/news-features/featured-images/thinking-ahead-new-york-communities-consider-risks-extreme-rainfall.

Lang, Kristin C. 1990. "Spencer-Peirce-Little House – Study of Mortar, Whitewashes, and Pargings Found on Exterior Wall Surfaces." Society for the Preservation of New England Antiquities.

Massachusetts. 2022. "2022 Massachusetts Climate Change Assessment." Executive Summary. https://www.mass.gov/info-details/massachusetts-climate-change-assessment.

Morrison, Hugh. 1952. *Early American Architecture.* New York: Dover Publications.

Nold, Carl R., Benjamin Haavik, and Julie A. Solz. 2012. "Collaboration and Preservation: Historic New England and the Proactive Preservation Interpretation and Planning (PPIP) Process." *The Artifact, Its Context and Their Narrative: Multidisciplinary.* ICOM-CC.

Pfeiffer, Brian. 2020. *1690 – Newbury Massachusetts – 5 Little's Lane – Spencer-Peirce-Little House.* http://www.archipedianewengland.org/1860-2/.

Phillips, Morgan W. 1990. "Spencer_Peirce-Little House, – Observations on the History of the Exterior Masonry Surfaces." Society for the Preservation of New England antiquities.

Pokwatka, Christina, Ben Haavik, and Bruce Blanchard. 2023. Telephone and email discussions with Christina Pokwatka, Ben Haavik and Bruce Blancahrd, Property Care, Historic New England regarding HNE maintenance records for Spencer Peirce Little.

Pranger, Susan. 2015. "Resiliency Assessment: Old State House." "Resiliency: Preparing Historic Buildings for the Impact of Climate Changes" Presented to the Faculty of the School of Design Studies in Partial Fulfillment of the Requirements for the Degree of Master of Design Studies. Boston Architectural College. Thesis.

Pokwatka, Christina 2021, Interview with Christine Pokwatka, HNE Property Care Team on site at Spencer Peirce Little Farm. (12).

Resilient, MA. n.d. *Rising Temperatures.* https://resilientma.mass.gov/changes/rising-temperatures.

Sasaki, 2014 "Designing with Water" https://www.bostonharbornow.org/wp-content/uploads/2017/02/PRT2-Designing-with-Water_Full.pdf

Selwyn, Judith, 2014. Telephone Interviews with Dr Selwyn, conservator and consultant re: renovations to Old State House.

Waterhouse, Campbell. 2023 Interview with Campbell Waterhouse, Director of Facilities and Preservation, Revolutionary Spaces regarding existing conditions and renovations to the Old State House and Old South Meeting House, on site at the Old State House (May 19)

weatherbase. 2021. *Newburyport, Massachusetts*. https://www.weatherbase.com.

CHAPTER 3.3

The San Antonio Missions

Introduction

In the 15th and 16th centuries, the Spanish built mission communities from Florida to California to further their goals of conquering and converting Mexico and the American Southwest. Five of these missions were established between 1718 and 1731 along an eight-mile stretch of the San Antonio River. These missions share a common history, climate, and construction techniques. A comparison of San José y Miguel de Aguayo (San José), built 1768–1782, and Nuestra Señora de la Purisima Concepción de Acuña (Concepción), built 1740–1755, illustrate how different use and maintenance have affected their current condition.

History

The Rise and Fall of the Spanish Missions

Spanish colonies included a religious mission, a nearby military presidio, and a pueblo or town. The "mission" refers to a religious community that included instruction in the Catholic faith, Spanish civic duty, and training in agriculture and construction. As resources became available, each mission built a church, cloister, convento (priest's residence), workrooms and courtyards, housing for the local Indians and a farm. Complex construction was achieved by a largely unskilled local population under the direction of the Jesuit or Franciscan missionaries and skilled Spanish or Mexican architects and master craftsmen. The mission's buildings, training, and resources were designed to assure long-term economic viability of the Spanish colony.

The mission system failed when the Spanish withdrew in the late 18th century. The Mexican government left each mission with local resources, but as the

DOI: 10.4324/9781003160489-17

missions lost their Indian converts, they also lost the plentiful labor required to maintain the buildings. Subsequent survival of the mission buildings depended on the strength of the remaining community and individual leaders who directed or championed maintenance and restoration.

History of the San Antonio Missions

By 1731, Franciscan missionaries had established five missions along an 8-mile stretch of the San Antonio River. The San Antonio missions flourished until official Spanish support began to dwindle in 1775. Between 1793 and 1824, ownership was transferred to the local diocese and then to local parishes, of the Catholic Church. By 1831, Texas was part of Mexico and all mission land, with the exception of the churches themselves, had been sold to private individuals. The buildings were abandoned; timbers and stone were often removed and reused nearby.

The Texas Army used each of the San Antonio missions, most in ruins and without roofs, as defensive fortifications during their fight for independence. After the United States annexed the Republic of Texas in 1845, responsibility for most of the missions was assigned to the Catholic Church. Work began at San Juan and Espada in 1858 but little work occurred at the missions at Concepción, San Valero (the Alamo) and San José until the beginning of the 20th century.

Restoration began with the creation of the National Monuments Program in 1906, the National Park Service in 1916, the San Antonio Conservation Society in 1924, and the Works Progress Administration (WPA) program in 1935. The historic preservation movement in the 1950s and 1960s brought renewed attention but mixed results depending on the available resources and shifting attitudes towards conservation and preservation.

With the exception of San Valero, the mission churches of San Antonio continue to serve as places of worship. Under a 1935 agreement, the Archdiocese of San Antonio under the name of Las Misiones, an umbrella non-profit, retained ownership and responsibility for the operation and maintenance of the churches at San José, San Juan, Espada, and Concepción, including decision-making authority of use, interpretation, and restoration. The remainder of each site is managed and controlled by the National Park Service, which only has decision-making authority for the grounds and secular buildings. The Texas Historical Commission has the authority to approve or reject any proposed treatment or modifications to both religious and secular buildings. All five missions became part of the San Antonio missions National Historic Park in 1978 and were designated as UNESCO World Heritage Site in 2015.

Each of the five missions are maintained and restored in response to different interpretive and preservation goals. All five missions now have roofs and some degree of exterior restoration, but missing lime plaster has not been restored in order to maintain the "ruined" appearance of the last century. Although all of the churches are now enclosed, the lack of the original plaster leaves them more vulnerable to the weather than originally intended.

Nuestra Señora de la Purisima Concepción de Acuña

Nuestra Señora de la Purisima Concepción was completed in 1755 and flourished as the administrative center of the missions. The mission was abandoned by 1797, but people continued to live in and around the mission, and it was used by the Texas army in 1835. After the Catholic Church regained ownership in 1841, a succession of religious orders occupied the buildings and completed work in the 1880s and 1930s and intermittent repairs when resources became available.. The general consensus seems to be that all five of the San Antonio mission churches were originally plastered with two or three coats. However, there is some disagreement about whether the rear and side walls of the church at Concepción were ever plastered. At least one report suggests that only the front façade and towers were originally plastered, while the mortar was struck flush with the surface of the stoneon the rear and side walls.

In the early 20th century, lime plasters were replaced with cement plaster that is now gradually being removed. Remants of the original lime plaster on the exterior and interior are being carefully retained, restored, or replaced. In spite of experiencing neglect and abuse similar to the other San Antonio missions, Concepción retains the most original fabric, including its original stone roof and extensive original interior and exterior plaster. Its survival has been attributed to its continued use, its proximity to San Antonio, and its secure foundation on a rock outcropping.

Figure 3.3.1 Nuestra Señora de la Purisima Concepción, San Antonio, Texas. (Photo by Daniel Thomas, 2021.)

San José y Miguel de Aguayo (San José)

The mission at San José y Miguel de Aguayo was first established in 1720 and moved twice before the existing complex was established in 1740. The church was completed in 1782 and intended to outshine the mission previously completed at

Concepción in size and ornament, but the size was reduced due to insufficient funds. The Churrigueresque style façade, however, is more ornate than most other missions including Concepción. There were efforts to restore San José after Texas established its independence, but the roof had collapsed by 1874 and the mission was in overall poor condition by 1900.

The south tower collapsed in 1920 and was rebuilt in 1929. The missing roof was replaced, and half of the walls were reconstructed in the 1930s. The dome and eastern two-thirds of the vault were rebuilt with a mixture of limestone and modern reinforced concrete construction. In 1948, missing portions of the carved limestone façade were replaced with cement or stone replicas, taking care to retain any original fabric.

Although much of San José is a reconstruction, the church, convento, and walls present a fairly complete picture of a mission complex. The church and grounds, with their aged appearance and exposed rough stone, are interpreted by the NPS as a ruin, with a focus on stabilization rather than reconstruction. Missing stones and mortar are patched and pointed, but only a few patches of original plaster and paint remain. The interior plaster and paint are remarkably intact even though it was exposed to the weather for more than a century. Assessment and restoration have been ongoing but intermittent since 2006.

Figure 3.3.2 San José y Miguel, San Antonio, Texas. (Photo by Daniel Thomas, 2021.)

Climate and Geology

San Antonio is located approximately 662 feet (202 meters) above sea level in Bexar County, Texas, an area that is bisected by the Balcones Escarpment. North of the escarpment are gentle hills; to the south are prairies and plains. The missions are all located on fairly level land within a half mile of the San Antonio River, an important source of water for the early missions. Concepción is located on a stone escarpment, which may contribute to its relatively sound condition.

The climate of San Antonio is considered Cfa – humid subtropical, under the Köppen-Geiger Climate Classification. Temperatures rarely fall below freezing, and even the dry months experience some rain. However, San Antonio has experienced a steady increase in all temperature metrics over the last half century including average summer and winter temperature, the number of warm and hot days with maximum temperature over 80, 90, or 100°F and the number of warm nights with minimum temperatures above freezing.

Overall, the total amount of annual rainfall in San Antonio has not increased significantly. However, there has been an increase in the number of dry days and in rainfall intensity – the amount of rain falling on wet days. In the future, San Antonio is expected to have more dry days and less total rain in the spring and fall but more individual days of intense rainfall. Longer periods of wet conditions are expected to contribute to biological growth.

Traditional Materials of the San Antonio Missions

Unlike the more typical adobe missions, the missions at San Antonio were built of a porous tufa stone covered with lime plaster and accented with carved limestone. Roofs were either flat timber or vaulted and domed masonry, both covered with lime plaster.

Tufa

The tufa walls of the San Antonio missions are almost 4 feet thick at the side walls, increasing to around 5 feet thick at the base of the towers and main façades. The thick walls could resist wind loads and earthquake movement even after their roofs were destroyed. A core of rough rubble set in lime or clay mortar was faced in uncoursed or roughly coursed, irregularly, or roughly dressed, tufa that was set in a thick lime mortar and covered with a thick layer of limestone stucco.

Tufa (not to be confused with volcanic tuff) is a tuffaceous limestone, a very soft and porous sedimentary rock quarried from the Karst formation on the banks of the nearby San Antonio River. Tufa has a large pore structure that easily absorbs water vapor, rainwater, and surface water, but also dries quickly and resists the expansion of salts or clay. However, rapid fluctuations in temperature and moisture, and prolonged saturation from intense rain or rising damp can deteriorate the stone. Water migrating through tufa can erode adjacent lime mortars and seep into adjacent limestone.

Frequent wet-dry cycles can transport and deposit salts deeper and higher in the wall and can expand the clay found in many sedimentary stones. This repeated expansion and contraction breaks the bonds within the stone and causes delamination and flaking on the exterior faces. Although infrequent, freeze thaw has created micro-cracks that leave the stone vulnerable to stresses.

Limestone Ornament

Figure 3.3.3 Detail of Retablo at San José y Miguel. (Photo by Daniel Thomas, 2021.)

San José and Concepción both have an elaborately carved limestone entrance façade or portal ("retablo") of a gray micritic limestone with bands of ochre. At San José, the two-story entrance retablo, the sacristy window surround, and other architectural details were carved with elaborate floral and shell shapes from a green-gray limestone in the Mexican Churrigueresque style. The carvings at the Alamo and at Concepción are more restrained, representative of the more austere late Baroque style of New Spain. Unlike the tufa, the dense, finely grained bioclastic limestone was never intended to be covered with plaster, although the carvings at Concepción were originally coated with a thin colored limewash or paint.

The limestone was quarried north of San Antonio in an area now known as Brackenridge Park, part of the narrow Austin Chalk Group formation. The limestone is bioclastic (containing fragments of planktonic marine and other shelled organisms less than 0.5 mm thick) and micritic (a recrystallization of fine-grained calcium carbonate mud). It has a high content of fine clays and non-carbonate fines but virtually no siliceous sand or silt. The stone is easily saturated due to a combination of the microporosity in the micritic mud and larger-scale porosity of unfilled bioclastic shells, and can be damaged by swelling clay or by frost.

Although the limestone is chalky and soft, it is also fine grained, dense, and homogenous with no naturally occurring cracks or cleavage veins and no distinct layers which makes it suitable for carving in all directions. It is soft enough to shape with woodworking tools when first quarried but becomes harder as moisture in the stone evaporates. The tool marks on the stonework are consistent with woodworking tools, reinforcing the belief that the stone was carved by woodworkers shortly after being quarried and before the stones case-hardened. However, there are fine microcracks parallel to the geological lamina (layers) of the limestone, and some cross grain cracks consistent with swelling of expandable clays that could explain the persistent scaling of the carved stones.

Lime Mortars and Plasters

The original mortars and plasters were a non-hydraulic mix of high calcium lime, sand, and clay. The lime was obtained by firing caliche soil or fragments of limestone on site. The sand was a coarse-grained natural carbonate sand that included naturally rounded shell fragments, limestone sand, and minor amounts of siliceous grains of quartz. Traces of organic matter have been detected in the plaster, but it is not clear whether these were added to make the plaster more durable or were absorbed from the wet paint.

Bedding and pointing mortars were thick to accommodate the porous, uneven tufa blocks in the walls, vaults, and domes and could have aggregates up to a half inch in size. Extruded bedding mortar was struck level to fill in the voids of the tufa stone. Most walls were then coated with at least one or two layers of lime plaster that keyed securely into the irregularities and open pores of the tufa.

At Concepción, a base coat of lime and sand, with aggregates less than 0.1875 inch thick, was applied to the towers, parapets, and the front (west) façade. It appears that the side and rear walls at Concepción were never plastered. The west façade received an additional finish layer with a high lime to aggregate ratio, using aggregates of medium to fine sand, to allow for a smooth finish. The plaster on the front façade and on the towers was incised to receive highly ornate paint patterns that were applied to the dried plaster.

Exterior Paints

Both exterior and interior plaster were stenciled and painted with bright polychrome geometric patterns using earth, charcoal, or calcium carbonate pigments in binders of plant or animal oils. The patterns were applied "al seco" to plaster that had already dried prior to being painted. Al seco paint is not impregnated into the plaster and was more likely to wear away than paints applied al fresco, to wet plaster.

At San José, only small fragments of painted plaster remain. At Concepción, there are multiple layers of lime plaster and paint on the west façade, indicating that the plaster and paint were originally reapplied periodically. There are also traces of colored limewash applied directly to the carved limestone entrance retablo at Concepción.

Figure 3.3.4 Traces of original paint at Concepción. (Photo by Daniel Thomas, 2021.)

Roof Construction

Flat roofs were built up in layers. Heavy timbers spanned the narrow nave and supported smaller branches. These supported a matt of still smaller fibers, which supported layers of soil or clay mud. The top layers of lime plaster were often treated with alum or salts to make it more waterproof.

Stone domes and vaults required more skill but were preferred due to the scarcity of large timber. The roofs of San José and Concepción included a vaulted roof with a stone dome at the crossing and sacristy, supported by simple stone pendentives. These roofs were also protected by cement plaster. The dome at Concepción was restored in 2020.

Wood

Exposed exterior wood was limited to the bell stocks, window lintels, and door and window jambs. At Concepción, the bell stock was made from small, knotty branches of hickory, a porous wood naturally resistant to decay. The ends of the stock were set into pockets that tended to wick water into the wood and stone.

Exposed wood weathered quickly from fungal, termite or UV damage, depending on exposure. UV damage is most severe on the south elevations, while fungal decay caused by wet wood is greater on the northwest, where the wood dries more slowly. There are traces of limewash or other white paint that originally protected the wood.

Projected climate changes are likely to increase UV intensity and related damage, and long durations of rain that will promote fungal decay. Traditional, permeable coatings such as limewashes can protect the wood from biological growth and UV damage while allowing it to dry.

Change, Weather, and the Future

The missions at San Antonio were abandoned and mostly without roofs for over a century. During that time, most of the protective lime coating was eroded by wind and rain, while the roofs and stone walls were dismantled by vandalism and damaged by war. As the original lime plaster eroded over time, the walls were either left exposed or coated with cement plaster. When restoration began, the most beneficial action was restoration of the roofs; the most detrimental was the use of cement plaster to replace the lime.

Erosion and Deterioration of Plaster and Stone

Wind and rain gradually eroded the plaster coatings that were intended to protect the underlying materials from water damage. Unprotected roofs and walls deteriorated and were more vulnerable to rain, wind, and earthquakes. (The flat roof at San Juan was partly destroyed by a hurricane in 1866.) Poor roof drainage can concentrate runoff, deeply eroding mortar joints and plaster coatings.

Deterioration of the original lime coatings and tufa is most pronounced on shaded elevations, or north and west exposures that do not dry as quickly. Deterioration will increase as longer events saturate the materials more frequently. Deterioration also frequently occurs in bands below chamfered, sub-horizontal surfaces at cornices, and other projections where cracks increase and concentrate damage from rain runoff from the roof and towers above.

Under normal conditions, rain would saturate only the surface of the limestone and sheet off in a heavy rain. However, moisture from the adjacent exposed tufa can migrate horizontally into the limestone pores and cause clay to swell and salts to accumulate. Although freeze-thaw conditions are not frequent in San Antonio, they have occurred enough times over the centuries to crack the stone.

Restoration of the churches at San José and Concepción has not included replacement of missing exterior plaster. This is partly due to limited resources but primarily due to a deliberate decision to maintain the familiar appearance of the exposed stone ruins. Unfortunately, this allows water to saturate the stone and move laterally in the wall, deteriorating the mortar joints as well as the adjacent carved stone. As rain events become more severe, the walls are likely to become more deeply saturated.

Original plaster is being reattached or sealed by injecting a lime-based grout with compatible permeability and physical properties. Friable plasters have also been consolidated with natural or synthetic materials.

Cementitious Repairs

At the beginning of the 20th century, Portland cement became popular and widely available, and the production of slaked lime almost disappeared. *The San Antonio Light* noted that modern materials, including cement, were used to repair the dome at San José in 1917 because the original "material is not to be had". Cement plasters and mortars were used at all of the San Antonio churches. Cement was also used to repair or replace eroded limestone at San José.

The high cement mixes did provide temporary protection but were impervious and rigid compared to the original lime materials. Impervious coatings trapped water; rigid plasters eventually cracked, channeling water behind the coatings. Rigid cement mortars transfer thermal and mechanical stresses to the adjacent stone.

The stone domes and vaults at Concepción are still covered with cement or concrete that has caused deterioration of the underlying tufa at pinnacles, cornices, molding, and other horizontal surfaces. The cement is gradually being replaced with lime plaster mixtures that match the composition of the original lime plaster.

Biological Damage

Epilithic growth on the surface of stone or plaster is unlikely to damage materials and may actually protect it from erosion and UV damage. High lime mortars resist biological growth, but endolithic growth of fungi, algae and lichen that extends into pores or cracks within plasters and mortar can enlarge cracks and support sequential growth.. This occurs more frequently on shaded elevations, as seen at Concepción. Biological growth that had been removed from original plaster in 2013 had regenerated by 2018.

Lichen damage could increase as severe rain events and more frequent wet-dry cycles enlarge existing cracks. Removing existing lichen can damage stone and plaster and be followed by new growth. A better option is to focus on water management.

Foundations and Drainage

Both San José and Concepción are on flat sites, with sandy soil over a dense clay or stone substrate. Surface and subsurface water flows toward the nearby river. San José is built on dense clay soil with modern drainage systems designed to carry stormwater away from the foundation walls. Concepción was built on a formation of karst stone, which provided good structural stability for the walls and domes.

Rising damp continues to deteriorate stone and plaster at the base of the walls at both Concepción and San José. At Concepción, the bottom foot of the tufa walls is eroding on all elevations from cyclical rising damp. It is interesting to note that throughout history, lime plasters were often considered sacrificial because it was preferable that salts accumulate and delaminate plaster coatings than damage the underlying stone.

Figure 3.3.5 Rising damp at base of wall of Concepción. (Photo by Daniel Thomas, 2021.)

Pilot Treatments at Concepción

Investigations in 2013 identified deterioration at Concepción's west façade, dome, and bell tower and led to a series of pilot restorations that used traditional lime plaster and mortar.

Investigations of the bell tower identified extensive damage where four different Portland cement stuccos had been used on horizontal and vertical surfaces. The cementitious patches did not bond well to adjacent materials and tended to leak at featheredges. Deterioration was most severe at vertical mortar joints and wherever water entered through open mortar joints or gaps in the plaster or cement. In 2018, the cement plaster on the bell tower was replaced with lime plaster identical to the original. A 2019 inspection confirmed that the repair was performing well.

Concepción retains its original vaulted stone roof including a drum and dome at the transept. The drum is constructed of small 2″ to 6″ tufa rubble, set in lime mortar. In 2019, the cement coating on the dome and drum over the transept at Concepción was found to be leaking and cracked. Water had saturated the underlying stone drum. The historic lime mortar had low compressive strength, comparable to other mortars of the same age, but was in good condition. However, cracks were found in the stones in the cornice at the base of the drum, which were covered with at least two layers of modern cement plaster. The lath from the 1997 installation was partly corroded and lath from previous installation was completely corroded. The cement plaster was carefully removed and replaced with lime plaster.

The original plaster on the front façade of Concepción was delaminating. In 2013 and 2018, the original stucco was reattached by injecting a lightweight grout of hydraulic lime, inert ceramic microspheres, and 5% acrylic polymer emulsion, and filling undercut edges with mortar of hydraulic lime and sieved caliche aggregates. Dry pigments and water-based paints were applied to the edge of repairs to visually blend new repairs with original material.

Figure 3.3.6 Lime plaster patch at Concepción. (Photo by Daniel Thomas, 2021.)

These pilot programs confirmed that original plasters could be carefully reattached, and that new lime plasters could be introduced and would perform well if the original stucco were stabilized prior to performing other pointing and restoration work. Cement stucco must be carefully removed to avoid delaminating any original plaster that is adjacent to or below the cement. The work also highlighted the importance of carefully coordinating and sequencing the work of different trades to protect the original fabric of the building.

The San José Retablo

The loss of lime plaster at San José continues to allow water to penetrate both the tufa and the adjacent never-plastered carved limestone. Damage is especially severe where cementitious plasters and patches prevented drying of the underlying stone and where cement mortar forced water to escape though the stone rather than through the joints.

An assessment of San José y Miguel in 2009 identified the following deterioration, all related to trapped moisture:

- Disaggregation of the original lime mortar
- Leaching of sulfates from cement mortars
- Expansion of clays inherent in the limestone
- Cracking of stone as trapped water freezes
- Deterioration related to migration and precipitation of salts
- Weathering of calcareous binders by sulfuric acid rain
- Migration of nitrates from ground water

Multiple phases of conservation work to the carved limestone of the main façade in 2010 through 2012 included replacement of incompatible parging with reversible and breathable products. Where there was risk that the original material would be lost to deterioration, the surface was consolidated with hydraulic lime slurry. Lime products are considered more appropriate than cementitious products because they are similar to the original materials in both the limestone and in lime mortars. They can bond with the original materials and are less likely to affect the porosity or permeability of the stone, mortar, or plaster. Cornices and overhangs were modified to divert water from projecting details, and the ground surface was revised to divert water from foundations.

However, the plaster coating on walls adjacent to the retablo has not been restored. Water diverted from the retablo is still absorbed by the tufa on either side of the retablo, where it can migrate into the retablo stone and joints. This migration of water is likely to increase during the projected heavy rain events.

Stone Details at San José Retablo

San José has more elaborate and projecting carvings and statues than any of the other San Antonio missions. By the 20th century, large segments of these original carvings had been lost to weather, deterioration, vandalism, and theft. In 1948–1949, carved stone details that were missing from the west wall were replaced with new carvings using a different limestone, likely sourced from Austin, Texas, or Indiana. Other carvings were roughly replicated with cement stucco over a bronze wire armature.

The bonding agents available to conservators in the mid-20th century were often not sufficient to adhere either replacement stones or cement reproductions to the remaining original stone, so conservators chose to overlap the original stone fragments with larger reproduction, changing the appearance of the overall composition. Although the replacements were crude and out of scale, they retained the original fabric and were often reversible. When stronger reversible bonding agents became available in the early 21st century, conservators were able to replace the earlier work with more appropriately sized stone carvings. However, the earlier patches did cause some damage by trapping moisture.

Conclusions

The history of damage and repair at the San Antonio mission highlights several familiar conundrums as new technologies become available. Traditional materials have a proven track record but may not meet the demands of a changing climate and economic and social priorities.

The installation and maintenance of lime plaster relies on the availability of appropriate materials and skilled labor and must function as an integral part of a larger system. At the San Antonio missions that system includes the foundations and subsoils, the tufa and limestone walls, and the design of roof. If any of these systems fail, other components are stressed.

The design of San Antonio's missions was adapted to a hot climate with intermittent rains. As the climate shifts to hotter days and more intense rainfalls, traditional systems may prove to be inadequate. Better maintenance of existing systems will help, but it may not be sufficient. New technologies may be required to avoid permanent loss of materials.

The rapidity and unpredictability of changes in the climate means that conservators may not always have the luxury of extensive testing of new materials. However, conservators must must consider whether treatments are reversible or correctable if they fail.

Bibliography

Building and Monument Conservation (B&MC). 2018. "Mission Concepción Pilot Project Treatment Documentation."

Bass, Angelyn, and Douglas Porter. 2016. *Condition Assessment and Conservation Treatment Planning for the Historic Stucco on the Facades and Exterior Elevations of Mission San José and Mission Espada in San Antonio, Texas*. Letter report, San Antonio: Conservation Associates.

Bell, Wayne, and Marsh Jackson. 1971. *Mission San Francisco de la Espada*. National Register of Historical Places Inventory – Nomination Form, Texas State Historical Survey Committee.

BMC. 2012. *Masonry Condition Assessment with Treatment Options and Preliminary Treatment Recommendations for the West Elevation and Towers of Mission Concepción*. Condition Assessment, Building Monument Conservation.

Browne, Byron. 2017. *Spanish Missions of Texas*. Charleston, South Carolina: The History Press. www.historypress.net.

Climate-Data.org. n.d. *Climate San Antonio*. https://en.climate-data.org/north-america/united-states-of-america/texas/san-antonio-417/#climate-graph.

Conservation Associates. 2019. "2019 Emergency Stabilization of the Historic Polychrome Stuccos and Conservation Treatment Planning for Mission Nuestra Senora De La Purismima Concepción De Acuma." Report Prepared by Conservation Associates For The Archdiocese of San Antonio, Ford, Powell, & Carson Architects & Planners, Inc. and Pugh Constructors.

Conservation Associates. 2011. "Mission Concepción Exterior Finishes Survey 2011." Report Prepared by Conservation Associates For Ford, Powell, & Carson Architects & Planners, Inc. and The Archdiocese of San Antonio (February, 2012)

Correia, Jennifer M. 2005. *Treatment Evaluation for the Micritic Limestone at Mission San José y San Miguel de Aguayo, San Antonio, Texas*. Thesis in Historic Preservation, University of Pennsylvania. https://repository.upenn.edu/hp_theses/23/.

Ford, Powell & Carson. 2019. "Mission Concepción-dome and Drum Investigation."

Ford, Powell & Carson. 2014. "Pilot Conservation Treatment and Emergency Stabilization, Mission Nuestra Señora de la Purísima Concepción de Acuña."

Hayhoe, Katharine. 2015. *Climate Trends in San Antonio and an Overview of Climate Projections for the South Central Region*. ATMOS Research and Consulting.

Las Misiones. 2022. *Las Misiones*. https://lasmisiones.org/welcome/history/.

McDowell, Katherine Ann. 1997. *Characterizations and Conditions Assessment of the Sacristy Window Mission San José y San Miguel de Aguayo San Antonio, Texas*. Master's Thesis, Philadelphia: University of Pennsylvania, Scholarly Commons. https://repository.upenn.edu/hp_theses/377.

Morrison, Hugh. 1952. *Early American Architecture*. New York: Dover Publications.

Myjer, Ivan. 2021. Interview of Ivan Myjer, conservator, BM&C on site at San Antonio Missions (San Valero, Concepción and San José, San Antonio).

Myjer, Ivan. 2010. *Mission San José y Miguel de Aguayo West Front Conditions Assessment Report*. Arlington: Building and Monuments Conservation.

NPS. (2023). San Antonio Missions. Retrieved August 8, 2023, from National Park Service. https://www.nps.gov/saan/index.htm

Ringenbach, Paul T. 2014. *San Antonio Missions, Nomination to the World Heritage List by the USA*. Los Compadres de San Antonio Missions National Historic Park, San Antonio: UNESCO. https://whc.unesco.org/en/list/1466/documents/.

San Antonio Light. 1917. October 14 as referenced by McDowell 1997

Testwell, Inc. 2009. "Stone Analysis Report." Report for BMC including in 2010 BMC report by Ivan Myjer.

CHAPTER 3.4

Missions of Southern Arizona

Introduction

San José de Tumacácori and San Xavier del Bac were built within a few years of one another and only 40 miles apart. They share an early history, but there are significant differences in their construction, recent history, and preservation approach.

San Xavier is in Wa:k, a village within the Tohono O'odham ancestral lands, south of Tucson, Arizona. San José de Tumacácori is farther south, in Santa Cruz County in an area originally frequented by the O'odham, Yaqui, and Apache. San José was constructed of a mix of fired and unfired adobe on soil foundations, while San Xavier, the earlier of the two, used only fired adobe and sits on bedrock. Both have rubble core walls with lime plaster coatings. Today San Xavier is almost fully restored and is actively used as a Catholic Church, while San José is stabilized as a partial ruin, similar to its condition when it was abandoned in the mid-19th century.

While it could be argued that the condition of San Xavier is due to differences in construction, other factors had equal or greater impact. Although both were abandoned for most of a century, San Xavier was always protected by the surrounding indigenous community, and has been fully restored through the collaboration of the Catholic Church, the Patronato San Xavier and the local Wa:k villagers.

The mission at Tumacácori was isolated, unsupervised, and already in ruins when the National Park Service (NPS) assumed control in 1916. The mission of the NPS continues to be to stabilize, maintain, and interpret Tumacácori as a ruin. Repairs and modifications are intended to preserve the existing original fabric, not to restore the mission to its original condition.

These two approaches to preservation – restoration versus stabilization – are both valid philosophical decisions with different implications. The full restoration of San Xavier has one set of challenges, including the substantial cost of complete

DOI: 10.4324/9781003160489-18

restoration, and modifications required to accommodate climate and safety. Stabilization of San José poses a different challenge, in particular that of maintaining exposed adobe in the face of increasing rain intensity.

History

The missions at Tumacácori and San Xavier del Bac were established by the Jesuits in the late 17th century. San Xavier was established at a thriving native settlement of Wa:k (amended to Bac by the Spanish) in 1700. The mission at San José de Tumacácori was established in 1753 when the "visita" of San Cayetano was relocated there after a Pima revolution. The Jesuits built adobe churches (which no longer exist) at both locations before being expelled by the Spanish in 1767.

A military Presidio was established at Tubac, 3 miles north of Tumacácori, in 1752, but was unable to defend the nearby missions against Apache attacks. The Franciscans assumed control of the missions in 1768. Although Tumacácori was poor and still under constant threat of Apache attack, its proximity to a source of water – the Santa Cruz River – made it more defensible than other villages. It became an important stop on the mission trail and a *cabecera*, a head mission for nearby villages.

The Franciscans completed San Xavier del Bac in 1797 and began construction of the church at San José de Tumacácori in 1799. San José was intended to be comparable to San Xavier but was not as well funded. Construction continued intermittently from 1799 to 1822 but was unfinished when the priest departed in 1828. A small group of local natives used the church until the Apache forced a complete retreat from Tumacácori and the Presidio to San Xavier in 1848. Farmers and ranchers intermittently used the buildings at Tumacácori, but the Apache were active in the area until 1890.

Southern Arizona was controlled by Mexico until the United States took control with the Gadsden Purchase in 1854. In 1874, the US government established a 72,000-acre reservation for the Papago (O'odham) Indians surrounding the San Xavier mission. By the end of the 19th century, both Tumacácori and San Xavier were relatively abandoned tourist destinations that were being vandalized by visitors who left graffiti or took architectural souvenirs. The local villagers began to control access to San Xavier, limiting damage, but the mission at Tumacácori remained unsupervised.

San Xavier del Bac

The Catholic church continued to serve the mission and even conduct repairs at San Xavier but had no official deed or control of the property until 1910, when 14 acres including the mission were granted to the Bureau of Catholic Missions.

Restoration work was intermittent throughout the 20th century with substantial campaigns from 1900 to 1912 and from 1949 to 1958. Mission San Xavier was designated a National Historic Landmark in 1960 and added to the National Register of Historic Places in 1966. Mission San Xavier was named to the "2016 World Monuments Watch", which identifies properties at risk from natural or social/political and economic forces.

Today the mission is under the control of the Catholic Church. Its care and conservation is managed by the church in collaboration with the Patronato San Xavier and the surrounding Wa:k community, which is part of the larger Tohono O'odham nation. A non-sectarian non-profit, the Patronato San Xavier assumed responsibility for maintenance and conservation in 1978 and secured millions of dollars in donations and grants for restoration that would have been unavailable to the Catholic Church. The Patronato has retained national and international experts and collaborated with the Wa:k (Bac) community to train locals in conservation techniques ensuring the preservation of both the church and traditional building techniques. In 2020, the Center of Architectural Conservation at the University of Pennsylvania completed a Conservation Management Plan that will guide future conservation efforts.

Figure 3.4.1 San Xavier del Bac, Tucson, Arizona. (Photo by Daniel Thomas, 2021.)

San José de Tumacácori

San José de Tumacácori was designated a National Monument under the Department of the Interior in 1908 and placed under the care of the nearest federal agency, the United States Forest Service. The NPS assumed control of Tumacácori in 1916 with the goal of stabilizing the ruins and preventing vandalism. The partially reconstructed ruins of the mission are conserved and operated as the Tumacácori National Historical Park by the NPS.

The NPS rebuilt the missing walls and replaced the missing roof in 1921. Stabilization work and modifications required to protect the original fabric have been ongoing throughout the 20th and 21st centuries, as well as extensive investigations, testing, and research by the NPS and others. Some deterioration of adjacent structures continued until only the church, a chapel next to the cemetery, and three convent rooms remained in 1978.

Today, the church is fully enclosed and rain runoff from the roof is effectively managed. The adjacent convent, chapel, and walls have been stabilized. Substantial areas of exterior adobe on the front façade of the church remain exposed to the weather. Conditions throughout the missions are carefully monitored and repaired or stabilized when required.

Figure 3.4.2 San José de Tumacácori, Santa Cruz County, Arizona. (Photo by Daniel Thomas, 2021.)

Climate and Topography

Existing Topography

San Xavier and Tumacácori are in a low desert plain of the Sonoran Desert, an area punctuated by volcanic outcroppings and surrounded by the Tucson Mountains to the west, the Santa Rita Mountains to the south, and the Rincon Mountains to the east. Elevations vary from with San Xavier at approximately 2,500 feet and Tumacácori at 3,260 feet to 9,000 feet in the mountains surrounding mountains.

San Xavier was built at the top of a slight hill, facing the south, where it is visible for many miles.

Both missions are on the west edge of the Santa Cruz flood plain, a broad valley running north to south. The native settlement at Wa:k (Bac) was established near a dependable spring, lush with "pine, ash, cypress, walnut, china trees, mesquite, alders, poplar, willow [and] tamarind" trees (Duell 1919). Tumacácori was close to the Santa Cruz River, surrounded by a mesquite forest. However, the river began to disappear when it was dammed in the 1800s and settlers began to pump groundwater in the 1890s. The mesquite forests lining the river had disappeared by 1950.

Existing Climate

Temperatures and precipitation in Arizona vary with elevation, from hot dry Sonoran deserts to snowcapped mountains. Phoenix was the hottest city in the United States in 2017, followed by Tucson, partly from the heat island effect of urban development. Although only 40 miles apart, Tumacácori is at a higher elevation, at the foot of the mountains, where it can be 10° F cooler than Tucson and at San Xavier.

Winter storms from the Pacific bring heavier rainfall during El Niño years and drier winters during La Niña when the jet stream moves farther north. Summer thunderstorms bring most of the rain, as well as strong wind, blowing dust, and lightning, as excessive heat lifts the moisture from the mountain ranges. Lightning storms are most frequent during the wet months of July, August, and September. Lightning strikes are naturally attracted to the iron cross and tall towers of the mission churches.

San Xavier mission and the city of Tucson are in the Köppen climate subzone BWh, dry and hot, while Tumacácori in the subzone Bsk, defined as "cold and semi-arid, a tropical and semi-tropical/semi-arid steppe climate". Overall, Tucson has less than 12 inches of annual precipitation, while Tumacácori gets more than 16 inches. The lack of cloud cover can produce dramatic temperature swings, from an average summer high of 101° F to a nighttime low of 68° F at San Xavier, with more moderate temperatures from 97 to 59° F at Tumacácori. In winter, temperatures at San Xavier drop from an average from 67 to 38° F at night, while Tumacácori is only slightly cooler at 65° F and 32° F.

Winds in southern Arizona come from the west or northwest in the winter and shift to come from the south or southeast from the Pacific Ocean or Gulf of Mexico in the summer. The moisture-laden summer wind combines with daytime heating of the desert to create the conditions for summer monsoons, called "las aguas," or "the waters" – intense wind driven thunderstorms. Summer heat also brings damaging updraft and downdraft winds close to the ground. Summer storms are short, intense, intermittent, and unpredictable from July to September. Winter rains are also intermittent, but more predictable as they cross the mountains from the west, usually providing short gentle rains, as equipatas ("little packages") of rain. Early fall can bring tropical cyclones from the Pacific. These violent storms, "Chubasco", are infrequent but are responsible for the record monthly rainfall amounts.

Future Climate

Arizona's temperature has increased 2° F over the last century, the fourth highest increase in average temperature in the United States. At Tumacácori, temperatures were 2° F to 8° F warmer than usual during the cool season and near normal in the warm season but lower than normal in January, creating the potential for frost events and freeze-thaw cycles.

San Xavier and Tumacácori both expect less annual rainfall in the future, but more extremes of drought, heat, and high intensity rainfall events. The Tumacácori area has experienced drought conditions since 2000, although the drought has been less severe in recent years. In 2017, rainfall was higher than normal in December, January, May, and July but lower than normal in the remaining months. A strong three-week monsoon in July represented 43% of the entire year's precipitation. Rainfall in May was 175% of the normal rainfall.

Climate change is making the summer monsoons ever more extreme and more unpredictable. The increase in winds can spread wildfires. July 2021 was the wettest month in Tucson on record with over 8 inches of rain, 70% of the yearly average.

Construction Techniques and Materials

Construction at Tumacácori and at San Xavier was directed by Spanish padres or master craftsmen but completed by local natives who lived in or near the mission. San Xavier was built first but used slightly more advanced materials and construction. Both churches were built with foundations of stone rubble, but San Xavier was set directly on bedrock; San José was built on poor soils, and partly over a buried drainage channel.

Both churches were constructed of adobe blocks made of local soil. San José used a mix of fired and unfired adobe, while all of the adobe in the church at San Xavier was fired.

Both churches were covered with lime plaster made locally from limestone gathered in the distant mountains. Limestone was heated over a wood fire in an open pit until the rocks began to swell and crack and could be hammered into powder. The lime powder was "slaked" for a few days, then mixed with river sand and water to make mortars and plasters.

San Xavier del Bac

The local soil at San Xavier is a dense, stable mix of sand and gravel with 10%–40% silt or clay, producing strong fired adobe blocks. Two feet below grade, the soil is denser, with more gravel and cobbles. The foundation walls are several feet thick, constructed of large volcanic basalt rocks in a thick bed of lime and sand that extends from the bedrock 5 to 6 feet below grade to 2 feet above grade.

Above grade, fired adobe blocks set in lime mortar line a core of basalt rubble and lime mortar. San Xavier also used brick arches, not previously seen in this area. The interior and exterior of the masonry walls, towers, and roofs, including the molded brick of the ornate portal, were finished with several layers of lime plaster.

Two bell towers flank an ornate entrance retablo of molded fired adobe bricks covered with plaster. The walls of the bell towers were thinner than the church below, constructed of several wythes of fired brick. The roof vaults and dome were only a single wythe thick. Typically, both bed and head joints were filled with lime mortar, but at the upper courses of the towers, only the bed joints were originally mortared. The west nave walls were reinforced over time with low stone buttresses set outside the stone foundations.

Figure 3.4.3 Volutes, canales, and stone foundation at San Xavier, Tucson, AZ. (Photo by Susan Pranger, 2021.)

The plaster surfaces were originally painted with local earth (mineral) pigments such as iron oxide. An oil binding medium is most likely but has not been identified.

The majority of the wood for the *vigas*, beams, balconies, doors, and windows are from pine trees from the nearby Santa Rita and Catalina Mountains, although mesquite and cottonwood were also used. The smaller latillas were made from the woody ribs of the saguaro cactus.

San José de Tumacácori

The plan for San José was intended to rival San Xavier but due to a lack of funds a flat roof was installed over the nave in lieu of the intended vault, and both transepts and the second bell tower were omitted. The masonry dome over the sanctuary, and a barrel vault over the sacristy were completed and the originals remain intact today.

Although rubble from a nearby riverbed was used for the foundation walls, San José was built over and with poor soil. A 1976 analysis determined that the soil was 66% sand and 34% silt and clay, which would produce an adobe with low strength. Later analysis of the adobe found even greater percentages of sand and silt, low compaction, and evidence of round river sand. Traces of sulfur, potassium, and chlorine were found, but these may have come from rising ground water.

Both unfired and low-fired adobe blocks were used at San José, reflecting a change in technique as the church was constructed. The lower walls of the nave are 6 feet thick, originally intended to support a masonry vault, and the walls at the base of the Baptistery are 9 feet thick to support the bell tower. Both are constructed of unfired adobe block set in clay mortar lining a core of large rocks set in clay mud mortar. The upper walls of the bell tower are thinner and lined with fired adobe block set in lime mortar. Low fired block in lime mortar was also used for the decorative retablo (portal), the sanctuary arch, the sacristy vault, arches, cornices, and other locations where higher strength was required.

Stones fragments extended out from the mortar joints to key the lime plaster coating that originally covered the interior and exterior. The lime was finished with a thin wash of burned gypsum that was polished by hand or using rawhide while still wet. Water from the roof ran through stone scuppers to open downspouts lined with lime plaster that was made more hydraulic by adding brick dust. Lime plaster on the dome of the church was later coated with a mixture of soap and alum.

Patterns of erosion indicate that the paint was applied to dried plaster over a white finish coat of gypsum. It is not clear what type of binder was used, and there is no evidence of animal proteins such as milk or eggs. With the exception of the indigo, the paints used at Tumacácori were all created with local materials, mostly minerals, suspended in an unknown binder. The blue is most likely a stain derived from the indigo plant, imported by Mexican artists.

The original wood used at Tumacácori would have been mesquite from the local area or pine from the nearby mountains. Velvet mesquite is most common in the

Tumacácori area, although honey mesquite and screwbean mesquite are also found. Traditionally, the wood was stained.

Figure 3.4.4 Interior painted plaster finishes at San José. (Photo by Daniel Thomas, 2021.)

Deterioration and Restoration

San José de Tumacácori

The roof of San José was reportedly leaking a year after the mission was abandoned in 1848. Either the roof collapsed or the timbers were removed for use elsewhere by 1889. The earthquake of 1890 cracked the interior west wall of church, the base of a façade column, pediment, and choir loft. Walls were vandalized by people searching for non-existent buried treasure, and the buildings became temporary housing for settlers, miners, or passing travelers. The nave remained unprotected for 30 years, until the roof was replaced in 1920.

Figure 3.4.5 Crack in east transept wall at San José. (Photo by Daniel Thomas, 2021.)

Damage to Tumacácori in the 20th century was attributed to the constant erosion of heavy winds, annual heavy rain events in the summer, and the continuous rains of the winter, punctuated by specific weather events: a cyclone in 1907, a cold snap in 1931, torrential rains in 1944, and lightning storms in 1939 and 1995. Tornados in 1964 and 1974 damaged nearby homes but missed the mission.

In 1918, the NPS started stabilization work by underpinning the bell tower and yard walls, and repairing the entrance arch and sacristy wall, using local workmen and matching materials wherever possible. Between 1920 and 1923, the walls and pediment were rebuilt, and a beam and plank roof constructed over the nave. However, large areas of plaster were lost between 1936 and 1939 to weather, vandalism, and poor bonding. Cement and other impervious coatings were being used as early as the 1920s and continued throughout the 1930s and 1940s. A campaign to remove the cementitious coatings began in the 1970s.

The roof was replaced in 1921 and in 1947 but roof leaks continued to cause the plaster to delaminate. In 1980, the existing beams were treated for a termite infestation and the nave sheathing and roof coatings were replaced. The roof slope was modified to direct rainwater to the canales, and a subsurface foundation drain was added along the west elevation. Scuppers on the east elevation were plugged to prevent water damage to the plaster below.

In 1985, the sanctuary dome was repaired with lime plaster and several coats of vinyl-acrylic-latex exterior masonry solution, which concealed structural flaws in the dome. Later, the roof was coated with several coats of limewash with alum (salt). In January 2010, a storm with 4 inches of rain caused roof leaks, failure of adobe, partial collapse of sanctuary window, and collapse of north wall of sacristy.

Circa 1976, the NPS established an extensive monitoring system at San José to study temperature and moisture content in the walls, soils, and atmosphere. Investigation determined that high levels of moisture were reducing the bearing capacity of the adobe to critical levels, especially where covered by impervious (cement) materials. Evaporating moisture was also depositing salts. After determining that the impervious coatings were forcing moisture higher in the adobe walls, NPS began to replace cementitious patches and coatings with more traditional adobe and lime materials. At the canales (open downspouts), a soft brick powder was added to make the lime more hydraulic, and the plaster was polished to make it more water resistant.

Figure 3.4.6 Plaster finish, fired adobe caps, and canales at San José. (Photo by Daniel Thomas, 2021.)

Attempts to divert surface water included excavation at the exterior perimeter of the church in 1965, installation of an elastomeric membrane in 1970, and removal of the membrane and installation of French drains in 1977. However, monitoring in 1976 suggested that settlement was still occurring. Most cracks opened and closed daily and seasonally in response to a change in temperature but one crack at the southeast corner of the nave was gradually increasing in width, while a crack of the west side of the same corner gradually decreased in width, indicating that the cracks were caused by ongoing settlement of the corner. Ground water was 25 feet below and unlikely to be the cause of settlement. However, an ancient natural arroyo (a drainage gully) extended southwest to the northeast, directly below the corner with the highest moisture and salt content and was identified as a possible source of rising damp during the summer and winter wet seasons and occasionally during dry seasons.

In 1984, paint conservators noticed that plaster damaged by torrential rains tended to reconsolidate with warmer, drier weather in the spring. Efforts to consolidate original lime plaster in the 1980s and 1990s with synthetics produced mixed results, often failed, and were usually irreversible. Materials incompatible with the chemical and physical properties of the original lime and adobe caused salt migration, shrinkage, and expansion from thermal or moisture fluctuation. Moving into the 21st century, conservators used traditional materials but experimented with the exact proportions of sand, lime, or adobe and tested injection grouting with a variety of materials. Preliminary tests in 2016 using ethanol-based nano-limes to consolidate friable lime plaster at San José were promising, with a 45% increase in mechanical strength, less than 1% decrease in porosity, and less than 2% decrease in permeability although the treatment did not improve frost resistance.

San Xavier

From 1837 until 1873, the mission buildings at San Xavier rarely had a priest, but the O'odham villagers continued to protect the mission church from theft or vandalism. Historical accounts reported that cracks were visible on the interior side of the dome with related water infiltration as early as 1850.

An American priest who arrived in 1881 led a campaign to replaster the church, and to repair the dome and foundation after the earthquake of 1887. After Portland cement was used to waterproof the roof in 1895, a single crack appeared in the roof but extensive cracking on the interior of the dome suggested that the cement coating was admitting and trapping water.

Even though the Catholic Church had no official control until 1910, they sponsored extensive renovations from 1900 until 1912 including a new coat of lime plaster, new balusters at the tower, reconstruction of the atrium and cemetery walls, and a new enclosure at the east courtyard. Low stone buttresses were added to reinforce the church walls that were eroded by salt migration.

The mission was in good condition when the Franciscans returned in 1913 but was showing the effects of weather by 1936. Heavy rain destroyed a deteriorated estípite, a short square column typical of the Churrigueresque style. In 1931, a section of the front façade and cornice fell during a cold snap in 1937, and half of the west bell tower was destroyed by lightning in 1939. A series of small projects between 1935 and 1958 included restoring the bell tower and adding a lightning protection cable.

Work between 1949 and 1958 included restoration of the sculpted plaster retablo with cementitious coatings, replacement of the wood balconies, and new brick and plaster volutes. The majority of the original brick and plaster estípite columns were missing and were replaced with cast concrete replicas of the two originals that remained in 1942. The Franciscans continued intermittent repairs, including patching cracks, replacing plaster on the lower level of the façade, and, in 1967, repainting the entire church. Unfortunately, much of this work incorporated cement, iron nails, and other synthetic materials that were not necessarily compatible with the original materials. Renovations as late as 1985 also used incompatible hard fired brick and a plaster mix of ordinary Portland cement (OPC), concrete additives, lime, and sand.

In 1988, biological growth was discovered where moisture was trapped under a peeling and delaminated elastomeric roof coating. The fired adobe below the coating was still sound except at the dome where there were multiple cracks and some deteriorated adobe.

The decorative scrolls, finials, and balustrades along the church parapet evolved in material and design over time. The missing finials were replaced in 1906 with French style octagonal plaster finials, in 1970 with plaster duplicates, and in 1985 with concrete casts based on grainy 1871 photographs. In the 1990s, the 293-pound concrete finials were replaced with lightweight Fiberglas replicas in response to concerns that even a mild earthquake might topple them. The finials on the tower were replaced in 2022.

Missing animal heads were replaced with thin plaster lion heads in 1906 and with hard brick and plaster scrolls copied from the church at Caborca in 1954. The deteriorated brick and plaster balustrades were replaced in 1951 with reinforced concrete replicas that remain today.

Restoration of the West Tower from 2002 to 2008 included removal of the cement plaster and concrete roofing, and repairs to brick damaged by a lightning storm. The East Tower was stabilized in 2017. Cementitious and elastomeric coatings were replaced with traditional lime-sand plaster mixed with cactus mucilage, finished with a traditional waterproofing coat of aluminum stearate, and burnished with river stones to close pores and increase resistance to water. The lime plaster is renewed periodically, although aluminum stearate is no longer used.

Nopal juice, a sticky substance pressed from the boiled pads of the prickly pear cactus, reduces the amount of water required, and increased workability, adhesion, water resistance, and flexibility. It also reduced shrinking and cracking as the plaster cured. Brick dust was added to the lime coating of the distinctive red canales, making the open gutters more hydraulic and water resistant.

Brick substrates were repaired with a custom low-fired brick, then prepared with a coat of limewash followed by a lime slurry with sand, and a thick brown coat of lime and sand to fill in gaps and imperfections. The brown coat is applied in several half-inch layers, built up to as much as 3 inches thick at carved joints. This is followed by another limewash or emulsion, a 1/8-inch finish coat to fill in all the divots and imperfections, a second "brown coat finish", and a third finish coat. The finish coat is followed by two or three coats of limewash applied over two to three weeks. The thickness and viscosity of the limewash is adjusted by the mason depending on the temperature and humidity conditions.

The walls and dome of the West Tower were reinforced with non-metallic fiberglass reinforcement. Fibers spun from basalt rock were used at the East Tower. Basalt fibers are stronger than Fiberglas, non-metallic and non-conductive, so it will not rust or conduct lightning. The basalt fibers were coated with epoxy resin to increase resistance to potential deterioration from moisture. Larger rods were embedded in a thin coating of liquid rubber (in lieu of epoxy) to absorb any differential movement between the rods and the plaster. The rods were also recessed up to 6 inches from the surface to minimize thermal stress. Basalt fibers were also used to restore integrity to plaster or brick that was cracked but not friable, to increase the bond or integrity of the wall itself, and to provide seismic restraints. Recent renovations at San Xavier installed lightning protection on both the towers and the scaffolding, along with a detection system capable of detecting lightning in a 5-mile radius.

Figure 3.4.7 Basalt fiber at San Xavier. (Photo by Susan Pranger, 2021.)

Figure 3.4.8 Lime plastered roof at San Xavier. (Photo by Daniel Thomas, 2021.)

Earthquakes

The worst earthquake in the area was the Sonora Earthquake of 1887. The magnitude 7.6 epicenter was approximately 140 miles southeast of San Xavier, on the western front of the Sierra Madre Mountains, south of the Arizona-Mexico border. Mission San Xavier would have experienced about 70% of that intensity, approximately 7.2 on the Richter scale; San José was even closer.

The 1887 earthquake demolished many of the adobe site walls at San Xavier, displaced one of the façade columns and finials on the upper west wall, cracked the west wall of the west transept, and aggravated an existing crack in the dome.

Two exterior walls that were previously damaged by the earthquake partially collapsed at San José in 2018. The wall was patched with new adobe and mortar, using fired adobe and lime mortar at the top three courses under the ledge, where future water infiltration was most likely.

Insect Damage

Honey mesquite is naturally resistant to termites, but not all mesquites are. Pine is especially vulnerable to termite damage, especially if it is wet. Wood boring beetles also attack mesquite but prefer fresh sapwood. Traditionally, mesquite was aged before being used, to avoid beetle infestation.

Termites plagued the roof timbers at San José between 1920 and the 1940, and again after the 1947 replacement. In 2017, termite-damaged pine beams in the east addition of Saint Xavier were restored with epoxy and treated with borates to prevent future damage. The beams in the adjacent sacristy arcade are mesquite and more resistant.

An Integrated Pest Management system, which monitors pest populations and intervenes at an early stage with the least toxic alternatives, is proposed by the NPS standards, and appropriate at all properties, historic or not.

Training

The Spanish missions were built and later maintained by the indigenous populations who were trained and supervised by Spanish or Mexican experts. San Xavier has continued this tradition by working with members of the local O'odham tribe, most recently under the supervision of Eric Means, an expert in adobe construction, and for the previous 75 years of work by the family-owned Morales Restoration and Builders. The interior plaster and artwork are being conserved by Tim Lewis, a local O'odham who started there as an apprentice and subsequently trained internationally, and his wife, Matilde Rubio.

Projected Impacts of Climate

Damage from climatic wear and tear can be exacerbated by lightning, earthquakes, fire, and biological and insect damage, not to mention deliberate vandalism. Earthquakes are not expected to increase in frequency or severity; however, San José's existing cracks are more vulnerable to future earthquake movement than San Xavier's fully restored and reinforced towers.

Although annual rain is expected to decrease across Arizona, the severity and duration of storms is expected to increase, along with the potential to increase erosion of the exposed adobe and protective lime coatings. The collapse of several adobe walls after unusually intense rain events in 2010, 2015, and 2018 prompted an interdisciplinary investigation by the Desert Research Learning Center and the NPS on the impact of rainfall intensity on adobe walls. They discovered that erosion of adobe walls was not just proportional, but exponentially greater with intensity of the rain. Doubling the amount of rain in a 30-minute period caused five times as much erosion. Increased intensity also causes deeper erosion.

Thermal changes drive the movement of moisture in adobe, including moisture that is trapped behind impervious coatings. Freezing temperatures do occur, but frost damage at San Xavier and San José is limited by the lack of moisture and large pores. However, repeated cycles of heat and cold, combined with intense sunlight, can cause cycles of thermal expansion and contraction that can break the bonds within and between materials. Cement and iron expand differently than adobe and lime.

Increased temperatures and wet conditions are both conducive to an increase in insect populations and damage to wood. Increased UV radiation can increase erosion of wood, making it more vulnerable to water and insect damage.

Conclusions

The church at San José de Tumacácori is in remarkable condition, given its unstable foundation and exposed adobe walls. Maintaining the building in its current condition as rain events increase in severity will require that the NPS maintain their current level of vigilance and stabilization. Additional protection might be necessary in the future to retain areas of exposed exterior adobe. The ongoing collaboration between the NPS, universities, and other researchers promises to continue advancing the body of knowledge regarding earthen buildings.

The restoration of San Xavier falls at the opposite end of the preservation spectrum, restoring missing elements, and replacing deteriorated materials with a mix of traditional systems and the latest technologies. The collaboration of local, regional, and international experts and local volunteers continues to train interns from the O'odham tribe, many of whom are now conservation experts in their own right. The ability to learn and adapt historic and modern techniques to conservation of materials will be critical while facing an uncertain climatic future.

The ongoing work at both missions highlights the need for a comprehensive and proactive plan of observation, maintenance, and repair. The projected increase in severe storms makes preventative action all the more imperative. Chapter 5 of the NPS *Management Policies 2006* includes the following criteria that should be considered when determining the extent of restoration or maintenance of a historic building:

- The nature and significance of a resource and its condition and interpretive value
- The research potential of the resource

- The level of intervention required by treatment alternatives
- The availability of data and the terms of any binding restrictions
- The concerns of traditionally associated peoples and other groups and individuals

Both approaches – restoration with new technology at San Xavier del Bac and stabilization/curated decay at San José de Tumacácori – have a role in responding to the future challenges of climate change. Equally important is the ongoing training, research, and sharing of knowledge.

Bibliography

Arizona-Sonora Desert Museum. 2023. *Patterns of Rain.* https://www.desertmuseum.org/books/nhsd_patternsrain.php.

ASU. 2022. *Monsoon & Desert Meteorology.* https://sgsup.asu.edu/basics-arizona-monsoon-desert-meteorology.

Bleser, Nicholas. n.d. *Tumacácori, From Rancheria to National Park.* Southwest Mission Research Center.

Bottemiller, Kitty. 2017. "On a mission: Preservation work underway at Tumacácori." Green Valley News. https://www.gvnews.com/news/on-a-mission-preservation-work-underway-at-tumacacori/article_022bccfe-8296-11e7-baf2-fffdbf499b52.html

Brean, Henry. 2021. *Close look at Mission San Xavier's in entrance reveals surprises.* https://tucson.com/news/local/close-look-at-mission-san-xaviers-intricate-entrance-reveals-surprises/article_6f0a1bee-5791-11ec-93ee-2ba53d07421a.html.

Britannica. 2020. *Ocotillo.* Inc Encyclopaedia Britannica. https://www.britannica.com/plant/ocotillo.

Browne, Byron. 2017. *Spanish Missions of Texas.* Charleston, South Carolina: The History Press. www.historypress.net.

Chan, Jocelyn Wai-Pui. 2015. *An Investigation of the Painted Finishes of Mission San José de Tumacácori's Facade.* Thesis, Philadelphia: University of Pennsylvania.

City of Tuscon. 2023. *The Santa Cruz River Heritage Project – A Brief History.* https://www.tucsonaz.gov/Departments/Water/Water-Projects-Programs/Water-Arts-and-Culture/The-Santa-Cruz-River-Heritage-Project

Crimmins, Mike. n.d. *Shocking Facts about Arizona Lightning.* https://cals.arizona.edu/backyards/sites/cals.arizona.edu.backyards/files/p16.pdf.

Crosby, Anthony. 1978. "A Preservation Monitoring System at Tumacácori National Monument." *Bulletin of the Association for Preservation Technology* (Association of Preservation Technology International (APT)) 10 (2, U.S. Park Service Issue). https://www.jstor.org/stable/1493612.

Crosby, Anthony. 1987. "Monitoring Moisture at Tumacácori." *APT Bulletin: The Journal of Preservation Technology* (Association for Preservation Technology International (APT)) 19 (4): 32–44. https://doi.org/10.2307/1494147.

Díaz, Declet, and M. Nicole. 2017. *Testing and Evaluation of Soil Based Grouts for the Adhesion of Consolidated and Unconsolidated Painted Lime Plaster at the Mission San José de Tumacácori.* Thesis, Philadelphia: University of Pennsylvania. https://repository.upenn.edu/hp_theses/619/.

Drain, Thomas A., David Wakely, and N. Scott Momaday. 1994. *A Sense of Mission, Historic Churches of the Southwest*. San Francisco: Chronicle Books. www.chroniclebooks.com.

Duell, Prent. 1919. *Mission Architecture as Exemplified in San Xavier del Bac Including a Complete List of the Missions in the Southwest*. Tucson: Arizona Archaeological and Historical Society.

Eatherly, Charles R. 2021. *Tubac Presidio State Historic Park*. https://azstateparks.com/tubac/about-the-presidio/park-history.

Eggington, Charles Morey. 1928. "Photo of Mission San Xavier." *Prent Duell Photo Album, compiled 1928*.

Fontana, Bernard L., Edward McCain, and David Carter. 2015. *San Xavier del Bac; Portrait of a Desert Church*. Tucson, Arizona: Southwestern Mission Research Center.

Google maps. n.d. google.com/maps.

Herr-Cardillo, Starr. 2021. Interview of Starr Herr-Cardillo of *Patronato San Xavier* regarding renovations to San Xavier del Bac mission at San Xavier, Tucson, Arizona.

Hopkins, Ross R., and Gordon Chappell. 1986. *National Register of Historic Places Inventory-Nomination Form*. Mission San José de Tumacácori.

Jang, Jean. 2016. "Performance Evaluation of Commercial Nanolime as a Consolidant for Friable Lime Based Plaster." Master's Thesis, University of Pennsylvania. https://repository.upenn.edu/hp_theses/597?utm_source=repository.upenn.edu%2Fhp_theses%2F597&utm_medium=PDF&utm_campaign=PDFCoverPages.

Kessell, John L., and Fray Bartholome Ximeno. 1964. "San José de Tumacácori 1773: A Franciscan Reports from Arizona." *Arizona and the West* (Journal of the Southwest) 6 (4): 303–312. https://www.jstor.org/stable/40167860.

Lewis, T. and Rubio, M. 2021. Interview with Tim Lewis and Matilde Rubio of *Tohono Restoration* regarding renovations to San Xavier del Bac mission at San Xavier, Tucson, Arizona.

Lim, Alex. 2018. "A Review of Changing Ruins Conservation Approaches at Tumacácori National Historical Park: A Case Study of the Convento Compound." *Crossroads in Science*. National Park Service, U.S. Department of the Interior, 56–66.

Lim, Alex. 2019. *Treatment Report, Emergency Repair of Tumacácori Church Wall Failures*. National Park Service.

Means, Eric. 2021. Interview with Eric Means of *Eric Means Constructions* regarding renovations to San Xavier del Bac mission on site at San Xavier, Tucson, Arizona.

Morrison, Hugh. 1952. *Early American Architecture*. New York: Dover Publications.

Moss, Jeremy. 2021. *Of Adobe, Lime, and Cement: The Preservation History of the San José de Tumacácori Mission Church, Part 1 – The Fabric*.

National Park Foundation. n.d. *The Story behind Tumacácori*.

National Park Service. 2006. *Management Policies 2006*. U.S. Department of the Interior. https://www.nps.gov/subjects/policy/upload/MP_2006.pdf.

NPS. 2023. *Culture, Nature, and Community*. https://www.nps.gov/tuma/index.htm.

———. 2019. *Measuring the Effects of Rainstorm Intensity on Adobe Walls*. National Park Service. https://www.nps.gov/articles/sodn_adobe_test_walls.htm.

NPS. 2018. *Status of Climate and Water Resources at Tumacácori Water Resources at Tumacácori National Historical Park*. National Park Service, U.S. Department of the Interior.

———. 2015. *Trees.* Edited by National Park Service. https://www.nps.gov/tuma/learn/nature/trees.htm.

———. 2020. *Tumacácori Virtual Tour - Mission Grounds.* U. S. National Park Service.

———. 2020. *Earthen Architecture – Preservation Techniques.* July. https://www.nps.gov/articles/000/earthen-architecture-preservation-techniques.htm.

Patronato San Xavier. 2020. *Patronato San Xavier.* https://patronatosanxavier.org/about/.

Penn Praxis. 2020. *San Xavier del Bac Conservation Management Plan.* Philadelphia: University of Pennsylvania.

PestKilled Team. n.d. *Termite Resistant Wood 101.* https://pestkilled.com/termite-resistant-wood-101/.

Pinkley, Frank. 2007. *Mission of San José de Tumacácori.* National Park Service.

States at Risk. 2015. *Arizona's Climate Threats.* https://statesatrisk.org/arizona.

Tohono Restoration. 2013. *Art Conservation and Restoration.* https://www.tohonorestoration.com/about_us.htm.

U.S. Climate Data. 2023. https://www.usclimatedata.com/.

USGS. 2020. *M 7.6 – The 1887 Sonora Earthquake, Mexico.* https://earthquake.usgs.gov/earthquakes/eventpage/official18870503221300000/impact.

Vint, Bob. 2021. Interview with Bob Vint, Vint & Associates, regarding renovations on site at to San Xavier del Bac mission, Tucson, Arizona.

weatherbase. 2021. *Tubac, Arizona Koppen Climate Classification.* CantyMedia. https://www.weatherbase.com/weather/weather-summary.php3?s=568820&cityname=Tubac%2C+Arizona%2C+United+States+of+America&units=.

Zamora-Reyes, Diana, and Christopher L. Castro. 2021. *Monsoons make deserts bloom in the US Southwest, but climate change is making these summer rainfalls more extreme and erratic.* https://theconversation.com/monsoons-make-deserts-bloom-in-the-us-southwest-but-climate-change-is-making-these-summer-rainfalls-more-extreme-and-erratic-166222.

CHAPTER 3.5

Lighthouses of the Pacific Northwest

Introduction

There are 87 lighthouses along the western coast of the United States, most built in the second half of the 19th century to support marine commerce. A comparison of four of these lighthouses illustrates how differences in construction, maintenance, and stewardship influence the durability of materials in a coastal marine climate.

Lighthouses were built at Cape Blanco, Heceta Head, and Yaquina Head in Oregon, and at North Head in Washington State between 1870 and 1898 by the U.S. Lighthouse Establishment. All four began service with full time caretakers but saw a decline in maintenance when they were electrified in the 1930s and automated in the 1960s. All have been challenged by the use of modern, impervious coatings that were intended to eliminate water intrusion but ultimately trapped moisture in the wall. The recent effort to use porous traditional materials is now being challenged by the need to endure more extreme weather events as the climate changes.

History, Location and Climate

History

The U.S. Lighthouse Establishment was created in 1789 to set standards for maritime navigational aids. Between 1870 and 1913, they directed the construction of lighthouses on both coasts of the United States, using standard plans adapted to the local topography and available materials. They also supervised the full-time caretakers who followed strict rules for the operation and maintenance of the light stations, which included the lighthouse, dwellings, and other support buildings. Whitewashing and painting were frequent activities, as were repairs to the roofs,

DOI: 10.4324/9781003160489-19

windows, and fences. In 1910, control was transferred to the Lighthouse Service (USLHS) under the Bureau of Lighthouses.

In the 1930s, the Lighthouse Service was replaced by the U.S. Coast Guard (USCG). By the end of the decade, most stations had been partially automated with the introduction of electricity. Fewer resident keepers were required to operate the lights, which lead to a decline in regular maintenance. Outbuildings became obsolete, were removed, or left to deteriorate. Full automation in the mid-century eliminated the need for full-time caretakers, exacerbating the deterioration of most light stations until the preservation movement in the 1970s brought attention to their condition and historic importance. Ownership of most lighthouses was eventually transferred from the Coast Guard to other private or public entities such as the Bureau of Land Management (BLM) or the State Parks Departments.

Cape Blanco is 59 feet high, the oldest existing and furthest west of Oregon's coastal lighthouses, projecting 1½ miles into the Pacific, where it often endures

Figure 3.5.1 Cape Blanco Lighthouse, Cape Blanco, Oregon. (Photo by Daniel Thomas, 2023.)

winds as high as 125 mph. Built in 1870, the lighthouse is on an exposed headland surrounded by 200-foot cliffs on three sides. The lighthouse was added to the National Register of Historic Places in 1992. Today the active navigation light is operated and maintained by the USCG, while the rest of the lighthouse is maintained by the BLM in cooperation with the Confederated Tribes of the Siletz and Coquille Indians and the Oregon State Park and Recreation Department (OPRD).

The site at Yaquina Head is significant as a Middle Archaic settlement that was abandoned more than 2,000 years ago, and more recently as ancestral territory used seasonally by the Yaquina and other Indian tribes. The lighthouse was completed in 1892 but not activated until the lens was installed in 1893. Yaquina Head Lighthouse is 93 feet high, set 83.2 feet above sea level on a narrow peninsula of Columbia basalt projecting a mile into the Pacific. In 1980, Congress designated the Yaquina Head Outstanding Natural Area to be managed by the BLM. The Coast Guard occupied the attached workroom until 1993. The lens is still the property of the USCG. The lighthouse was listed on the National Register in 1993.

Figure 3.5.2 Yaquina Head Lighthouse, Newport, Oregon. (Photo by Daniel Thomas, 2023.)

Heceta Head Lighthouse, the last coastal lighthouse built in Oregon, was completed in 1894. It is 56 feet high and built on a cliff 205 feet above the ocean, sheltered by higher cliffs to the northeast. It was placed on the National Register in 1978 and transferred from the Coast Guard to OPRD in 2001.

Figure 3.5.3 Heceta Head Lighthouse, Florence, Oregon. (Photo by Daniel Thomas, 2023.)

In 1852, a presidential order designated the entirety of Cape Disappointment, Washington State, a military reservation, and set aside land for two lighthouses. Cape Disappointment Lighthouse was built at the tip of the cape facing the Columbia River in 1854. North Head Lighthouse was built in 1898, 65 feet tall and perched on a 190-foot-tall rocky basalt bluff on the north boundary of the cape district, facing the Pacific Ocean. The Coast Guard transferred ownership of the reservation and the North Head Lighthouse to Washington State Parks in 2012. However, the Cape Disappointment Lighthouse is still owned and operated by the USCG. Both lighthouses are part of the Cape Disappointment Historic District, which was added to the National Register of Historic Places in 1975.

Figure 3.5.4 North Head Lighthouse, Cape Disappointment, Washington. (Photo courtesy of Washington State Parks and Recreation Commission, 2022.)

Location and Climate

Weather patterns tend to move west to east across the Pacific Ocean, raising temperatures but bringing lots of moisture to the coasts of Washington and Oregon. The summer mean temperature is 66°F, and rarely exceeds 90°F; the winter averages 38°F, and rarely falls below 16°F. Annual rainfall is 75 to 90 inches along the coast, half of which falls in the winter months of December through February, with very little rain in the summer. Coastal snowfall rarely exceeds 3 inches per year.

The marine climate is naturally harsh, subject to wind-driven sand and salt water, and high levels of UV exposure. Hurricane force winds occur several times a year, usually from the south or west. The sun shines as little as 20% of the time in the northern coastal area of Oregon.

The lighthouses at Cape Blanco, Yaquina Head, and North Head are located on particularly exposed peninsulas, while Heceta Head is sheltered on the north side. Deterioration is a result of wind-driven salt, water and sand, corrosion of dissimilar materials, and lack of ventilation and maintenance, especially after the lighthouses were electrified and automated. Cape Blanco and Heceta Head both report winds in the 100 to 125 mph range; North Head has reported gales with winds up to 160 mph. Yaquina Head also suffered from wind-driven gravel, prompting construction of a fence along edge of the bluff to deflect some of the airborne gravel.

The most severe erosion occurs on the southwest exposures, while moisture, algae, and fungi are most severe on the shaded and damp north exposures. High winds draw moist air into the tiniest joints, so ventilation and the use of porous coatings are key to allowing interior masonry to dry.

Changing Patterns in Climate

The temperatures in Oregon are expected to increase 0.5°F in the next decade or two. The frequency and intensity of storms are already increasing. While the average annual precipitation is not expected to change significantly along the Oregon coast, winters are expected to get wetter and summers drier. Storms will be longer and more intense.

Construction Techniques and Materials

Lighthouse Construction

Traditionally, a light station includes a lighthouse with an attached workroom and separate buildings for living quarters and storage. Early lighthouses such as Cape Blanco and Heceta incorporated a dedicated oil room, but light stations built after 1880–1890, such as Yaquina Head and North Head, used kerosene that was stored in separate buildings because it posed a greater risk of fire.

Although configuration and materials varied slightly, all lighthouses were designed with a common function: to house and operate a lamp and Fresnel lens to aid navigation. Every lighthouse included a lantern mounted on a tower and one or two attached rooms. Windows on four sides of the tower provided views of exterior conditions and allowed light into the stairwell and watch room.

The metal and glass lantern, lens, and lamp, and cast iron stair and landings were the standard components furnished by the U.S. Lighthouse Establishment and assembled on site. However, there were variations in the construction of the masonry towers and in the use of stone, cast iron, and other metals. The height of the lantern was determined by the height above sea level and the typical level of fog.

The lantern room is enclosed by a glass wall in a metal frame that protects the lamp and lens. A series of cast iron plates are mechanically connected to form the floor of the lantern and cantilever outward to form the exterior gallery floor.

A metal roof is capped by a round vent-ball with a lightning rod, a bronze pinnacle with a platinum point. Initially these components included an assortment of dissimilar metals, including copper, bronze, cast iron, and wrought iron, that led to electrolytic corrosion. By the 1890s, the standards changed to a more simplified collection of cast iron and bronze components. Cape Blanco and Yaquina Head both have a copper roof and more mixed metals; later construction at Heceta Head and North Head had roof plates of painted cast iron, although some copper was still used for cornice trim and bronze for the lantern framing.

The windows and doors provided ventilation when the lantern was not in use and reduced moisture inside the tower but were closed when the lantern was operating. The ducted louvers were adjusted by the light keepers to adjust the airflow through the lantern room and control the combustion of the oil or kerosene light. The vent-ball exhausted air and combustion fumes. After the lantern was electrified, the vent-balls were often closed to reduce leakage.

Figure 3.5.5 Ventilation louver at Cape Blanco Lighthouse. (Photo by Daniel Thomas, 2023.)

Cape Blanco, Heceta Head, and Yaquina Head Lighthouses were each constructed with an interior and an exterior brick wall, separated by air spaces but connected by ribs. The exterior wall typically varied in thickness, getting thinner as it got higher, while the interior wall was one or two brick wythes thick. North Head Lighthouse, however, was built with a single masonry wall of varying thickness, lined on the interior with 1½″ thick 12″ × 12″ ribbed clay tiles. Although there were voids between the tile and the brick, drainage was obstructed by mortared ribs that secured the tile to the brick wall.

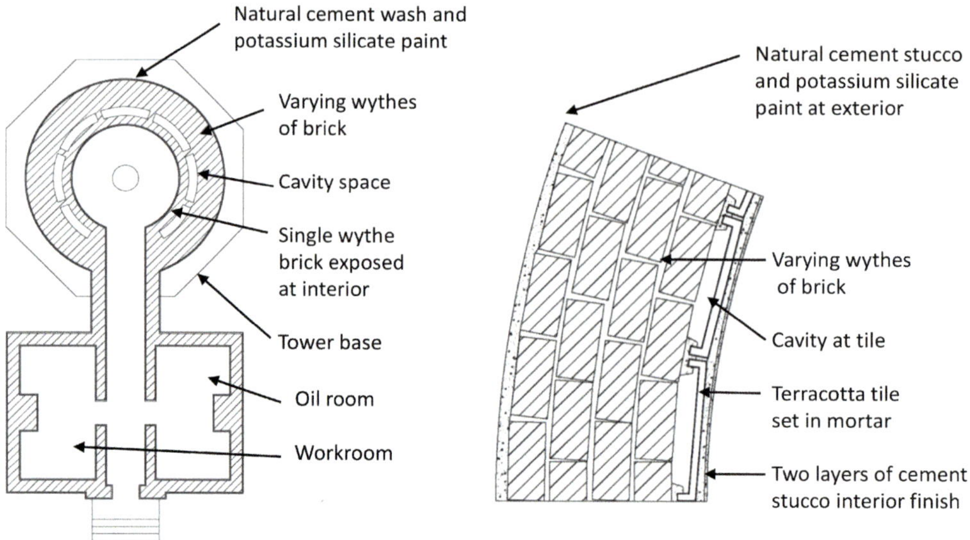

Plan of Yaquina Lighthouse

Detail of North Head Lighthouse

Figure 3.5.6 Typical Construction at Yaquina and North Head Lighthouses. (Drawing by Susan Pranger, Daniel Thomas, 2023.)

North Head and Heceta were originally finished with a lime or natural cement stucco and a limewash; Yaquina and Cape Blanco were finished only with a thinner wash that allows the profile of the brick joints to be seen. The quality of the brick varied: some were made locally and others were shipped in from Portland or San Francisco. The windows at Cape Blanco were cast iron; those at Heceta, Yaquina, and North Head were wood. All four lighthouses were constructed with concrete foundations.

Circa 1870, natural cement slurries were introduced and in 1890, the U.S. Government began a policy of covering their forts and lighthouses with a half inch coating of cement stucco. Originally, the Light House Service directed their personnel to paint exteriors with a mix of turpentine, boiled linseed oil, white lead, and a small quantity of Japan dryer. The use of elastomeric paint was introduced in the 20th century.

Deterioration, Restoration, and Other Changes

Deterioration of Lighthouses

The lighthouses began to deteriorate most dramatically in the 1930s after the transfer to the Coast Guard and electrification of the lanterns led to a decrease in staffing and maintenance and other unfortunate modifications. Although intended to exclude water, the use of ordinary Portland cement (OPC) coatings trapped moisture that penetrated the stucco by wind-driven rain or trickled down through from leaks in the lantern or windows.

When the airflow was not required for combustion, the vent-balls were often plugged to eliminate water leaks. However, the resulting lack of ventilation increased humidity and accelerated deterioration of both metal and interior masonry. Dehumidification was added to many of the lighthouses in the early 21st century but proved to be counterproductive, at least in the damp winter months, when it drew exterior moisture into and through the walls.

Chlorides are common in salt-laden air, and in the local sand used in the original mortars. Chlorides in the mortar and on the surface react with the metal to create rust that decreases the structural capacity of the iron and causes surrounding brick to spall. Salt water and acid rain can also cause graphitization of unpainted cast iron, a process that can weaken the metal without visible exterior deterioration. Water trapped in the cast iron can continue to cause corrosion underneath paint coatings.

Figure 3.5.7 Interior efflorescence and corrosion at Heceta Head Lighthouse. (Photo by Daniel Thomas, 2023.)

The use of dissimilar metals in the lighthouse lanterns, especially in this high salt environment, caused galvanic corrosion of less noble metals, such as cast iron and bronze. Cast iron is less vulnerable to corrosion than wrought iron because cast iron develops a patina that resists further corrosion. However, high humidity, salt, and lack of paint maintenance accelerated damage to the cast iron and allowed water to leak through the roof and gallery floor plates into the masonry below. Elastomeric coatings introduced in the 1960s were unsuccessful in keeping out wind-driven rain and were also not breathable enough to let moisture escape.

Thermal stresses can also cause differential movement and cracks in the masonry. These stresses can cause the vertical cracks often seen in the masonry lighthouse tower. OPC mortars are harder than the historic brick and will transfer damaging stresses from moisture or thermal expansion to the brick.

Figure 3.5.8 Vertical crack and replacement corbels at Heceta Head Lighthouse. (Photo by Daniel Thomas, 2023.)

While deferred maintenance certainly took its toll, the harsh coastal environment accelerates deterioration even when the buildings are regularly maintained. Algae and lichen thrive in on the sheltered north exposures and wind-driven rain permeates any available crack.

Figure 3.5.9 Lichen on North elevation of Cape Blanco Lighthouse. (Photo by Susan Pranger, 2023.)

Cape Blanco Lighthouse

At Cape Blanco, the water table, windowsills, window frames, and necking course were all made of cast iron. The roof and vent-ball were copper sheet metal over a wrought iron frame, capped with a cast iron crown ring, set on an unpainted bronze base, cornice, and gutter. The lighthouse retains its original ironwork, decorative brickwork, brick sills, and cast iron window frames but the original cast iron casement sashes were replaced with wood sometime prior to

1990. The doors, windows, and roof details of the workroom were modified in 1932 and in 1988.

The lighthouse was partially renovated by the USCG prior to 1990 when management was transferred to the BLM and operation to the OPRD (the USCG retains ownership). Undated specifications indicate that the interior walls were coated with two coats of lime, OPC, and sand, followed by a lime whitewash. In 1992, the interior and exterior metal was painted and interior walls of the tower were aggressively sandblasted to remove OPC coatings, damaging the faces of the bricks.

In 1996, researchers from the University of Oregon reported water infiltration, deteriorated glazing, caulking, and flashing and a lack of ventilation. It also reported mildew and efflorescence on the building exterior and spalling of the sandblasted interior brick. Subsequent repairs in 1996 through 1999 included painting all metal with epoxy paint, painting exterior masonry walls with an elastomeric acrylic resin paint, reglazing windows, and repairs to the ventilation system.

In 2001, the lantern ventilation system was still leaking, draining water into the masonry below. Restoration of the lantern in 2003/2004 included reinforcement of the wrought iron ribs with stainless steel, replacement of the copper roof with a heavier copper, and restoration of the original cast iron ventilator ball to reopen the vents. The masonry was repointed and repaired with epoxy injections. Epoxy paint was used on metal surfaces and elastomeric coating on exterior masonry and concrete. However, in 2003, the lantern glass was replaced with a thinner glass and glazing tape, which caused new leaks that seeped through rusted joints in the cast iron floor plates to penetrate the masonry below. Damage was most severe on the southwest, south, and southeast elevations.

A 2020 assessment reported that the exposed exterior brick had continued to deteriorate from moisture and biological growth, a general lack of maintenance and poor restoration practices. Saturated masonry contributed to corrosion of the embedded iron framing and windows and rust jacking of adjacent masonry. Although the brick and embedded iron was most damaged at the top of the tower, the cast iron windows, and their wrought iron anchors also corroded at lower levels. Air bubbles inherent in cast iron collect water and corrode the cast iron from within. Clear sealants have increased corrosion by trapping moisture against the cast iron. Dehumidification helps dry the masonry during the dry summer season. However, dehumidification in the wet winter months drew additional moisture from the exterior through the masonry and has been discontinued.

In 2023, brick dust accumulating on the stairs and floor and patches of efflorescence and algae on the watch room walls are evidence of ongoing deterioration. Whatever remained of the original exterior limewash is now covered with lead paint, at least five layers of elastomeric paints, and a final layer of acrylic latex paint added in 2011. The lighthouse looks surprisingly good upon casual inspection. However, the thick impermeable coating is lifting and peeling. Algae and fungi have created a green patina with orange accents on the north elevation, and bare brick is exposed at window surrounds.

Figure 3.5.10 Latex paint over multiple layers of impermeable elastomeric coatings at Cape Blanco Lighthouse. (Photo by Daniel Thomas, 2023.)

Yaquina Head Lighthouse

Yaquina Head Lighthouse has a copper roof and a double gallery. The lower walkway is supported by a series of decorative cast iron brackets. The tower is accented with a combination of brick corbels and granite.

Yaquina Head has a fixed lens supported by a pedestal between the lantern room and the service room. Prismatic lenses in the floor plates allowed light from the lantern to illuminate the service room so that the keeper could confirm that the lens was still shining.

The tower was completed in 1873 and whitewashed by 1875. Breakage of glass and shutters by windblown gravel from the beach was frequently reported. A heating system was installed in 1900. Lightning damage was reported occasionally, although damage was minimal because the tower was protected by a copper strip that extended from the vent-ball spindle to an iron tube of coke buried into the ground.

The few records of work during the USCG era indicate that the lighthouse was frequently painted. Other work completed by the USCG prior to 1960 included filling the windows with brick to exclude water, sandblasting the cast iron gallery and base, and replacing or covering the exterior cast iron upper collar with cast-in-place concrete. In 1984, the gallery railings, door pendants, and parapet were repaired, the exterior of the tower was painted, and the interior was sandblasted. A 1992 assessment recommended improvement to site drainage and fencing, repointing and damp proofing the brick foundation, restoration of the cast iron, and replacement of the windows.

The lighthouse oil room would have been heated to keep the lard oil from solidifying. The fireplaces in the workroom and oil room were closed during the USCG era; they have been restored to their original appearance but are not functional. Electric unit heaters added in 2006 are also used to control humidity. Dehumidifiers have never been used at Yaquina Head.

The BLM has completed extensive assessments and restoration since they assumed control. Extensive work in 2006 included stripping 134 years of coatings from the interior and exterior metalwork, stone, and masonry as well as asbestos abatement, lightning protection, masonry repair, painting, reglazing, metal repair, upper belt course and cornice work, and lantern work. The brick was repaired with pressure injection of epoxy, repointed with a lime and sand mortar, and then finished with an elastomeric coating. Lantern wall repairs included replacing the glass, repairing cracks in the cast iron, adding splice plates, adding gaskets to isolate dissimilar materials, replacing vent screens, and roof and pinnacle work.

The exterior elastomeric coating was removed in 2014 and replaced with a natural cement wash and a potassium silicate mineral paint. The mineral paint quickly became soft and chalky and washed off in the rain. In 2020, any remnants of the 2014 cement wash and mineral paint were removed and replaced with a three-coat potassium silicate finish by a different manufacturer. The cornice and roof were repaired in 2016, followed by ongoing maintenance.

A 2021/2022 restoration of the workroom included abatement of lead paint and refinishing the walls with a historically accurate lime-based mortar wash followed by a three-coat silicate-based coating. This coating "bubbled" in a few spots after two weeks but stabilized after another two weeks.

The permeable exterior and interior coatings are holding up well, and interior humidity is well controlled. However, moisture continues to be an issue today due to the severity of the weather. The lantern glazing and cast iron deck are leaking, and algae are growing on the north elevation. BLM has applied for funding from the Great American Outdoors Act (GAOA) to complete another restoration based on a 2020 Condition Assessment.

Figure 3.5.11 Corrosion of cast iron at Yaquina Head Lighthouse. (Photo by Daniel Thomas, 2023.)

Heceta Head

Heceta Head Lighthouse is constructed of brick masonry with basalt trim at the base, window frames, and cornice. The lantern dome is constructed of cast iron plates with a copper ventilation ball.

Modifications prior to 1977, during the USCG era, included infilling several windows with masonry and applying exterior cement stucco. The Coast Guard also used iron inserts, chicken wire and gunite to cover basalt stone corbels that had been previously damaged by the corroded cast iron clamps. The interior face of the brick had been sandblasted to remove a previous cement coating.

In 2001, the masonry was sound but there was evidence of wind-driven rain and chloride damage on the interior surface of the brick, including corrosion of embedded metal components. The joints between the cast iron plates of the galley floor had corroded due to loss of the original lead sealants. The corroded vent-ball was another source of water infiltration.

In 2012–2014, OPRD completed a comprehensive restoration of the entire lighthouse, including the masonry tower and the lantern metalwork. Cement stucco was removed from the exterior brick and stone masonry and replaced with a mix of natural cement, lime and sand, and a finish coat of potassium silicate. Basalt stone corbels that were damaged by rust jacking of iron inserts were repaired or replaced with new basalt corbels. Missing and non-original windows and doors replaced with historically accurate handmade wood replicas.

The Heceta Lighthouse has not received regular repairs or maintenance since the 2014 restoration. The attached workrooms are heated in the winter and dehumidified in the summer to reduce humidity. The interior appears to be in good condition except for a few areas where efflorescence and rust are evidence of moisture infiltration from the lantern or windows. However, the exterior paint coating shows extensive wear and biological growth, and new cracks have appeared in the exterior masonry.

Figure 3.5.12 Crazing of Potassium Silicate and Lime on Heceta Head. (Photo by Daniel Thomas, 2023.)

North Head Lighthouse

The North Head Lighthouse has a cast iron roof, ventilation ball, and gallery floor plates. The tower is constructed from a single unventilated load bearing masonry wall and interior clay tile lining, coated with stucco inside and outside.

The brick walls incorporate a variety of cornice, window, and other details in sandstone and cast iron, and sits on a sandstone base over a solid concrete foundation. The exterior brick was originally protected with an unpainted alum and potash cement stucco mix. The original sandstone was a local fine-grained feldspathic lithic wacke sandstone with a high content of porous volcanic grains, clay, and mica. The grains are only lightly cemented, which limited its resistance to expansion from increased moisture, freezing, or salt.

After the lighthouse was fully automated in the 1970s, six of the original windows were removed and the deteriorating sandstone belt courses, window hoods and sills were covered with stainless steel or cut flush with brick and covered with stucco. Additional work in 1978 included patching, caulking, and painting masonry, stucco, and metals and reglazing the lantern. In 1983, the interior tile walls were aggressively sandblasted to remove interior lead paint, removing some of the cement coating.

Additional layers of cement stucco, impermeable elastomeric paints, and sealants were applied on the exterior until 1992, when the Coast Guard removed the stainless steel cladding and patched the sandstone details with an acrylic patching compound. By 2011, hidden cracks in the joints between sandstone, brick, and infilled windows were allowing substantial water infiltration and causing efflorescence on the interior clay tile walls.

Figure 3.5.13 Moisture escaping at joints in interior tile at North Head Lighthouse. (Photo by Susan Pranger, 2021.)

An assessment in 2011 found that areas of the cast iron, stucco, and masonry were deteriorated, and the structural steel was in critical condition. Water was infiltrating the tower through the lantern glazing, the inoperable vent-ball, and corroded joints in the cast iron, and the circa 1960 cement coating was trapping moisture and causing further deterioration and spalling. The saturated masonry had caused corrosion of the support beams only at the interface with the masonry but not where the beams were completely exposed or concealed. The surface of the sandstone was cracked and eroded by abrasion and salt crystallization wherever it was saturated by lantern leaks and wind-driven rain. The base was completely eroded; the cornice was sound on the north exposure but easily fractured on the south.

Figure 3.5.14 Deterioration of original sandstone base at North Head Lighthouse. (Photo courtesy of Washington State Parks and Recreation Commission, 2013.)

Washington Parks and Recreation began planning restoration work in 2010 but did not start work until the property was transferred to them in 2012. A series of projects from 2012 until 2022 completely restored the exterior metals, masonry, wood, and stucco.

The original sandstone was replaced with Amherst Gray Berea Sandstone from Ohio, selected to match both the appearance and physical properties of the original stone which was no longer quarried. Similar properties of water transmission, thermal expansion, and elasticity were desired because the new stone would be keyed into the remaining original sandstone. Berea also performed slightly better than the original in salt crystallization testing. The Berea stone has more quartz and less feldspar than the original sandstone, and a higher percentage of open pores, but the particles themselves are less porous, and the stone overall is more coherent and durable. The stone is performing well, although there is evidence of minor pyrite staining. Algae continue to be a chronic issue and are treated annually with a quaternary cleaner.

The replacement cement stucco and mortars were a mix of natural cement, hydraulic lime, and sand: more hydraulic than pure lime but not as hard as OPC. In 2016, a finish coat of natural cement stucco was applied in a single day, using one of four mixes tested in different weather conditions. The final mix used sand and natural cement and was misted with water for three days after installation to control the rate of curing.

The stucco was covered with a coat of potassium silicate paint, which failed within two years of installation, apparently due to abrasion from windblown sand. The coating was removed and replaced with a permeable elastomeric coating. Although previous layers of elastomeric paint trapped moisture, the careful use of a single layer of permeable elastomeric appears to be performing well.

The cast iron floor plates of the tower are painted annually but continue to rust, possibly due to water within the cast iron that is still working its way out. The lime-based paint on the interior started to fail within months as moisture in the masonry was drawn through the finish by dehumidifiers. The use of dehumidifiers has been discontinued.

Projected Impacts of Climate

Increases in UV degradation and an increase in the severity and duration of storms are likely to increase the stresses placed on the lighthouse materials. Despite ongoing restorations and maintenance, the salt content, high humidity, and high winds force moisture into any opening. A plan for ongoing inspection and maintenance is especially appropriate to this climate. Ongoing maintenance must pay special attention to maintaining the ability to dry through the continued use of porous coating and appropriate ventilation.

Figure 3.5.15 Interior biological growth at Cape Blanco Lighthouse. (Photo by Daniel Thomas, 2023.)

Heceta Head, Yaquina Head, and North Head Lighthouses have all received major restorations in the last two decades. Yaquina Head and North Head are being actively maintained, but Heceta Head is showing severe wear. Although the permeable coatings are best suited to the harsh environment, they must be regularly replenished if they are to withstand an even more intense environment in the future.

Conclusions

A comparison of four lighthouses along the Oregon and Washington Coast illustrates the impact that humidity and rain have on traditional stone, brick, lime, and metal and on modern cementitious and elastomeric coatings, and the importance of ongoing maintenance.

The collaboration of multiple governmental and private non-profit organizations has diluted clear responsibility for maintenance and repairs of the lighthouses in the past. However, effective collaboration between government entities and non-profit advocates can shed light (no pun intended) on the needs of historic lighthouses and provide multiple opportunities for obtaining resources for their maintenance and repair.

Lessons learned include not just the need for good design and maintenance, but that changes in materials and operation may be required for the buildings to be practically preserved. However, short cuts that attempt to simply block water rarely work. The unrelenting power of water to move through a building must be understood and acknowledged.

The "Cape Blanco Lighthouse Historic Preservation Plan" is a good example of an approach that recognizes the need for on-going observation and maintenance. It documents history, existing conditions, and establishes recommendations and priorities for future maintenance and repairs.

While the priority is to address structural and safety issues, the plan also prioritizes a series of actions that collectively address the need to dry out the masonry and provide ongoing ventilation of the entire lighthouse:

- A dehumidification plan and system that responds to seasonal changes.
- Restoration of the vent-ball assembly and operable windows at various levels.
- Removal and replacement of existing lead, elastomeric, and acrylic paints with a permeable limewash, polymer-modified limewash or similar breathable coating on the masonry.
- Frequent maintenance of protective paints and sealants on metal surfaces.

Bibliography

Achterman, Gail. 2009. *Coastal Climate Change Effects*. https://oregonexplorer.info/content/climate-change-oregon?topic=44&ptopic=38.

Alex Klahm. 2016. "Report by Alex Klahm, Architectural Metal and Design, Inc."

Anderson, Kraig. 2023. *Heceta Light House*. https://www.lighthousefriends.com/light.asp?ID=131.

BLM. 2014. *Yaquina Head Lighthouse Painting.*

———. 2020. *Specifications for 2020 Yaquina Head Lighthouse Painting.*

———. n.d. "Cape Blanco Lighthouse Interior Painting Contract."

———. 1996. "Paint Specifications." *Division 9*. August 9.

———. 2003. "Part VI – Specifications for Cape Blanco Lighthouse Preservation."

BLM. 2022. "Restoration Work for the Yaquina Head Lighthouse: Work Room 2022."

———. n.d. "Yaquina Head Outstanding Natural Area Historic Lighthouse."

Chambers, Cullen. 2001. "Heceta Head Light Station." Overview of Existing Conditions.

Collins, George. 2006. "Restoration of Yaquina Head Lighthouse."

Crawford, Mike. 2023. Specification information for natural cement and "Project Summary: Historic Preservation Heceta Head Lighthouse Restoration, Award Date 10/10/23" furnished by Mike Crawford, President, Pioneer Waterproofing Co., Inc by email to Susan Pranger 9/8/2023.

Donovan, Sally. 1992. "Yaquina Head Lighthouse." National Register of Historic Places Registration Form.

Edwards, Deb. 2023. Interview of Deb Edwards, Park Ranger, OPRD *on site at Heceta Head Lighthouse in Florence,* Oregon by Susan Pranger. *sit*, (February 28).

Friends of Yaquina Lighthouses. n.d. *Yaquina Bay and Yaquina Head Lighthouses*. https://www.yaquinalights.org/about-yaquina-head-lighthouse/.

Hansen, David M. 1975. *Cape Disappointment Historic District*. National Register of Historic Places Inventory – Nomination Form, National Park Service.

Heceta Lighthouse B&B. 2020. *History*. https://www.hecetalighthouse.com/history.

Highbridge Materials Consulting, Inc. 2016. "Petrographic Examination Reports (Existing Stone; Berea Sandstone)."

Howell, Scott. 2010. "North Head Lighthouse Architectural Metals Assessment."

LaFollette, Cameron. 2022. *Yaquina Head Lighthouse*. November 9. https://www.oregonencyclopedia.org/articles/yaquina-head-lighthouse/#.Y-HD3bMK8c.

LaLande, Jeff. n.d. *Cape Blanco*. https://www.oregonencyclopedia.org/articles/cape_blanco/pdf/.

Lambert, John. 2016. "Report on Original and Possible Replacement Stone (Berea Sandstone) for North Head Lighthouse." In *Summary of Results of Petrographic Analysis by John Walsh*, Highbridge Materials Consulting.

Licht, Susan, interview by Susan Pranger. 2023. (January 12).

Licht, Susan. 2020. "Cape Blanco Lighthouse Historic Preservation Plan."

Licht, Susan. 2023. "Cape Blanco Lighthouse Masonry and Steel Investigations."

Licht, Susan. 2011. "North Head Lighthouse HSR." Historic Structures Report.

Licht, Susan. 2014. "North Head Lighthouse, Materials Investigation Report."

Light-House Board. 1871. "Summary of Light House Board Reports." Summary of Light House Board reports.

LightHouse Board. 1873–1920. "Summary of reports by the Light House Board."

Lighthouse Friends. 2023. *Cape Blanco Lighthouse*. https://lighthousefriends.com/light.asp?ID=125.

———. 2023. *Heceta Lighthouse*. https://www.lighthousefriends.com/light.asp?ID=131.

Lighthouse Service. n.d. "Instructions for the Personnel."

Lighthousefriends. 2023. *Yaquina Head Lighthouse.*

Masko, Dave. 2012. *Famous Lighthouse on Oregon Coast now closed for Renovations.* March 7. https://huliq.com/10282/famous-lighthouse-oregon-coast-now-closed-renovations.

McMurry, Alex, interview by Zoom call by Susan Pranger. 2023. (January 11).

McMurry, Alex, interview by Susan Pranger. 2023. (February 27) *Site Visit* Caper Disappointment, WA.

Minor, Rick, and the US BLM, Oregon State Office Debra C. Barner: Heritage Research Associates. 1991. *Yaquina Head: A Middle Archaic Settlement on the North-central Oregon Coast.* Heritage Research Associates.

NPS, USCG, DOD Legacy Resource, management Program and US Lighhouse Society. (1997). *The Historic Lighthouse Preservation Handbook.* Retrieved August 8, 2023, from https://www.nps.gov/maritime/nhlpa/handbook.htm

Neblett, LCDR N.P., and LT R. J. Williamson. 1977. *National Register Nomination Form Heceta Head Lighthouse.* Nation Register of Historic Places Inventory – Nomination Form, National Park Service.

OPRD. 2013. "Heceta Head Lighthouse Restoration." PowerPoint presentation.

———. 2005. *Heceta lighthouse.* https://web.archive.org/web/20060321141624/http://www.oregonstateparks.org/images/pdf/heceta_lighthouse.pdf.

Otac. 2001. "Cape Blanco Lighthouse Assessment Report."

Penberthy, Bryan. 2017. *History of the Cape Blanco Lighthouse.* https://www.us-lighthouses.com/cape-blanco-lighthouse.

Pranger, Susan. 2023. Site Visit by Author to Yaquina Head Lighthouse, Newport, Oregaon. (February 28).

———. 2023. Site Visit to Cape Blanco Lighthouse, Oregon.

Ruud, Cassie. 2016. *Video/Illuminating: Yaquina Head Lighthouse.* https://www.thenewsguard.com/news/video-illuminating-yaquina-head-lighthouse/article_e2621434-b3eb-11e5-b488-abba81c882f6.html.

Staehli, Alfred. 1992. "Condition Assessment of Yaquina Head Lighthouse." Condition Assessment.

University of of Oregon. 1996. "Cape Blanco Condition Assessment." Preservation Field School, University of Oregon.

———. 2004. US Department of Labor, Bureau of Land Management, Salem District "Yaquina Head Lighthouse Preservation – Specifications and Drawings."

US Lighthouse Service. 1910. "Lighthouse Inspection Reports Database – North Head Light Station." https://uslhs.org/national_archives/north_head.htm.

US Lighthouse Service. 1910. "Lighthouse Inspection Reports Database Yaquina." https://uslhs.org/national_archives/yaquina_head_1910_files/lantern.htm.

USLHS. 1881. "Lighthouse Inspection Reports Database, Yaquina Bay."

Wood, Stephen. 2020. *North Head Lighthouse Celebrates 122 years.* https://www.adventureawaits.com/352/Commute-to-a-park-6-19-20.

WPRC. 2015. "Cape Disappointment State Park, North Head Lighthouse Restoration."

———. n.d. "Project Manual for North Head Lighthouse Restoration at Cape Disappointment State Park."

WRCC. 2023. *Climate of Oregon.* https://wrcc.dri.edu/Climate/narrative_or.php.

Yaquina Lights. 2018. *A new LED Lamp and the Evolution of Yaquina Head Lightsources.* https://www.yaquinalights.org/news/a-new-led-lamp-and-the-evolution-of-yaquina-head-lightsources/.

CHAPTER 3.6

Michigan Central Station

Introduction

The Michigan Central Station (MCS) in Detroit, Michigan was one of the largest railroad stations at the peak of rail activity in the early 20th century. After decades of abandonment and deterioration, the building is now being restored by the Ford Motor Company for their new Innovation Center. The restoration will retain most of the exterior stone, brick, terracotta, cast iron, and copper that managed to survive decades of weather and neglect, while introducing new technologies and materials that are compatible with the historic materials.

Figure 3.6.1 Michigan Central Station (MCS), Detroit, Michigan. (Photo by Tim Pranger, 2022.)

DOI: 10.4324/9781003160489-20

History, Location and Climate

History

The Michigan Central Railroad Station was the tallest train station in the world when it was built in 1912. Fifteen stories of offices towered over a three-story base with a dramatic waiting room, concourses, two mezzanines, and a covered carriage house. Travel by train gradually declined after World War I, and even more sharply in the 1950s as travel by car and air became popular. Many areas of the building were closed in 1967, although train service continued. Amtrak took over the building in 1971 and reopened the waiting room and main entrance in 1975, but closed the building and ceased all train service in 1988.

There were many changes in ownership and attempts to restore the building over the following decades. Various enterprises owned by the Mouron family acquired the building in 1996, removed the train shed in 2000, and used the remaining tracks for an "intermodal freight facility" until 2004. New uses were proposed, including a headquarters for the Detroit Police, a new Trade Processing Center, and a convention center or casino. A 2009 resolution by the city council to demolish the building was stopped by a lawsuit citing the National Preservation Act of 1966, and MCS was nominated to the National Register in 1975. By 2011, there was significant vandalism and water damage; most of the original roofing and windows were missing or damaged, and the concourse was open to the sky. From 2011 through 2014, efforts were made to stabilize the building by installing temporary power and lights and new windows and louvers that allowed the interior to begin to dry. In 2018, the Ford Motor Company acquired the complex and began renovations to convert the building to the Ford Innovation Center.

Location and Climate

Detroit is located in the Köppen climate zone "Dfa", hot summer continental climate, and has hot, humid summers and cold winters. Average summer temperatures reach 74°F in July, although the highest recorded temperature of 104°F was in June. The average low is 26°F in January, but the lowest recorded temperature was –21°F. The average annual precipitation is almost 30 inches, with the most precipitation in December and the least in August.

Detroit has already experienced increased temperatures and precipitation. A comparison of the 30 year average from 1981 to 2010 to the average from 1961 to 1990 reveals an increase of 1.4°F in average temperatures, 4.3°F in nighttime temperatures, an increase of 11% in precipitation. The number of hot, dry days and hot, wet days both increased by at least three days in the summer, while the number of cool, dry days decreased by more than ten days.

Construction Techniques and Materials

The building's exterior is a beautiful composition: a two-story base of granite and Indiana limestone supporting a brick and terracotta tower. Cast iron and steel windows, and copper roofing have since been replaced. The steel structure is supported by concrete columns and foundations below grade.

The exterior construction is unusual because there are large expanses of masonry without relieving angles or expansion joints. The masonry is self-supporting for the full height of the building, tied to, but not supported by, backup masonry or steel structure. Flexible lime mortar and vertical bands of window and spandrel absorbs most thermal movement.

Building Construction

A concrete foundation supports a steel structure with concrete floors. Most of the steel is encased in concrete. The exterior of the two-story concourse is clad with 8- to 12-inch-thick Indiana limestone above a few courses of granite base, both keyed into a brick backup wall. Arched masonry openings surrounded the original cast iron windows or wood doors.

The tower walls are a single wythe of brick with a generous cavity and a non-load bearing backup wall of either brick or clay tile (speedtile). The backup walls are supported by the structure at each floor level and stabilize, but do not support, the exterior brick veneer. The upper floors of the tower, including the Juliette balconies, are clad with unglazed terracotta veneer that imitates stone. The masonry bays of the tower alternate with a strip of windows and spandrels that extend almost the full height of the building. The cast iron spandrel panels are supported by the steel structure at each floor. The spandrels were installed prior to the masonry and brace, but do not support, the masonry. The windows at the tower and base were originally steel with cast iron spandrels or trim. A few windows at the base were wood.

Materials

Laboratory analysis of the exterior mortars existing in 2019 revealed a range of types, including hydrated lime mortar at the limestone, a type N mix of Portland cement and hydrated lime at the granite, and a type O mix of Portland cement and hydrated lime at the brick. All were in good condition.

Analysis of the exterior masonry identified a coarse-grained granite of feldspar, quartz, and biotite mica. Microcracks throughout the stone appeared to be inherent, but there are also microcracks within 2 mm of the surface from freeze-thaw damage.

The limestone is a medium density dolomite, a calcium magnesium carbonate with 1% quartz crystals, 1% white clay (kaolinite or illite), and traces of manganese oxide. The clay is dispersed in thin layers mixed with small dolomite crystals. Pyrites were not observed in all samples, but up to 2% iron was found in different layers of the stone, including hydrous iron oxides-hydroxides in the beige stone and hematite in the darker, reddish bands. Hairline cracks occur in the clay layers and scaling occurs along the boundaries between layers. Freeze-thaw cycles have caused flaking and crumbling and loss of up to an inch of surface.

The terracotta is a hard, dense fired clay product with low water absorption, 3% quartz and 97% grog (ground fired clay) and binder.

Deterioration, Restoration, and Other Changes

By the time Ford acquired the MCS, it had experienced severe deterioration of the exterior and interior due to vandalism, lack of maintenance, and decades of weather exposure. Most of the copper roofs had been stolen, leaving both exterior walls and building interiors vulnerable to water penetration. The unheated and unprotected building was saturated with water, with 4 feet of standing water in the subbasement.

Water traveling downward in the interior had saturated plaster walls and ceilings, and interior and exterior masonry, exiting through exterior masonry joints. The plaster initially appeared to be sound, but quickly disintegrated after temporary enclosures and heat were installed. In spite of this, both the concrete and steel structure were in good condition.

Roofs

The original roofs of MCS were deteriorated or missing by 2011, with significant deterioration of the underlying concrete slabs. The temporary roofing was quickly stripped and blown away by heavy winds, as MCS is the tallest structure in the area.

Many of the concrete roof slabs were replaced; the built-up gravel roofs were replaced with membrane roofs, and the missing concourse roofing is being replaced with new copper batten seam roofing, and temporary roofing membranes were installed on the floors below.

The carriage house originally had a copper roof but will be restored with a glass enclosure designed to reveal the dramatic riveted wrought iron structure.

Exterior Masonry

The masonry exterior was in remarkably good condition given the lack of maintenance and roof protection.

The original lime mortar was intact and had continued to absorb the expansion and contraction of the building during both occupancy and vacancy.

However, face mortar was typically missing at bands of terracotta, brick, and limestone immediately below parapets or other projections where water was allowed to infiltrate from above. All of the joints in the granite and limestone and approximately 20% of the mortar joints in the brick, mostly around window openings or along floor lines, were repointed with new lime mortar to match the existing.

Granite, being non-porous, was generally in good condition and only needed cleaning and minor repairs. The brick and terracotta were generally sound except where the brick was damaged at the corners of the tower from expansion of the brick and mortar. Although some steel was discovered at the corners, it appeared to be a later insertion and not the cause of cracking. In these areas, the brick is being replaced with new brick and lime mortar to match existing.

Figure 3.6.2 Brick cracking at corner of MCS tower. (Photo by Angela Wyrembelski, 2018.)

The terracotta was cracked or spalled, with rusting steel support structure and anchors, primarily along unprotected roof cornices, at the Juliette balconies and other projections. Biological growth was also significant at the Juliette balconies. Rusted steel and anchors were cleaned and coated or replaced. The cornice and Juliette balconies were replaced on the full perimeter of the tower with new fiberglass (glass fiber reinforced polymer) panels to reduce the weight on the existing steel structure but had to be secured to resist significant wind uplift. Other deteriorated pieces were replaced in-kind with new terracotta. Minor chips and spalls were patched and reglazed.

Figure 3.6.3 Terracotta repair at cornice of MCS. (Photo by Angela Wyrembelski, 2018.)

The two-story limestone walls and columns were in good condition except where individual stones were saturated over a long period of time. Unprotected cornice stones at the top of the wall were both saturated with water and displaced by rust jacking of the exposed steel and mortar deteriorated by water through the masonry joints. Repairs included cleaning or replacement of corroded steel, replacement of individual stones, dutchman repairs, and patching of small areas with proprietary patching compounds. Abandoned limestone blocks salvaged from the Dark Hollow Quarry, the quarry originally used to supply the limestone, were more beige in color than the original, but blended well when used for full height dutchmen or full blocks. Damaged stones were removed so that the texture and surface treatment of dutchman repairs and patches could be matched in a controlled shop environment. The patching mortar was selected to blend in with the range of existing limestone colors; they are slightly grayer than the adjacent limestone but blend in as they weather.

Figure 3.6.4 Deteriorated limestone cornice at waiting room of MCS. (Photo by Angela Wyrembelski, 2018.)

Figure 3.6.5 Dutchman repair at limestone capital at waiting room, MCS. (Photo by Tim Pranger, 2021.)

Windows

The majority of the original wood and steel windows in the tower were replaced in 2011. Replacement of the aluminum windows was recommended in 2018 because they did not match the operation, profile, or sightlines of the original windows but the replacement windows were in good condition and were retained due to the cost of replacement. The original cast iron spandrels were rusted but sound, and were repaired, cleaned, and repainted.

The remaining steel windows in the concourse were replaced but the steel windows in the waiting room and a few wood windows in the concourse were restored. Ford produced replicas of the cast iron ornaments with a proprietary resin-based 3-D process that picked up all the details and imperfections of the original cast iron and are indistinguishable from the original.

Figure 3.6.6 Resin replica of ornamental grilles in MCS waiting room. (Photo by Tim Pranger, 2021.)

Insulation

Hygrothermal WUFI analysis showed that it was safe to add batt insulation on the interior of the brick walls and terracotta walls of the tower, using a detail that provided a gap, metal studs with batt insulation, a vapor barrier and gypsum wall finish. However, the analysis indicated that comparable insulation would trap moisture in the limestone and granite walls at the base, so those walls were not insulated.

Projected Impacts of Climate

The combination of freezing temperatures and heavy precipitation creates ideal conditions for freeze-thaw cycles that would be very damaging to wet porous stone, which explains the damage at horizontal features of the limestone and terracotta. The combination of rising temperatures and increased rain combined with intermittent freezing temperatures will increase the fluctuations in temperature that cause cycles of freeze-thaw.

Conclusions

The original masonry and mortar at MCS are all moderately dense and should continue to perform well with appropriate water management. Drainage systems should be sized to accommodate increased volumes of rain and maintained to avoid future blockages.

Bibliography

CTL. 2019. "Petrographic Examination and Chemical Analysis of Mortar Samples from the Michigan Central Station Project in Detroit, Michigan." Technical Report.

CTLPC. 2019. "Petrographic Examination of Stone and Plaster Samples from Michigan Central Station." Technical Report.

Detroit Historical Society. 2023. *Michigan Central Station*. https://detroithistorical.org/learn/encyclopedia-of-detroit/michigan-central-station.

Kniffell, Leonard. 1974. "Michigan Central Railroad Station." National Register of the Historic Places Inventory-Nomination Form. https://catalog.archives.gov/id/25341108

Quinn Evans Architects. 2018. "Project C Conditions Assessment Report (MCS)." Conditions Assessment Report.

Quinn Evans. 2018. "Project C Appendix C Window Types." Conditions Assessment Report.

Wikipedia. 2023. *Michigan Central Station*. https://en.wikipedia.org/wiki/Michigan_Central_Station.

Wyrembelski, Angela, Ron Staley, and Richard Bardelli. 2022, Interview with MCS project team regarding restoration of Michigan Central Station on site at MCS, Detroit Michigan.

CHAPTER 3.7

Early Chicago Concrete at Unity Temple and Baha'i Temple

Introduction

It is not surprising that the first uses of exposed concrete for religious structures were in the Chicago suburbs of Oak Park and Wilmette, Illinois. Chicago was a center of concrete innovation, producing more cement than any other region in the country and distributing more than any other city in the world. From 1892 to 1906, the annual production of Portland cement production increased in the United States from 500,000 to 46,500,000 barrels, accelerated by a shortage of steel in 1897 through 1898, and a decline in lime production. Exposed concrete was beginning to emerge as a feature of modern architecture and as an economical alternative to stone or masonry veneers.

Frank Lloyd Wright used concrete at Unity Temple to express the material's character and to create an organic form. Jean-Baptiste Louis Bourgeois (Louis Bourgeois) conceived of the ornate Baha'i temple as an expression of faith, independent of a material choice, and later discovered that concrete was the only material capable of executing the intricate design.

Unity Temple was constructed within three years of its 1906 design, when Wright later acknowledged that they "knew much less of reinforced concrete construction than we know now" (Siry 1996).

Construction of Baha'i Temple did not begin until 1920 and work continued intermittently until its completion in 1953. This long overlapping period of design and construction, and collaboration of designers and contractors, gave Baha'i the distinct advantage of incorporating, as well as driving advances in concrete and precast technology.

Both projects were conceived and executed by small congregations, although Baha'i had access to support from an international community from the beginning,

DOI: 10.4324/9781003160489-21

and Unity Temple was not supported by the broader preservation community until the 1970s. Today, both are internationally recognized; both have undergone major renovations with the goal of preserving these iconic buildings for the future.

History, Location and Climate

History of Unity Temple

Unity Temple was the first structure that Wright built entirely of concrete, but it aligned with his general philosophy of the "truthful expression of materials" (Siry 1996). Early sketches suggest that Wright may have briefly considered a brick veneer for the temple, but he clearly desired, and promoted, the use of exposed concrete. This was partly to be economical, but also in response to express the "authentic nature" of concrete.

Figure 3.7.1 Unity Temple, Frank Lloyd Wright, Architect. (Photo by Daniel Thomas, 2022.)

The complex was constructed from 1906 to 1908, with an exposed "bird's eye" pebble finish. Wright collaborated closely with Paul Mueller, who had extensive prior experience with concrete and solved many of the technical issues required to execute Wright's design, but the character of the materials and the complex design quickly led to leaking roofs and cracks in the walls. A series of roof repairs in the following decades did not solve the underlying issues. In 1961, the entire exterior was resurfaced with a stucco-like coating. A decade later, the cracks were more widespread and exposed corroded reinforcing steel.

In 1973, the Unity Temple Restoration Foundation (UTRF) was established to oversee repairs and restoration. In addition to ongoing roof and crack repairs, the stucco coating replaced with a layer of shotcrete (gunite), a concrete spray applied with pneumatic hoses. Although neither the 1961 stucco nor the 1973 shotcrete exactly replicated the original appearance of the concrete, they did prolong the life of the building by reducing moisture, freeze-thaw, and carbonation.

Delaminated concrete and exposed rebar were once more visible on the underside of the cantilevered slabs by 1987, and large areas of the fascia were unsound. Repairs to the roofing and exterior continued throughout the 1990s. Unfortunately, the complex configuration of roofs and skylights continued to leak. Unity Temple was listed on Landmark Illinois' Most Endangered list in 2000. A comprehensive master plan in 2006 identified interior and exterior structural, environmental, and decorative damage and the need to dry the saturated concrete.

In 2009, there was a renewed sense of urgency after a section of the interior ceiling fell in the unoccupied auditorium. Nominations to the World Heritage Status in 2008 and the National Trust's List of Most Endangered Properties in 2009 brought renewed efforts and resources, including donations from the Alphawood Foundation, the Tawani Foundation, the congregation, and the UTRF. Emergency repairs to the south roof were completed in 2010, field mockups were used to update the master plan in 2014, and a comprehensive renovation began in 2015.

The 2015 project included extensive repairs to the interior and exterior concrete structure and finishes, roofing systems, skylights, windows, and doors. The mechanical, electrical, fire alarm and lighting systems were upgraded, new geothermal wells and air conditioning were added, and interior decorative finishes were restored.

In 2015, ownership was transferred from the Unity Temple Unitarian Universalist Congregation (UTUUC) to an independent limited liability corporation, UTP, LLC. In 2022–2023, management and maintenance of the building was shifted from UTUUC to UTRF. It is hoped that this shift in responsibility will allow for more proactive maintenance and long-term preservation planning.

History of the Baha'i Temple

The Baha'i community purchased land in Wilmette in 1907 and placed a symbolic cornerstone in 1912 but World War I delayed their call for submissions until 1920. Louis Bourgeois' design was not based on a particular material; stone, terracotta, metal, and concrete were still being investigated when the foundations were poured in 1921. Precast concrete was selected only after Earley's test panel survived a year in the harsh Chicago climate, but the processes required to execute Bourgeois' design did not yet exist; they were developed over the next three decades, concurrent with fundraising, design, and construction.

Figure 3.7.2 Baha'i Temple, Louis Bourgeois, Architect. (Photo by Daniel Thomas, 2022.)

Allen McDaniel, a structural engineer and member of the Baha'i community, proposed that the initial phase construct a complete weatherproof structure including three tiers of reinforced concrete galleries capped by a steel frame dome sheathed in glass and aluminum. The second phase would clad the exterior with a layer of ornate precast concrete panels, while the third phase would provide an interior layer of ornamental precast panels. Each of the three layers was perforated with openings, aligned to allow sunlight to filter into the building during the day and the interior light to shine out at night.

The weatherproof building structure was completed between 1921 and 1931. The Phase 2 cladding was not awarded to John Earley's Studio until shortly before Bourgeois' death in 1930, but the two men had already spent almost three decades exploring ways to adapt Earley's pioneering work in precast to Bourgeois' designs. The exterior cladding was fabricated in Earley's workshops near Washington, D.C., shipped by train to Wilmette and installed by 1942. Interior cladding was completed in 1948.

The concrete materials and workmanship were of the highest quality. However, the design of the flashing interface between ornamental concrete façade and cast-in-place structure trapped water. Extensive deterioration of the saturated concrete required a major restoration in the 1980s. The brilliant quartz concrete was also discolored by a combination of dirt, gypsum deposits, and biological growth. Developing a cleaning method that did not destroy the concrete's reflective quality required more investigation and innovation in the late 1980s. A second round of restoration and cleaning occurred between 2002 and 2012.

While the design was well documented, the specific materials and processes were closely held trade secrets, lost when Earley's studio closed in 1973. True to their methodical approach, the Baha'i community assembled a multi-disciplinary team of Architects, Engineers, master craftsmen, scientists, and contractors. They collaborated from 1983 to 1990 to reverse engineer, and sometimes improve, the materials and techniques required to restore the concrete exterior and aluminum skylight. The results were thoroughly documented and archived as part of a 1,000-year conservation plan. Specialized tools and stockpiles of quartz and other materials were stored in the building or nearby for future use.

Today the House of Worship is in almost pristine condition, under the watchful eye of its conservation team. However, the continuing effects of weather will require constant vigilance and maintenance to meet the Baha'i's goal of 1,000-year life.

Location and Climate

Chicago is in the Midwest of the United States, 800 miles from the Atlantic Coast and 2,000 miles from the Pacific Coast, and lacks moderating effects from either ocean. The area also lacks nearby mountains that block the movement of arctic air. As a result, Chicago experiences extremes of cold and heat. Their snowy winters average temperature between 16°F and 31°F in January, but the lowest recorded temperature was –27°F with a windchill of –83°F. Summers are hot and wet with 18 to 21 days over 90°F each summer, and a record high of 105°F in 1995. In the hottest month of July, the average temperature swings from 66°F at night to 84°F during the day. An average of 38 inches of snow falls in December through February, with frequent but short rain showers in the summer.

The Chicago area is already noticing changes in weather that are outside the normal cycles. Average temperatures have increased steadily since the 1970s, and more rapidly in the last four years. Average temperatures between 1997 and 2006 were 5°F higher than temperatures recorded between 1961 and 1990. In the four years prior to 2006, annual temperatures increased by 2°F to 4°F with more dramatic increases of 7°F in the winter. There were fewer cold waves in the 1990s compared to the prior two decades but more extreme heat waves and record temperatures. Heavy rain events, tied to flooding, have doubled in frequency.

Unity Temple and Baha'i Temple both experience the extreme weather conditions of Chicago area although the microclimates of the two buildings are slightly different: Unity Temple is nestled into a relatively level suburban site, sheltered by other buildings and trees, while Baha'i Temple is an imposing structure surrounded by gardens overlooking lake Michigan. Although the prevailing winds in Chicago are from the northwest, the prevailing wind at Baha'i Temple is generally from the northeast, off the lake. However, it is the differences in the execution of the two temples that best illustrate the impact of climate and weather on concrete. The composition of the concrete and the detailing of the building determine the effect of Chicago's extreme weather on these two buildings.

Materials and Construction

Both Unity Temple and Baha'i Temple used Portland cement concrete, exposed aggregate, and a mix of cast-in-place and precast components to achieve a specific aesthetic. However, Baha'i used a sharp, hard quartz aggregate, while Unity Temple used a smooth, porous bird's eye pea gravel, which absorbed more moisture, and had a weaker bond. Both temples had a combination of cast in place concrete and precast ornament, and both reduced shrinkage by controlling the water to binder ratio in the cement. However, Unity Temple used a dry cement process similar to rammed earth, whereas the Baha'i Temple used a wet mix with a secret process to extract excess water as the concrete cured. Neither building incorporated expansion joints, but Baha'i exterior was a series of panels; Unity relied upon vertical windows and building geometry to control movement.

The quality of ordinary Portland cement (OPC) improved in the decades between the construction of Unity Temple (1906–1908), the construction of the Baha'i structure (1920–1930), and the Baha'i exterior cladding (1931–1942), and even more so by the time restoration work began on both buildings. During and after World War II, the cement particles were ground more finely to produce a quicker setting concrete, water-reducing additives were developed to control shrinkage, and air entrainment was added to increase resistance to frost damage.

Unity Temple

Materials and Construction

Unity Temple was constructed of monolithic mass concrete, using different aggregates and techniques to meet different aesthetic and structural goals. All concrete was cast-in-place on site except for the precast fascia panels and precast ornament on columns and flower boxes. The exterior concrete had an exposed aggregate finish except where covered with roofing membranes. The interior surfaces were covered with painted plaster with wood trim.

The unreinforced concrete foundations contained a strong limestone aggregate that increased compressive strength. The remaining walls, floor, and roof slabs were reinforced with steel reinforcing rods (rebar) because they used pea-gravel or coal cinder aggregate with a lower compressive strength. The steel beams and trusses required for longer spans were also encased in concrete. The concrete in the roof slabs had a high water content, which further decreased the strength of the binder.

Wright designed the overall structure without expansion joints. The vertical windows between the corner stairs and the flat walls of each elevation serve as expansion joints to some degree but do not continue through the roof or foundation. Vertical cracks can be seen above and below these windows.

The combination of weak binder, porous aggregate, and lack of expansion joints meant that the reinforced concrete roofs and walls were vulnerable to expansion and contraction caused by thermal, wet-dry, and freeze-thaw cycles.

Roofs

The exterior roof slabs were poured-in-place reinforced OPC concrete, topped with a lightweight concrete fill sloped to drain. Simple spans used a lightweight coal cinder aggregate, while cantilevered slabs used bird's eye gravel. There is speculation that the formwork on cantilevered slabs was removed to expose the aggregate before the binder was fully set and may have contributed to the creep that later developed.

The intersections of 16 separate roofs, many with skylights, led to constant roof leaks but the sequence of repairs is unclear. One report indicates that the roofing was repaired in 1921 and replaced or recoated in 1924. Other sources claim that the roofing was not replaced until after the Great Depression of the 1930s. In any case, water penetrating the relatively porous cinder concrete slabs eventually led to deterioration of the concrete and corrosion of the reinforcing steel. In 1938, the roof facias had cracked and the concrete of the corners of the cantilevered slab had spalled, exposing corroded rebar. It is not clear that any repairs were undertaken at the time.

The roofs were repaired in 1962/1963, re-roofed in 1971/1972, and repaired or replaced in 1982, 1986/1987, and 1999. However, the roof slabs continued to leak and the interior concrete was frequently saturated. By 2006, most roofing dated to 1982 or 1986, although a few new membranes were installed at Unity House in

2000 and 2004. WJE's assessment in 1987 found that large areas of the west overhang were deteriorated from a combination of related factors including a weak binder, fully carbonated concrete, freeze-thaw damage, and significant levels of chlorides (often associated with coal cinder). The west cantilever of Unity Temple was repaired in 1992 and in 2003/2004, six of the cantilevered roofs were "restored" and additional roofs "re-roofed".

Despite ongoing leaks, a 2000 report indicated that most freeze-thaw damage was in the lightweight concrete topping, not in the underlying structural roof slabs of the auditorium. However, there was significant deterioration on the undersides and vertical fascia of the exposed cantilevered slabs, and some deterioration of interior concrete structures. A series of repairs to the cantilevered slabs and supporting ornamental columns were completed between 2000 and 2013.

Exposed Exterior Architectural Concrete Walls

The exposed concrete walls were cast-in-place with bird's eye pea gravel in a very dry mix, compacted in 6-inch lifts, with a maximum of three lifts in 24 hours "to ensure proper setting." The dry mix and generous setting time reduced the size and number of shrinkage cracks. Exposed surfaces were washed with an acid solution to dissolve the binder and expose the pebble aggregate, creating a surface like coarse granite.

The exterior walls were protected by deep overhangs, and damage was initially limited to shrinkage cracking. However, freeze-thaw cycles enlarged cracks and loosened aggregates. In 1960, the cracks were repaired, and the walls were covered with a stucco-like latex coating that adhered the pebbles but obscured the color and texture of the original wall. The coating was removed by light sandblasting in 1973 and replaced with a pneumatically applied 1-inch layer of shotcrete. The shotcrete was lightly sandblasted to expose the aggregate and then sealed with linseed oil. There is no record that the linseed coating was ever reapplied.

The 2000 CTL report noted that there was substantial cracking and related delamination at the parapet walls and at the shallow fascia of the stair towers, generally at construction joints or where shrinkage cracks existed prior to the 1973 repairs. Repair work varied from crack repair to complete replacement of some parapets. Although the underlying "parent" walls remained structurally sound, cracks in the shotcrete appeared in 2000, and had substantially increased by 2014. The shotcrete did not delaminate at the original shrinkage cracks, which had rounded edges from the initial sandblasting. However, there was significant delamination at cracks from thermal or moisture movement, which were deeper, wider, and had sharp edges. Horizontal expansion cracks appeared at cold joints where walls met the roof slab; vertical cracks occurred offset from building corners by the thickness of the wall.

The North Façade Project in 2015–2017 focused on restoration of the 1973–1974 shotcrete coating. Open cracks in sound concrete were enlarged to a V-shaped groove to receive a bond breaker and two-part polyurethane sealant, then covered with coarse aggregates matching the shotcrete. More aggressive repairs were required where corroded rebar had to be exposed and thoroughly cleaned or replaced with stainless steel rebar. Stainless-steel mesh was added at crack repairs to prevent telegraphing and at new shotcrete patches to reduce shrinkage.

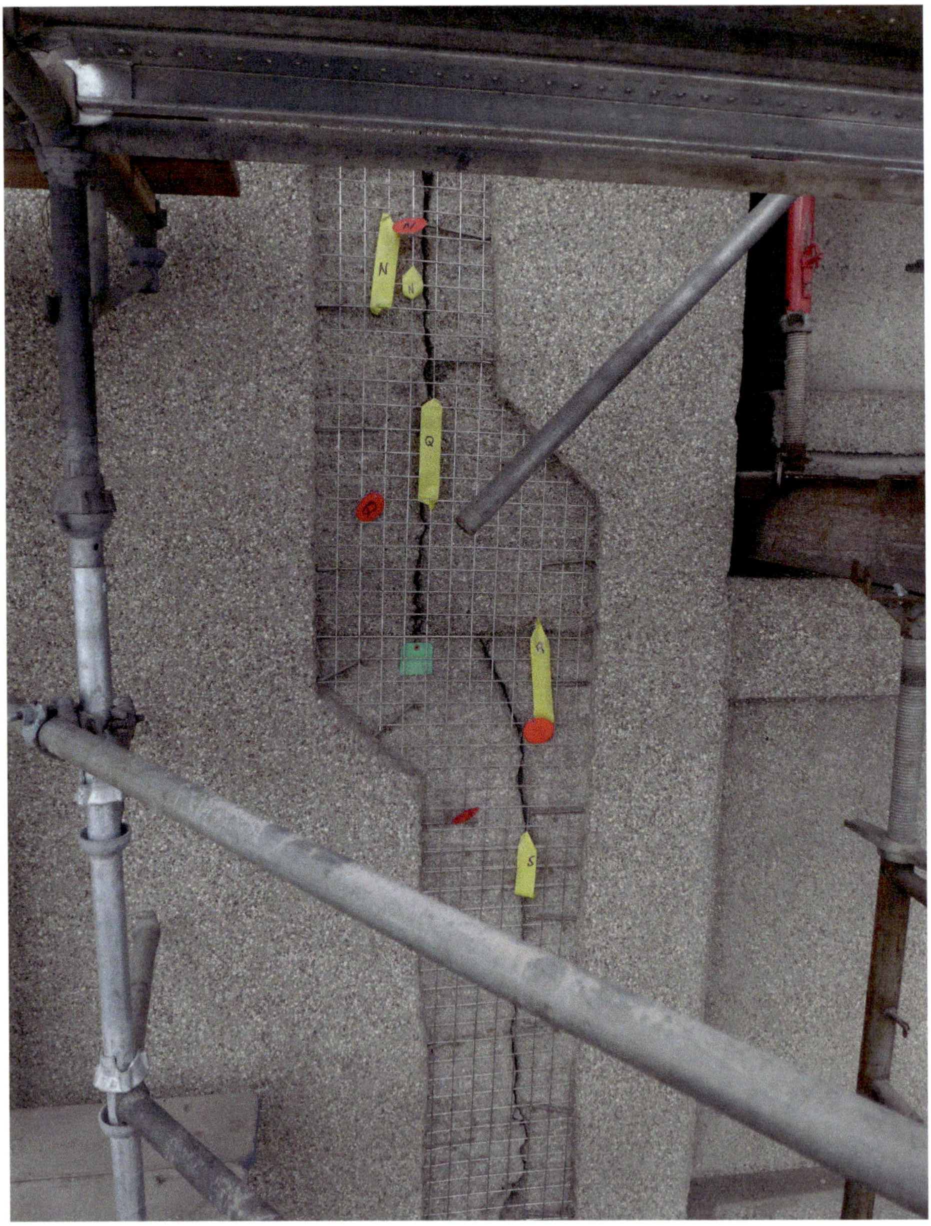

Figure 3.7.3 Unity Temple – concrete patch at Unity Temple. (Photo courtesy of Harboe Architects, 2016.)

The ornamental precast columns are protected from the weather by substantial overhangs and remained in good condition They were not covered with shotcrete in 1973 but did receive a penetrating silane sealer in 2001 and a water repellant, Wacker BS crème, in 2015–2017. This Wacker repellant reacts with the cement binder to form a polymeric silicone resin that reduces water absorption but does not block pores or reduce the permeability.

Baha'i Temple

Materials and Construction of Baha'i Temple

Phase 1 construction included a complete weatherproof structure, using reinforced cast-in-place concrete for the foundation and the lower, middle, and upper galleries, with windows set into the concrete. The top tier gallery cornice, often referred to as the "crown", was capped with a steel truss structure with secondary trusses enclosed by a skylight of aluminum frame and wire glass. The Phase 2 precast cladding of the dome was offset about 5 feet from the skylight to accommodate a catwalk for maintenance.

The Phase 2 cladding was a puzzle-like combination of precast panels and cast-in-place reinforced concrete. Precast sections shipped from Earley's workshops were hung off the structure except at the piers, base of the building, and base of the crown, where precast panels were set into cast-in-place concrete. Concrete was cast on site into detailed formwork based on Bourgeois' designs or used to backfill filigreed precast panels as thin as 2½ inch.

The largest difference between Phase 1 and Phase 2 concrete was the type of aggregate. The Phase 1 structural concrete was mixed with local sand and limestone aggregates. The Phase 2 exterior cladding used custom ground quartz sand and quartz aggregates to achieve a luminous appearance and exceptional durability. The aggregates were carefully selected, ground, and graded to specific proportions so they would pack well and minimize binder, provide a dense, strong concrete and maximize the appearance of the quartz on the surface. Earley experimented with a variety of grinding machines to achieve a coarse cubical or spherical shape that added strength and maximized the desired sparkle.

Equally important were the processes developed to cast and cure the precast concrete. The panels varied in thickness from 1 to 3 inches. Galvanized steel reinforcing bars were bent to follow the shape of the panels and set an inch below the surface in the thicker bands of the precast. Exposed anchorage steel was made of corrosion resistant "copper bearing steel" and anchorage bolts were stainless steel. Earley used a wet mix but minimized shrinkage by hand tamping small amounts of a wet mix into the molds, striking the form to settle the aggregate, and using burlap to draw excess water out of the concrete. When the first test panel cracked, Earley reduced the size of the sand by 0.0015 inch to reduce the total surface area of the aggregate. This reduced surface tension and increased the capillary action needed to draw moisture out of the concrete mix as it cured.

The precast panels were less likely to crack because they were removed from the mold 16 to 18 hours after casting and not constrained while curing. The exposed surface was immediately brushed by "all hands" in the workshop to remove surface cement and sand, and then the exposed aggregate was cleaned with a final rinse of dilute acid and water.

Detailing

Despite the durable concrete, extensive damage occurred where defective flashing detailing directed water into the joints between the precast and underlying structural concrete. The most widespread damage occurred to the crown at the base of the dome and at the monumental stairs. However, there was also damage at the first story cornice and at the hollow caps on top of the piers. Subsequent repairs at each of these locations have eliminated the underlying source of water, which should prevent future problems.

At the base of the dome, water drained to a flat 3 feet wide copper gutter with only three drains along the entire perimeter that were easily clogged by bird debris and corroded by acidic droppings. The original overflow flashing was covered by Phase 2 precast, which forced overflow into the joint between the precast and the structural concrete, saturating both. The precast initially resisted damage but was displaced by as much as 3 inches when freeze-thaw action deteriorated the structural concrete and sheared off anchors. Extensive renovations were required to replace both precast and cast-in-place structure. The gutter was modified to slope to a series of new drains, one in each of the nine bays.

Lack of drainage was also the culprit in the deterioration of the original composite precast treads and risers installed over foundation hall. Freeze-thaw damage was caused by deicing salts, water trapped below the treads, and aggravated by the difference in movement between the quartz topping and limestone substrate of the precast steps.

Figure 3.7.4 Separation of quartz concrete surface from limestone concrete substrate at precast stair at Baha'i Temple. (Photo by Robert Armbruster, 1997.)

Corrosion occurred in two conditions where the rebars were close to the surface but occurred only recently because of the slow carbonation rate of the quartz concrete. In the first condition, part of the rebar in a precast panel was placed too close to the surface in the mold. The high level of consistency in the Earley Studio meant that this error was repeated in all 27 locations with the same mold. Rebar also corroded where Earley reduced the thickness of a Phase 2 panel to as little as 3/8 inch to offset an error in the layout of the Phase 1 structure.

A 27-year Restoration (1983–2010)

In 1983, Baha'i began a 27-year restoration program by first assembling a restoration team to identify the source of the damage and replicate the original construction materials and process. This included investigating improvements in technology and materials, ensuring that they were compatible with the existing fabric and that new materials and methods could be obtained, repaired, or replicated in the future. A phased plan was implemented in a series of projects over the next 25 years beginning with structural and waterproofing issues.

The new concrete mix had to match the appearance at conditions from small patches to full size precast panels, have low water content to avoid shrinkage cracks and high flowability to fill the intricate formwork. Craftsmen and engineers collaborated to develop a mix of white Portland cement, crushed quartz aggregate, a crushed quartz sand, a clear water reducing additive, and a clear air-entraining additive. A custom retarding agent applied to the molds produced the exposed quartz aggregate finish. Patches were treated to match the adjacent weathered condition, while new panels were left unweathered.

After decades of thermal exposure and bird droppings, almost all the original wire skylight glass was broken but intact, and the aluminum caps were corroded. The wire glass was cleaned and covered with new laminated glass with neoprene gaskets and new aluminum caps. Netting was installed at the rear face of the exterior dome cladding to reduce future intrusion from birds.

New precast steps were fabricated in a workshop nearby over a ten-year period – 2002–2012. A durable membrane was installed over Foundation Hall, and gaps were incorporated to allow expansion and contraction of the treads along with drainage of the water. A hot water snow melt system was incorporated to reduce the need for de-icing salts.

Cleaning and Consolidation

The temple was first cleaned in 1942, but soon covered again with dust, carbon, grease, calcium carbonate, algae, and lichen. By 1969, leaks and pollution had contributed to gypsum deposits. Chemical cleaners removed stains and biological growth but were ineffective against the hard black crust of gypsum and left a yellow residue. It quickly became apparent that sandblasting the gypsum crust destroyed the surface of the concrete and dulled the quartz aggregate.

Dilute acid successfully killed and removed algae but did not prevent its re-growth. In 1989, the temple was cleaned with dilute ammonium bifluoride and gentle water wash to remove biological growth, and then with water misting to remove the gypsum. Cycling a water mist for 5 minutes on, 20 minutes off, for 4 to 6 hours gradually softened the gypsum until it could be washed off with a mild pressure wash. Although successful, the mist cleaning method was labor intensive, and was last used in 1991. Chemical cleaning was repeated in 2006.

Since 1993, the maintenance program has included regular rinsing of the monumental stairs and the more accessible areas of the building. The schedule varies from an annual cleaning of the most accessible first floor to cleaning every four or five years at the least accessible dome.

Figure 3.7.5 Water misting at Baha'i Temple. (Photo by Robert Armbruster, 1989.)

Consolidation of the concrete was considered in 2006 to reduce erosion of the surface and increase resistance to freeze-thaw and acid rain, but tests indicated that it had little penetration and changed the appearance of the concrete. Additional tests determine that silane could reduce the rate of carbonation by 25% without changing the exterior appearance; a silane repellant was applied in 2006–2007 in the hope of extending the life of the concrete, reducing carbonation, and reducing the frequency of cleaning.

Ongoing Maintenance

While Robert Armbruster has retired from a 27-year effort overseeing the restoration and protection of the House of Worship, the Baha'i community continues to take a proactive approach. Their conservation plan includes extensive documentation in a computerized, cross-referenced database. A photogrammetric survey completed in 1990 documents the three-dimensional surface within 1 mm so that details can be replicated in the future.

Spare materials, samples, and customized tools are kept in or near the temple where they will be available for future repairs. There are regular routines for inspection and cleaning. The gutter at the crown is cleaned regularly, with a more aggressive maintenance every four years to inspect joints and rinse all surfaces.

Deterioration and Weather

Deterioration at both temples was caused by environmental conditions but influenced by the character of the materials and by the methods of construction and subsequent repair. Understanding the relationship between these factors can provide insights into the ways that current changes in temperature and rainfall could affect durability in the future.

Deterioration of the exterior concrete at Unity Temple began with cracks but was aggravated by porous concrete and a lack of water management, all of which contributed to carbonation and corrosion. When Unity Temple was built in the early 20th century, designers and builders did not fully understand the impact of weather on carbonation of concrete or the effect of that carbonation on the corrosion of steel. They also underestimated the ability of shrinkage and freeze-thaw cracks to accelerate carbonation and corrosion.

The dense concrete used at Baha'i Temple resisted damage and carbonation except where design flaws saturated the concrete and caused deterioration of the binder, early carbonation, and corrosion. The surface of the concrete was discolored by a combination of biological growth, gypsum, and other atmospheric deposits and chemical stains. Other sources of deterioration were alkali-silica reaction and electrolytic reaction between the aluminum skylight and adjacent steel framing. These conditions were aggravated by excess moisture.

Rain, Heat, and UV Exposure

Although there were no expansion joints, Unity Temple initially experienced only minimal shrinkage cracks due to the use of a well-compacted dry mix. Over time, however, the porous roof slabs absorbed moisture from leaking roofs and interior humidity. The large thermal mass of the concrete walls expanded and contracted with changes in heat, moisture, and solar exposure but was restrained at corners. Over time, the concrete walls gradually developed more cracks, and existing cracks were aggravated by years of repetitive daily and seasonal fluctuations in humidity and temperature and freeze-thaw cycles. When structural cracks did eventually develop in the temple's exterior, they corresponded with the location of earlier shrinkage cracks. At both Baha'i and Unity Temples, thermal movement caused vertical cracking at restrained corners.

Figure 3.7.6 Expansion crack at corner at Baha'i Temple. (Photo by Daniel Thomas, 2021.)

At Baha'i, moisture damage was a result of poor detailing but was aggravated by variations in solar exposure. The east, west, and south elevations experienced greater temperature fluctuations and more cycles of freeze-thaw because those surfaces are heated by the sun, while the underlying concrete structure remains cold. Damage appeared first on the east elevations of the crown and lower gallery facias, while there was less damage at the north elevation.

UV damage is also seen in the drying and deterioration of the bituminous skylight glazing at Baha'i Temple. Although it was not reported, it is likely that UV also contributed to the deterioration of the skylights at Unity Temple.

Wind

The wind has had no significant impact at Unity Temple although some erosion is visible on the original precast columns. However, Baha'i Temple has lost as much as 3/8" of binder due to a combination of freeze-thaw damage and erosion by wind and rain. Deterioration is most severe on the east facades that experience both high winds and daily solar fluctuations but little loss of binder on surfaces that are protected from the weather.

Figure 3.7.7 Concrete aggregate is more exposed by weathering above sill than below. Baha'i Temple. (Photo by Daniel Thomas, 2022.)

Biological

Photographs from 1930 show the concrete walls of Unity Temple covered with climbing plants and ivy (Siry 1996, 133). There is no historical reference to damage attributed to this vegetation; however, roots of climbing plants can create and enlarge cracks; foliage can hold moisture against the surface and block drying rays of the sun.

Figure 3.7.8 Ivy growth on Unity Temple. (Photographer Unknown, c. 1920, photo courtesy of UTRF.)

At Baha'i, algae, moss, and fungus did not adhere to the hard quartz aggregate but quickly formed on the sand and cement binder. Over time, the spots developed into black streaks and areas of "crud", especially on wet north elevations. The ornate crevasses provided opportunities for birds to perch, and acidic bird droppings contributed to the corrosion of the concrete, copper gutters, and aluminum skylight. Although provisions were made for access, drains and gutters were not cleaned often enough to prevent accumulation of nest materials and windblown debris.

Carbonation and Corrosion

Carbonation and corrosion damage at Unity Temple was limited during the first half century of its life and slowed by coatings applied in the 1960s and 1970s. However, by 1987, the roof slabs were fully carbonated, resulting in poor compressive strength. Future carbonation of the walls can be mitigated by careful attention to crack repairs, but the combination of porous concrete and increasing heat and humidity is likely to expedite carbonation.

Fortunately, the depth of carbonation in the precast at Baha'i Temple was ¼ inch or less when tested in 2008. Nonetheless, corrosion occurred where moisture was high (as at gutter or roof leaks). The recent renovations that eliminated sources of water leaks and applications of silane should ensure that carbonation continues to move at a slow rate, but future vigilance is necessary to meet the goal of a 1,000-year life.

Figure 3.7.9 Corrosion Damage at Baha'i Temple. (Photo by Daniel Thomas, 2022.)

Chemical Damage and Atmospheric Pollution

Tests have indicated that the slabs at Unity Temple contain significant levels of chlorides. Chlorides are often associated with cinder concrete and can accelerate corrosion. Pyrites, which create local corrosion, were found in the bird's eye aggregate.

At Baha'i, chemical attack has been from external rather than internal sources. The brilliant white surface of the temple was covered with dust by a dust storm in 1934, before it was even complete. Gypsum deposits continue to form on the underside of the soffits of each gallery, especially where leaks have carried calcium carbonate deposits to the surface and edges alternate between being wet and dry. Rain washes gypsum deposits off some surfaces and concentrates them on crevasses, horizontal surfaces, and porous areas. It is very difficult to remove gypsum deposits once they have dried. Gypsum deposits are not just cosmetic; they destroy the integrity of the concrete binder and can trap water behind the crust, causing additional deterioration and spalling of the surface.

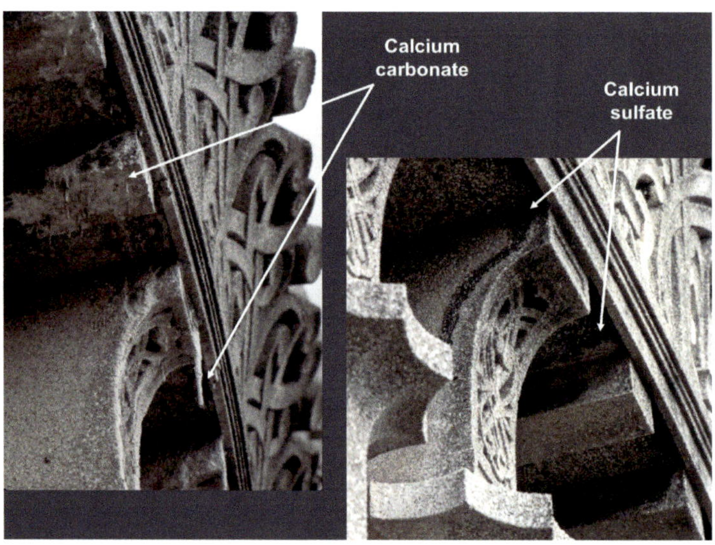

Figure 3.7.10 Calcium carbonate and calcium sulfate (gypsum) deposits at Baha'i Temple. (Photo by Robert Armbruster, 1987.)

Projected Impacts

The climate in Chicago is expected to gradually become warmer and more humid, with an increase in severe weather events, including heat waves and intense rainstorms. The increase in overall winter temperatures will increase the number of freeze-thaw cycles.

Severe weather events, wet-dry and freeze-thaw cycles increase water saturation of porous materials and cause sudden changes in surface temperatures, all of which contribute to the expansion and contraction, cracking and erosion of surface binders and aggregates. Cracks, combined with increased heat and humidity, can accelerate the rate of carbonation and corrosion.

Higher humidity, more frequent rain events, and warmer temperatures are likely to increase the rate of biological growth. Frequent cleaning may accelerate biological growth by eroding the surface, making it easier for bacteria, algae, and fungi to bond.

The global reduction of pollutants in the air has slowed the formation of gypsum in recent decades, and more intense rainfall might be expected to wash areas where gypsum can accumulate. However, weather and leaks that bring calcite to the surface will continue to form gypsum.

Conclusions

Both Unity and Baha'i Temples have invested millions into comprehensive renovations to improve water management and repair damage. However, concrete is not impervious to the weather, and it will continue to weather and naturally carbonate. Both institutions are now faced with the challenge of continuing a high level of surveillance, maintenance, and repair.

Figure 3.7.11 Delaminated shotcrete at Unity Temple. (Photo by Daniel Thomas, 2022.)

Although Unity Temple does not have the depth of financial resources available to Baha'i Temple, they share a strong commitment to their history and future. The full breadth of Baha'i Temple's maintenance planning may not be feasible for Unity Temple, but the vulnerable nature of Unity Temple's concrete makes it equally critical that UTRF collaborate with their consultants and funding sources to implement an ongoing maintenance and repair program that addresses ongoing cracking and delamination in a timely manner. Baha'i is a much larger facility but must still address small problems as they arise.

Unity Temple's porous and easily cracked concrete is more susceptible to carbonation than other denser concretes, but carbonation will continue at both locations and only be accelerated by projected climate changes. Both Baha'i and Unity Temples may eventually need to consider options to decarbonize concrete through cathodic protection or chemical treatment.

The uncertainty of future climate conditions makes it impossible to predict future needs. However, all the existing types of deterioration are likely to accelerate. A collaborative and proactive response is required; one that combines historical knowledge with new innovations. Planning must be nimble enough to address small problems as they arise, and forward looking enough to anticipate larger problems such as the continuing carbonation of the concrete.

Bibliography

Armbruster, Robert F. 2022. Interviews of Robert Armbruster, regarding the construction and restoration of Baha'I Temple, by zoom and on site, Wilmette, Illinois.

———. 2011. "Concrete Heritage Conservation – The Baha'i House of Worship." *Construcao Magazine.*

———. 1997. "The Fair Face of Concrete, Conservation and Repair of Exposed Concrete." *DOCOMOMO Preservation Technology Dossier 2: The Fair Face of Concrete: Conservation and Repair of Exposed Concrete. 1997* http://www.armbrusterco.com/Armbruster_Company/Robert_F_Armbruster_Articles_and_Videos_files/Armbruster%20-%20Restoring%20Brilliant%20Ornamentation%20-%20DOCOMOMO%201997.pdf.

Armbruster, Robert F. 2014. "Secrets of John J. Earley's Mosaic Concrete on the Baha'i Temple." "Celebrating 100 Years of the Work by John Joseph Earley and the Earley Studio Symposium, the American Concrete Institute Fall 2014 Convention, Washington, DC https://www.youtube.com/watch?v=buMySF2mHyE https://www.researchgate.net/publication/315090193_Secrets_of_John_J_Earley's_Mosaic_Concrete_on_the_Baha'i_Temple.

Armbruster, Robert F. 1993. "Weathering of Architectural Concrete on the Baha'i House of Worship." Presented at the American Concrete Institute Fall Convention, Minneapolis, Minnesota. http://www.armbrusterco.com/Armbruster_Company/Robert_F_Armbruster_Articles_and_Videos_files/Weathering%20of%20Architectural%20Concrete%20-%20Armbruster%201993.11.pdf.

Armbruster, Robert, and Jack Stecich. 1993. "The Baha'i House of Worship Restoration." *Concrete Repair Bulletin. (September/October issue)* http://www.armbrusterco.com/Armbruster_Company/Robert_F_Armbruster_Articles_and_Videos_files/Armbruster%20-%20Restoration%20of%20Baha%27i%20House%20of%20Worship%20ICRI%20Bulletin%201993.pdf.

Chicago. 2007. *Chapter Two – Climate.* November 7. https://www.chicago.gov/city/en/progs/env/climateaction.html.

Chilsen, Liz. 2018. *Unity Temple.* https://www.landmarks.org/unity-temple/.

CTL. 2000. "Report on an Evaluation of the Concrete Structure of Frank Lloyd Wright's Unity Temple." Performed for Unity Temple Restoration Foundation. Performed by Construction Technology Laboratories, Inc., Skokie, Illinois.

CTL. 2015. "Summary Report for the Concrete Structure of Unity Temple." CTLGroup Project No. 231126, Report to Unity Temple Unitarian Universalist Congregation & Unity Temple Restoration Foundation.

Franek, Dave (2022) Tour by Dave Franek, *UTRF Docent* on site at Unity Temple, Oak Park, Illinois.

Harboe Architects. 2015. "Unity Temple Restoration Package 1 – Exterior." Drawings issued for Permit and bidding. Chicago.

Harboe Architects. 2017. "Unity Temple – Final Outcomes Report." https://www.getty.edu/foundation/pdfs/kim/unity_temple_final_outcomes.pdf

Harboe. 2014. "Unity Temple Restoration-Phase 1 – Report." Basis of Design. Prepared for Unity Temple Restoration Foundation.

Harboe, T. Gunny. 2022. "Restoring Unity Temple." In *Unity Temple. A Good Time Place Reborn*, edited by Patrick F. Cannon, 106–108. Oak Park, Illinois: Unity Temple Restoration Fund.

Joyce, Bob, 2022. Phone interview with Bob Joyce, *President, Quality Restorations, Inc, regarding his work on Unity Temple and Baha'i Temple.*

Kelley, Stephen J. 1991. "The Role of the Engineer in Preservation." *APT Bulletin: The Journal of Preservation Technology* 23 (1): 6–21. https://doi.org/10.2307/1504392.

Koziarz, Jay. 2017. *Curbed Chicago*. June 7. https://chicago.curbed.com/2017/6/7/15751444/frank-lloyd-wright-architecture-unity-temple-restoraton.

———. 2019. *Curbed Chicago*. October 10. https://chicago.curbed.com/2019/10/10/20908624/unity-temple-frank-lloyd-wright-restoration-preservation-award.

MatWeb. 2022. *Wacker Silres BS Creme C Silicone Masonry Water Repellant*. https://www.matweb.com/search/datasheettext.aspx?matguid=83fb74e7fc60419ab6c5a2a592d484da.

McClier. 2006. *Unity Temple Restoration Master Plans*. McClier Preservation Group.

Ruehle, Heide, interview by Susan Pranger. 2022. *Director, UTRF* (December 13).

Seal Green. 2020. *Rust Remover MSDS 2022.pdf*. https://sealgreen.com/content/Rust%20Remover%20MSDS%202020.pdf.

Sika. 2022. *Sika Albitol Concentrate*. https://can.sika.com/content/dam/dms/ca01/9/SikaAlbitolConcentrate-pds.pdf.

Siry, Joseph M. 1996. *Unity Temple*. New York: Cambridge University Press.

Sokol, David M. 2008. *The Noble Room*. Oak Park: Top Five Books, LLC.

UTRF. 2015. "Getty Submittal."

———. 2019. *Unity Temple Restoration*. (May 3). https://www.flwunitytemple.org/unity-temple-restoration/

———. 2020. "Unity Temple Posters."

———. 2022. *History of Unity Temple*. https://www.flwunitytemple.org/history/

W. R. Hasbrouck, AIA. 1970. *National Register Nomination*. National Park Service. https://npgallery.nps.gov/NRHP/SearchResults?view=list.

Whitmore, Bruce W. 1984. *The Dawning Place, the Building of a Temple*. Wilmette: Baha'i Publishing Trust.

Wikipedia. 2022. *Unity Temple*. https://en.wikipedia.org/wiki/Unity_Temple.

Wright, Frank Lloyd. 1906. "Specifications for Unity Temple."

PART 4

Lessons from Abroad

CHAPTER 4.1

Scotland

Lime Past and Present

Introduction

Scotland has a rich heritage of stone and lime construction. Many organizations, including Historic Environment Scotland, the Scottish Lime Centre Trust, and the Ridge, are working to restore the physical fabric in Scotland, and also to uncover and share a deeper understanding of lime and its use. Projects vary from the more prominent Stirling Castle to small residential projects in the towns of Culross and Dunbar.

History, Location and Climate

History

Clay was used as mortar in Scotland when lime or the timber used to burn lime were not available or too costly. Lime was used on high status churches and castles as early as the 12th century. As the lime industry grew and lime became more available, clay bedding was used less often, or was protected with lime pointing mortar or lime render. Shells, rather than limestone, were burnt to produce quicklime in coastal communities.

Render highlighted the original architectural character of buildings by highlighting the windows, doors, cornices, and other details rather than the joints of the individual stones. However, in the Victorian era, the rustic look of exposed stone masonry was popularized, and the lime renders were either not restored or actively removed at the end of the 19th century. Restoration of lime coatings is now preferred for reasons of both authenticity and durability, but the education and craftsmen required to restore and maintain lime render are lagging behind the demand.

DOI: 10.4324/9781003160489-23

Location and Climate

Scotland is located between the Atlantic Ocean and the North Sea, at the north end of Great Britain, and is classified by as Koppen subtype "Cfb Marine West Coast Climate". Temperatures are moderate due to the influence of the sea, with cool summers and mild winters. There are few extremes in temperature or rainfall. The temperature varies from 59 to 60°F in August to 39 to 40°F in December–January across the county.

Rainfall, however, varies more widely across the country, due to the influence of southwest winds and the North Atlantic Drift. Mountains also create a rain shadow effect that reduces rainfall in the east. Annual rainfall varies from 118 inches per year in the western highlands to only 31 inches on the east coast, but it is typically frequent and of low intensity. Stirling sees an average of 147 days of rain per year, compared to 115 days in Dunbar. Thunderstorms are rare, but gales and high winds can occur in the winter.

The weather in Scotland is already changing, with a year-round increase in average temperatures, dryer summers, and wetter winters. Previously rare intense rainfall events will increase in the summer and winter. Frost and snowfall will be reduced, although this can increase cycles of freeze-thaw. The increase in total annual rainfall differs by region, with a 24% increase in the east and a 45% increase in the west.

Training

There are many organizations, charities, and craftsmen using or providing training in traditional lime in Scotland.

Historic Environment Scotland (HES) is a non-departmental government organization with charitable status responsible for the care of 300 historic properties. They provide financial support and advice, training, and research on conservation and climate change to individuals and organizations. A wide range of publications are available, as well as apprenticeships, a Conservation Summer Camp, skills training centers in Stirling and Elgin and other opportunities to learn conservation skills.

The Scottish Lime Centre Trust (SLCT) was established in 1994 to provide practical advice, training, and education in the use of traditional lime, based on the premise that it was not practical to promote the use of traditional lime mortar and finishes if there were no craftsmen or contractors competent to do the work. A large percentage of their work is troubleshooting to identify the cause of stone or lime failures. They operate out of a small workshop in Charlestown, Fife, Scotland, and are supported primarily by revenue from training and providing building assessment and advice. The remaining revenue is from small grants, such as one from HES to support scholarships for training or educational programs for schoolchildren.

The Ridge Foundation is part construction company and part social service program in Dunbar, a small town on the east coast of Scotland. The program restores historic properties in Dunbar and provides technical assistance to property owners while offering training and job opportunities for local youth, almost all from disadvantaged backgrounds. Over the past 10 years, staff, youth, and volunteers have developed expertise by testing and experimenting with both traditional techniques and new materials. In addition to advising other property owners, the Ridge has acquired and is in the process of restoring the buildings of Black Bull Close. They have gained a more nuanced understanding of the materials and simultaneously reduced costs by doing as much material preparation "in-house" as possible; for example, slaking their own lime. The Ridge Foundation is part of the larger charitable organization, the Ridge SCIO (Scottish Charitable Incorporated Organization) that provides social support, financial training, and education, access to gardening programs, and holistic support for those with addiction and other social issues.

Materials

Traditional Stone Construction

Traditional stone walls in Scotland were approximately 2 feet thick, load-bearing stone with earth mortar, bonded as if dry laid. A small percentage of lime might be added to the mix. Early renders successfully used sharp, angular local sand, often beach sand that would be considered inappropriate today.

Lime mortars were originally used only at vulnerable corners, window, and door openings and for lime render or limewash. As it became more available, lime was used to point or replace clay bedding mortar. If rendering the entire surface was not feasible, the pointing mortar was struck flush with the face of the masonry. By the 19th century, buildings in cities used lime exclusively.

Figure 4.1.1 Sandstone hard cement mortar and lichen, Dunbar, Scotland. (Photo by Daniel Thomas, 2022.)

Renders of lime and sand were generally gray or the color of the local sand. Limewashes were white or colored with earth pigments. Yellow ocher and burnt sienna produced the color "king's gold" used at the Great Hall at Stirling Castle. Iron sulfate (copperas) was used to produce a reddish limewash, like that used at the Palace (manor home) in Culross. The mixture was green when wet but turned orange-red as it dried and oxidized (essentially becoming rust). Green and blue pigments were not used because they were not stable in lime.

Lime

Several projects in Scotland have experimented with the use of "hot lime", the traditional method of preparing lime mortars. Mortar mixes were applied while the hydrated mix was still hot or shortly after it had cooled. Hot limes pull moisture out of the existing masonry or plaster, resulting in a mottled finish as it dries to the interior. If the surface is too dry, failure or delamination can occur. If the underlying surface is too wet, drying and carbonation can be delayed, increasing the risk of frost damage.

Delamination of harling has been associated with using mixes that are too hard and impervious, too thick, or used too close to the ground.

Some, but not all, organizations modify traditional mortar mixes in response to the climate. SLCT uses a range of additives to improve performance:

- Air entrainers, for better freeze/thaw resistance
- Water retainers, to ensure full hydration
- Water repellants to reduce water without blocking vapor.

Projects

Stirling Castle

Stirling Castle was constructed on a dolerite outcropping in the 15th and 16th centuries, with a few structures from the 14th century and the outermost defenses dating to the early 18th century. The royal castle has always been under the control of the state and is now managed by HES. The Renaissance-influenced Great Hall was completed in 1503 and was modified for military purposes from the 18th century until 1964.

Most, if not all, of the buildings would have originally been covered with a lime render, tinted to a yellow-orange that became known as "king's gold". By the end of the 19th century, the render had either worn away or been removed to reflect the Victorian taste for exposed stonework. Restoration of the Great Hall, completed in 1999, included replacement of the hammerbeam roof and parapet and a controversial restoration of the exterior lime and colored whitewash. SLCT had discovered evidence of the original lime render and color and developed a lime mix and color to match. The restored finishes are a mixture of harled lime, pressed back harl, and limewash applied directly to ashlar stonework. The surrounding community, who were accustomed to the exposed stone, were not initially happy with either the lime coating or the yellow color, but quickly came to appreciate it as more authentic.

Figure 4.1.2 The Great Hall at Stirling Castle, Scotland. (Photo by Daniel Thomas, 2022.)

More than 20 years after the restoration, only one elevation was redone in c. 2011. The impact of the gold render is still dramatic, but closer inspection reveals substantial deterioration. This is not entirely unexpected given the exposed location, a largely unheated building, and the lack of substantive maintenance. The cost of scaffolding and the reluctance to remove the hall from public view both make the prospect of renewing the lime wash unlikely, even though frequent renewal is more cost effective in the long run. If the limewash is renewed frequently in the first decade, the time between subsequent coatings can be gradually increased. Renewing the limewash protects the underlying stucco coating.

Figure 4.1.3 Stirling Castle: Great Hall deterioration. (Photo by Daniel Thomas, 2022.)

The Lime Kiln at Charlestown

In 1760, four large "draw" kilns were constructed at Charlestown to produce lime on a commercial scale. Eventually, another 10 kilns were added, each with the latest technical improvements. The complex included limekilns, a foundry, coal production, and a model town constructed to support the lime works.

The kilns were built of local sandstone against the side of a sandstone cliff and lined with hard-fired refractory brick. The kilns were continuously filled from the top of the cliff with alternating layers of limestone and coal and fired from the bottom. It took 14 days for the lime at the top to burn its way to the bottom. Limestone vaults provided access to draw holes on each side of the kiln, used to pull out burnt lime and control air feeding the fire. The kilns ran continuously, with one of the 14 kilns taken out of production each month so that the lining of refractory brick could be replaced.

The local limestone is very hard, about 80% calcium carbonate, but high in minerals from the fossils of crinoids. It produced a high-quality, highly hydraulic quicklime that was shipped from Charlestown nationally and internationally until the kilns were closed in the 1950s. The kilns were filled with rubble and cement and left to ruin. The Charlestown Limekilns are legally protected as a Scheduled Monument, and privately owned by the Broomhall Estate.

In 2014, HES began a four-year conservation effort to clear vegetation and stabilize the front face of the kiln by repairing and replacing stonework, repointing, replacing selected lintels, and adding protective iron grilles. The project included the development of a maintenance plan, including monitoring and advice from HES to the owner, and the development of connections with the community.

Figure 4.1.4 Charlestown limekiln, Charlestown, Scotland. (Photo by Daniel Thomas, 2022.)

Culross

The National Trust of Scotland owns and maintains 27 historic white-harled, red-tile-roofed houses in Culross, an intact 17th–18th century burgh. Believing that occupied homes are best conserved, the NTS rents the houses to tenants, and conducts an annual maintenance program to check the gas, clean roofs and gutters, and sweep the chimneys. They also provide expandable flood gates for doorways to respond to increased flooding due to climate change without physically altering or attaching to the buildings. The unusual level of care is evident, as is the constant challenge of the coastal environment.

Figure 4.1.5 Biological growth on cottage, Culross, Scotland. (Photo by Daniel Thomas, 2022.)

Most of the houses in Culross are modest white homes with white paint over cement render from the 1960s and 1970s, but the Culross Palace, a "palatial house" built from local sandstone was restored with a traditional harled lime finish.

Figure 4.1.6 Culross Palace, Culross, Scotland. (Photo by Daniel Thomas, 2022.)

Dunbar

Dunbar, located on the coast of the North Sea just south of the Firth of Forth, was designated a royal burgh in 1369. The main street of Dunbar, High Street, runs parallel a few hundred meters from the shore. High Street is now lined with 19th- and 18th-century buildings but retains its medieval layout of "burgage plots," narrow plots perpendicular to the street. The buildings form a continuous street wall of shops with a series of "close," narrow passageways perpendicular to High Street that lead to small residences on either side of the passageways, with enclosed courtyards or gardens beyond.

According to several sources, this configuration of narrow buildings with minimal access has confounded several 20th-century efforts to redevelop High Street and served to preserve remnants of traditional 17th- and 18th-century construction in the side closes of the town. Buildings and walls throughout the town are constructed with a local red sandstone, which is not very durable but workable in any direction. Lime render is gone from most facades and walls facing the streets, but remnants of clay bedding mortar and lime render can be seen on the neglected rear facades.

Figure 4.1.7 Dunbar: Rear wall of 74/76 High Street, Dunbar, Scotland. (Photo by Daniel Thomas, 2022.)

The Ridge began restoring the courtyard walls and ruined buildings at the rear of 92/94 and 96 High Street in 2017. An archeological evaluation in 2018 confirmed that the rear buildings dated to mid-18th to mid-19th century, but included some 17th-century elements, including wood molding, reused ship timbers, and clay-bonded wall footings.

HES sponsored the Ridge's first project, restoration of the rear garden wall, to test the durability of hot lime mortar and the feasibility of working with inexperienced workers. Work was scheduled in three phases over the course of the winter and spring (November 2016–May 2017) to test the materials and workmanship under harsh conditions. Local volunteers, supervised by more experienced craftsmen, experimented with different sands and local shell aggregate, salvaged stone, and quicklime from Cumbria to rebuild the wall. Wood ash, a mild traditional pozzolan, was added to accelerate setting. The mortar performed well except in one section where the existing historic mortar absorbed too much moisture from the new mortar and prevented proper curing. The project successfully matched historical materials and confirmed that inexperienced labor could produce good work when properly trained and supervised. Working through the winter did not affect the quality of the finished work but did slow curing and production.

The second project was a more ambitious restoration of the South Building/ Building 4 (named Roger's Land in recognition of technical and financial support from Roger Curtis and HES) that again served as a training ground for local youth and a place to experiment with traditional crafts and emerging technology. Renovations included reconstruction of the second floor and roof framing, restoration or rebuilding of the lime-rendered sandstone walls and chimneys, and reconstruction of the pantile roof. About 25% of the masonry walls were rebuilt using a mix of salvaged earth mortar and nearby soils. Dimensional red sandstone from a local stone yard was shaped and textured by hand on site. The floor and roof framing and clay pantiles were new. Experimental products included the use of permeable wood fiber insulation (with fire retardant coating) in the floor framing, and insulating hemp plaster covered with a topcoat of lime and sand. Wood windows and linseed oil paint were handmade by the Ridge joiners (carpenters).

The North Building is now being restored with techniques revised to reflect their experience on the South Building. The chimney and end walls are being rebuilt with lime mortar, and then harled with lime mortar and washed with lime wash. The mortar used on the chimney restoration includes salt, which is purported to help the lime carbonate.

Figure 4.1.8 Restored Chimney at North Building 74/76 High Street, Dunbar, Scotland. (Photo by Daniel Thomas, 2022.)

The Fisherman's Monument

The Fisherman's Monument was built in Dunbar's Old Harbor in 1856 to house a barometer for the local fishermen. A 1998 restoration of the Class B listed structure replaced the marble frieze and the barometer. The stonework was restored and repainted in 2012 with mineral paint, but wind and rain quickly destroyed the finish.

It is now being recoated with limewash on an annual basis to prevent additional deterioration, experimenting with different compositions to improve durability. A wash of lime slacked with seawater and applied hot was used annually for 2 years. An additive of Palm Jaggery, a type of sugar commonly added to lime mortar in 19th-century India, is being considered for the next application.

Figure 4.1.9 Fisherman's Monument, Dunbar, Scotland. (Photo by Daniel Thomas, 2022.)

Figure 4.1.10 Lime wash after one year of wear on Fisherman's Monument. (Photo by Daniel Thomas, 2022.)

Conclusions

The call for education, research, and training has been repeated wherever there are efforts to revive traditional materials and techniques. In spite of a centuries-long history of using lime, craftsmen and preservationists in Scotland have also had to reinvent the knowledge and skills required to restore traditional homes and national landmarks.

Their efforts have proven that lime mortars and renders can be durable in harsh climates, given appropriate materials and execution. However, lime is a temperamental material susceptible to weather conditions. Successful restoration requires that craftsmen understand the impact of temperature and humidity while performing their work and the impact of future weather conditions. Further exploration and research are required to increase to ensure a level of durability that responds to the reality of a more intense climate and the practical constraints of labor and material costs.

Bibliography

Addyman, Tom, and Tim Meek. 2019. "HES Technical Paper 31, Historic External Lime Finishes in Scotland." Historic Environment Scotland.

Artis, Roz, interview by Susan Pranger. 2022. *Scottish Lime Centre Trust*.

BARR. 1993. *Black Bull's Close, 72–74 High Street*. http://canmore.org.uk/event/891093.

Curtis, Roger. 2018. *Refurbishment Case Study 26, Black Bull Close*. Historic Environment Scotland.

Docent, interview by Susan Pranger. 2022. *Stirling Castle* (September).

Dunbar Harbour Trust. 2022. *Fishermen's Monument*. https://dunbarharbourtrust.co.uk/.

Fotheringham, Mitchell, interview by Susan Pranger. 2022. *Charlestown Kilns*.

Frew, Craig. 2018. *Technical Paper 29, Review of Hot-Mixed External Lime Coatings in Scotland 1997–2016*. Historic Environment Scotland.

HES. 2019. *Buildings to rear, 72 and 74, High Street, Dunbar*. https://www.buildingsatrisk.org.uk/details/891093.

HES (2014). *Scotland's Environment*. Retrieved 2023, from Environment Scotland: http://www.environment.scotland.gov.uk/get-informed/air/air-quality/

HES. 2004. *Statement of Significance Stirling Castle*. Historic Environment Scotland.

IFLI. 2018. *A5.1 The Chalestown Limekilns*. End of Project Report, Inner Forth Landscape Initiative. https://www.innerforthlandscape.co.uk/.

Met. 2014. *How will Scotland's climate change in future?* Environment Scotland. https://www.environment.gov.scot/our-environment/climate/changing-climate/#:~:text=How%20will%20Scotland's%20climate%20change,will%20be%20warmer%20and%20drier.

Morrison, J. 2019. "Black Bull Close, Dunbar, Evaluation." *Discovery Excab Scot, New*. http://canmore.org.uk/event/1104713.

Parker, Geoffrey N. 2022. *Battle of Dunbar*. https://www.britannica.com/event/Battle-of-Dunbar.

Pugh, R. J. M. 2018. *Dunbar, The High Street Closes*. self published.

Ridge. 2022. *The Ridge Training*. https://the-ridge.org.uk/.

Sked, Phil and David Learmont. 2022. *Culross*. Edited by Hilary Horrocks. Edinburgh: National Trust for Scotland.

Scottish Lime Center Trust. 2019. "25 Years of Excellence 1994-2019, Scottish Lime Center Trust." Booklet, Charlestown, Fife, Scotland. www.scotlime.org.

Stewart, Cara. 2022. *Preserving the traditional buildings of Culross*. https://www.nts.org.uk/stories/preserving-the-traditional-buildings-of-culross.

2023. *Stirling Castle*. https://www.undiscoveredscotland.co.uk/.

Stockdale, Andy. 2022. Interview with Andy Stockdale of The Ridge on site at Black Bull Close by Susan Pranger.

Undiscovered Scotland. 2023. *Stirling Castle*. https://www.undiscoveredscotland.co.uk/.

weatherbase. 2023. *Dunbar; Stirling*. https://www.weatherbase.com/.

Yeoman, Peter. 2018. *Stirling Castle, Official Souvenir Guide*. Edinburgh: Historic Environment Scotland.

CHAPTER 4.2

Wood Coatings in Norway

Introduction

The availability of good pine and the ease of construction made wood the primary building material in Norway; however, it must still be protected from the weather and related biological decay. In lieu of the lime and linseed coatings used in more southern climates, Norwegians used the materials at hand: tar coatings derived from pine stumps, cod liver oil along the coast, and paints made from minerals from slag heaps in mining communities. These coatings were labor intensive to produce and had to be renewed frequently but were cost-effective where labor and materials were readily available.

Wood in Norway tended to survive because buildings were maintained as long as they were in use. When they were no longer useful, they were often disassembled. Reuse of original components was common, not only in the stave churches. Stave churches could be disassembled and reassembled with a mix of new and old components.

History, Location and Climate

History

Tar has been used since prehistory to protect wood structures in Norway. Tar production expanded in the 8th century to waterproof the fleets of Viking longships, including their sails and ropes. In the 13th century, tar was used to protect the medieval stave churches that translated the architecture of Europe's stone churches into traditional Norwegian wood, using complex wood arches similar to inverted Viking ships.

 DOI: 10.4324/9781003160489-24

In the 16th century, industrial-scale copper mining began to produce huge piles of iron-rich slag, a byproduct of copper smelting. This led to the development of a mineral-rich flour-based distemper paint. The mines and smelters closed in the mid-20th century, but enormous slag heaps remain, and production of the paint continues on a small scale.

Figure 4.2.1 Heddal Stavkirk. Heddal Stavkirk. Notodden, Norway (Photo by Daniel Thomas, 2022.)

Location and Climate

Norway has a maritime climate, with mild winters and cool summers moderated by the North Atlantic Ocean. The climate varies greatly from east to west as well as north to south, with coastal areas generally wetter but with less variation between heat and cold. Large areas of Norway are classified as Dfc, Cool continental climate/ subarctic climate, but the more heavily populated southern areas are Temperate Continental (Dfb) or Tundra climate (ET). Many of the stave churches are in areas classified as Dfb, with more moderate temperatures and substantial precipitation, where tar coatings were appropriate.

Roros is classified as Dfc, cold with little precipitation, most of it between June and September, and therefore has no tradition of using tar. Average temperatures vary from −12°F in January to 53°F in July, although there is a large variation in extremes. The coldest recorded temperature in January was −50°F, and the warmest was 51°F. In the summer, the temperature has gone as low as 28°F and as high as 83°F.

Climate Projections

Over the past 35 years, temperatures have increased, and winters are shorter. The annual average temperature in Norway increased by almost 2°F since 1900 and is expected to increase by up to 8°F by the end of the 21st century, with the largest increases farther north, inland, and during the winter. Temperatures in the artic areas of Norway are rising twice as quickly as the global mean. Annual precipitation in Norway increased by 20% between 1900 and 2014 and is expected to increase by 30% by 2100, with most of the increase occurring in autumn, winter, and spring.

The combination of higher temperatures and higher precipitation is expected to cause more rain floods, and fewer snowmelt floods. Snow season will almost disappear in lowlands, but still occur in the highlands, where an increase of winter precipitation could be as high as 25% by 2100. Average wind conditions are not expected to change, but there may be an increase in the frequency of high winds.

There are more periods of rapid temperature change and more freeze-thaw fluctuation, increased snow, and periods of winter rain. Summer temperatures have been more extreme, with periods of high heat and drought. The combination of higher temperatures and high humidity is noticeably increasing decay and the growth of brown rot fungi.

Organizations

Norwegian Directorate for Cultural Heritage

In 1978, the Cultural Heritage Act established protections for all cultural assets dated prior to the Reformation, specifically including churches built prior to 1536 CE. The Norwegian Directorate for Cultural Heritage (Riksantikvaren) is responsible for oversight of these assets, including the remaining 28 stave churches. The Directorate requires the use of traditional tar coatings on all remaining stave churches, increasing the demand for the production of kiln-fired tar and for craftsmen who understand how to apply and maintain the coatings.

The Nordic Tar Network

Norway, Sweden, Finland, and Germany have developed successful strategies for producing or collecting and distributing locally produced kiln-fired tar through cooperatives to stave churches.

The Nordic Tar Network is one such clearing house for research and craft.

Rorosmuseet and the Preservation Center

Rorosmuseet (the Roros Museum) owns and manages the Sylthytta World Heritage center and museum, the Olav Copper works, the Preservation Center workshops, and 80 historic buildings. The Preservation Center maintains the buildings owned by the museum, advises private owners, and occasionally works on private historic projects. The center also provides education, resources, and technical expertise on the traditional construction and maintenance required to maintain the town and similar buildings, including training courses during their annual Craftdays.

Craftdays is an annual event with a dual purpose: to educate the community about the resources and activities of the Preservation Center, and to provide specific education and hands-on training in preservation crafts. Training is led by Preservation Center staff and national experts; participants include curators and staff of cultural assets and museums, craftsmen in related or other fields, and students and homeowners seeking to learn more about traditional buildings. Experts and novices work together to explore and share expertise regarding techniques, educational resources, preservation approaches, climate, and related challenges. The 2022 Craftsdays included multi-day training on traditional masonry, paint coatings, carpentry, and metal forging.

Projects

Stave Churches

Norway's medieval churches mimicked the form of stone churches in Europe but were constructed of wood protected with tar. Over 1,000 Catholic stave churches were built in the 12th–15th century, but they all became property of the king during the Reformation in 1536. Many were sold and fell into disrepair, in spite of maintenance efforts by owners and other locals. As of 2023, only 28 stave churches remain of the original 1,000 stave churches in Norway.

Industrially produced tar is good quality, but it is not allowed on the stave churches because it does not have the same color, consistency, or smell as traditional kiln-fired tar. Traditional tar is formed by the dry distillation of pine wood, usually pine stumps in a carefully constructed kiln. Dry distillation requires carefully controlled heating with limited air access so that liquid terpenes and resin are excreted and collected without combusting. Chunks of stump wood ("fat wood" rich in oil) are tightly stacked, covered first with firewood and then with peat to hold in the heat and limit access to oxygen. The trick is to burn the firewood so that the heap gets hot enough for the sap in the stump wood to flow, without having the stump wood or tar catch fire. The tar is drained to the center of the heap, then drained away in a wood pipe. The tar is initially light-colored, but gets darker and thicker as the wood burns.

The tar is boiled at about 108°F to reach the correct consistency and applied in thin layers. Correctly applied and cured, several thin layers will harden into a thick, hard, often knubby, coating that can survive indefinitely. The tar must adhere to the wood, but not be absorbed, to cure on the surface. If it is applied too thickly, or gets too hot before it cures, it will literally run off. A too thin coating will be absorbed by the wood but not protect the surface. Properly cured, the tar will become hard and eventually brittle. Crushed coal or charcoal, a by-product of producing tar, was sprinkled onto the soft, tacky surface to protect it from UV light.

Temperature and humidity affect durability and hardness of tar coatings, so they are usually applied in the autumn. Tar coatings on south and east exposures and roofs are more rapidly deteriorated by UV exposure or melted by solar heat.

Figure 4.2.2 Hard cured tar has a knubby appearance similar to charred wood, Heddal Stavkirk. (Photo by Daniel Thomas, 2022.)

Heddal Stave Church

Heddal Stavkirke, Norway's largest stave church, was erected in the 13th century, but some ore-pine timber dates to approximately 1000 CE. The church fell into disrepair after being sold in 1723. A reconstruction in 1849 replaced most of the original materials and was not historically accurate but did restore the church to active use. From 1952 to 1954, the church was completely disassembled and restored to a more historically accurate configuration based on historical records, guided by details from comparable medieval stave churches. Corner staves and plank walls were reused, but sleepers, stave beds, and most of the roof were replaced due to moisture and fungal damage. Perimeter drains were set in gravel to reduce splashing and carry water away from the foundations to a drywell.

The exterior cladding is local pine, coated with traditional kiln-fired pine tar. Except for the original door carvings, all of the cladding was replaced with machine made pine shingles and galvanized nails in the 1950s. The shingles on the south roof were replaced in the 1950s with machine sawed shingles, and shingles on the south gallery roof were replaced in 1998 and 2002 with more authentic hand-split wood shingles attached with wood nails hand split from fresh, but not green, heartwood of local pine.

The shingles wear more heavily where they are exposed to rain and sun. The south walls and main roof were black when tarred two years ago but are now brown, with a loose but protective layer that is easily worn by abrasion, including by people touching it. The roof is still black, but the tar temporarily softens for a short period in the spring and runs off on hot sunny days. Loose tar particles are washed away by rain.

Figure 4.2.3 Loss of tar on vertical surface at Heddal Stavkirk. (Photo by Daniel Thomas, 2022.)

Garmo Stavkirke

The Garmo stave church was built in the early 13th century, extended several times through 1730, and disassembled in 1880. It was reassembled in 1921, largely from

17th- and 18th-century components, at the Maihaugen museum in Lillehammer, Norway.

It was not possible to establish the dates of shingle and tar replacement during the author's visit in August 2022; however, the existing conditions very clearly illustrate the impact of different solar exposures. The tar on the southern exposures is excessively worn. The north roof and walls are older, but in good condition with a hard tar coating, except where concentrated water run-off from a valley had worn the tar coating.

Figure 4.2.4 (left) Tar at SE corner, (right) tar at NW corner, Garmo Stavkirke, Lillehammer, Norway. (Photos by Daniel Thomas, 2022.)

Ringebu Stavkirke

Ringebu was built in 1220 on the site of an earlier 12th-century pole church. The external gallery, side aisles, and apse were removed after the reformation, and north and south transept wings, a choir, and a distinctive red tower were added in 1631. It is actively used as a parish church.

Although the history of recent work was not available, observations by the author in 2022 confirmed the impact of solar orientation. Tar on the south exposure had been recently replaced; the north roof and wall were heavily coated with an older but very hard tar in good condition. The coating on the northeast was thinner. Areas directly below roof overhangs were generally in better condition.

Figure 4.2.5 Roof overhangs protect tar coatings at Ringebu Stavkirke, Ringebu, Norway. (Photo by Daniel Thomas, 2022.)

The Mining Town of Roros

In 1646, the King of Denmark and Norway granted the Roros Copper Works a lease that included control of all timber, ore, water, and people within a circumference of four Norwegian miles (roughly 25 miles) of the mine. A town was purposely built to support the mine and smelt works, and unofficial workers built small log houses south of the smelting facility. The town was rebuilt in 1679 after it was destroyed by Sweden's attempt to disrupt Norway's source of copper. Zoning in 1970 began to regulate street patterns and the types of repairs and materials to preserve the character of the town. Roros became a UNESCO World Heritage site in 1980. 80 buildings are individually listed under the Cultural Heritage Act.

About 60% of the residential structures are privately owned, with the rest owned by the Foundation Rorosmuseet. Although the mine closed in the 1950s, the town has maintained its total population by shifting to tourism, manufacturing of textiles and furniture, and other activities.

Figure 4.2.6 Town of Roros with slag heap in foreground. Norway (Photo by Susan Pranger, 2022.)

The original houses were constructed with stone foundations, log or heavy timber walls, and sod roofs. Vertical wood paneling was added to many houses to achieve a "Swiss chalet" aesthetic in the mid-19th century, and traditional sod roofs were replaced with slate in the late 19th century. (Borgus, 2020) Both log construction and the later wood panels are maintained with the traditional ore paint, which gives the town its distinctive coloring.

Figure 4.2.7 Traditional paint finishes in Roros. (Photo by Daniel Thomas, 2022.)

A traditional emulsion paint in Norway and other European countries with a mining tradition is made with a water-soluble reddish-brown pigment that is extracted from iron-rich waste slag. The pigment is mixed with rye flour, iron vitriol, and water, then stirred and boiled until it reaches the desired consistency. The permeable paint penetrates and protects the wood, and it wears off rather than peels as it ages. The ore pigment was also added to the linseed oil used to protect the more expensive doors and windows.

Figure 4.2.8 Roros Craftdays: Creating emulsion paint. (Photo by Daniel Thomas, 2022.)

Figure 4.2.9 Roros: Biological growth under overhangs and above water table. (Photo by Susan Pranger, 2022.)

Conclusions

Norway's traditional protective coatings for wood already show accelerated deterioration from solar UV and heat and from abrasion from rain and wind. An increase in UV, heat and rain events will reduce the durability and life of existing coatings and increase the decay of unprotected wood.

An increase in sunny days can lead to an increase in photodegradation. Higher humidity and temperatures can increase biological growth and wood decay. Severe storm events and heavy winds both increase abrasion and erosion of coatings. Walls oriented to the east and south suffer the most photodegradation from solar exposure and heat, but western exposures often experience more wear from wind and storms.

Higher interior temperatures, a lack of insulation and ventilation under traditional sod roofs, increased precipitation, and more fluctuations in winter temperatures are creating conditions where interior heat is melting snow, creating leaks, freeze-thaw damage, ice dams, and other moisture related problems. One response is to make minor modifications to traditional construction, such as adding waterproof membranes and increasing ventilation. Another approach is to increase the maintenance cycle for protective coatings.

Bibliography

Anker, Leif. 2020. *Heddal stave church*. https://snl.no.

Bakosgjelten, Birgit, interview by Susan Pranger. 2022. *Rorosmuseet Preservation Center.*

Brady, Michael. 2020. *Tar truth revealed by Viking Age finds*. https://www.norwegianamerican.com/tar-truth-revealed-by-viking-age-finds/.

Brekke, Arnstein. 2021. "What can we learn from cultural heritage?"

Daugstad, Karoline. n.d. *The Climate and Weather Conditions*. https://verdensarvenroros.no/en.

Egenberg, Inger Marie, and Jens Glastrup. 1999. "Composition of Kiln-Produced Tar."

Foss, Arild S. 2012. *Bronze Age people with a mine of their own*. https://sciencenorway.no/archaeology-forskningno-norway/bronze-age-people-with-a-mine-of-their-own/1373362.

Fox, Felicity. 2019. *Urnes and Norway's Stave Church Preservation Programme*. Cathedral Communications Limited. https://www.buildingconservation.com/articles/urnes-stave-church/urnes-stave-church.html.

Gothenburg University. 2023. *Nordic Tar Network - network about Nordic tar*. https://www.gu.se/hantverkslaboratoriet/nordic-tar-network-natverk-om-nordisk-tjara.

Heddal Stavkyrkje. 2022. *The Stave Church*. https://www.heddalstavkirke.no/om-stavkirken-en/.

Hennius, A. 2018. "Viking Age tar production and outland exploitation." *Antiquity*. https://doi.org/10.15184/aqy.2018.22.

Hennius, Andreas. 2018. "Viking Age tar production and outland exploitation." *Antiquity*. https://doi.org/10.15184/aqy.2018.22.

IEA. 2022. *Norway Climate Resilience Policy Indicator*. https://www.iea.org/articles/norway-climate-resilience-policy-indicator.

Kaste, Anders Moen. 2021. *Norway's wooden stave churches are a demaning heritage to maintain*. https://sciencenorway.no/architecture-church-medieval-history/norways-wooden-stave-churches-are-a-demanding-heritage-to-maintain/1800129.

Kroken, Ingmar, interview by Susan Pranger. 2022. *Heddal Stavechurch* Notodden, Telemark, Norway.

Nagel, Anne-Hilde. 1994. "Norwegian Mining in the Early Modern Period." *GeoJournal* 32 (2). http://www.jstor.org/stable/41146719.

Riksantikvaren. 2021. *Climate strategy for cultural environment management 2021–2030*. https://www.riksantikvaren.no/en/climate-strategy-for-cultural-environment-management-2021-2030/.

Rørosmuseet. n.d. *About the Roros Museum*. https://rorosmuseet.no/.

Stavekyrkjeeigarforum. 2023. *Heddal Stave Church*. https://www.stavechurch.com/heddal-stave-church/?lang=en.

Stoltenberg II Government. 2013. *Climate change adaptation in Norway*. Meld St 33 Report to the Storting (white paper), Norwegian Ministry of Climate and Environment.

Tarlach, Gemma. 2021. *To Save Norway's Stave Churches, Conservators had to relearn a lost art*. https://www.atlasobscura.com/articles/stave-church-tar-conservation.

visitlillihammer. 2023. *Ringebu Stave Church*. April 13. https://www.visitnorway.com/listings/ringebu-stave-church/8016/.

CHAPTER 4.3

Lahore Fort

Khalid Ibrahim and Zahid Usman

Introduction

The ancient city of Lahore, Pakistan, is believed to have been founded in 1000 BCE. The Lahore Fort was originally constructed in mud masonry in 1200 CE. This case study focuses on Kala Burj and Jahangir's bedchamber, constructed in 1633 CE. Previous restorations, variations in temperature and rainfall, and increased threats from chemical and biological agents will likely impact the durability and maintenance of burnt bricks, plaster coatings, marble, and sandstone.

Figure 4.3.1 Jahangir's bedchamber view from Southwest. (Photo by Ibrahim and Usman, 2023.)

DOI: 10.4324/9781003160489-25

History, Location and Climate

History

Lahore is an ancient city and the second-largest city in Pakistan. The old maps of Lahore show a water channel flowing near the boundary of Lahore Fort and the Badshahi Mosque called Budha Darya or Old River by the locals. This river used to flow along the city walls, making it a logical defense against invaders from the north. It is believed that the Hindu prince Loh, son of Rama Chandra, laid the city's foundation in 1000 BCE. The city has witnessed prosperity, invaders and yet survived after the invasion of Mahmood of Ghazni in 1000 A.D., the sultanate period (1206–1524), the Mughals (1524–1712), the Sikh regime (1764–1849), and the British captivity (1849–1947). It ultimately became a provincial capital after the independence.

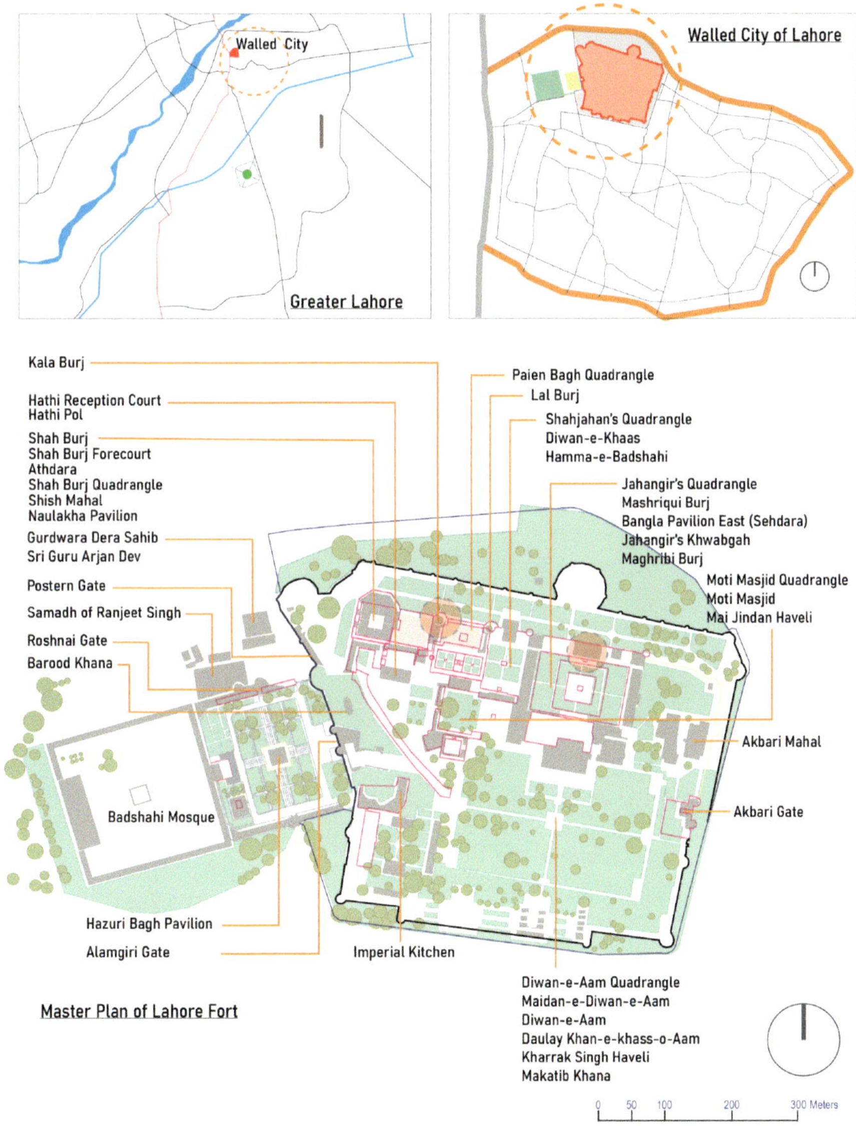

Figure 4.3.2 Master Plan of Lahore Fort. (Diagram by Ibrahim and Usman, 2023.)

The Lahore Fort, locally referred to as Shahi Qila, is a citadel in Lahore located at the northwestern corner of the walled city of Lahore. The exact date of construction of the Fort is not certain; however, archaeological evidence indicates that it was built long before 1025 CE. Some scholars attribute the footprint of the Fort to Mahmud of Ghazni when he conquered the city. The fort depicts a consolidated history of Mughal architecture and was declared a world heritage site by UNESCO in 1981. Lahore Fort is a large complex with an irregular plan spanning 1400′ × 1200′, with strongly fortified brick walls and the same formal organization as the c. 1565 CE Agra Fort in India. Jahangir's Quadrangle was constructed in 1622 and Kala Burj in 1618 by Emperor Jahangir. The fort was severely damaged by the bombardments in 1841 and 1842 in Sikh battles just before the British occupation of 1849, which continued until 1923, when it was handed over to the Archaeological Survey of India. After the partition of the sub-continent in 1947, its control was transferred to the Department of Archaeology, and remained under their control till 2014. The first conservation master plan of the Fort was developed by the Department of Archaeology in 1973, and the second by UNESCO in 2005. Since 2014, a comprehensive master plan and conservation of all the buildings in the Fort have been carried out by Walled City of Lahore Authority.

Location and Climate

Lahore Fort is situated near the banks of the Ravi River on the northern boundary of the walled city of Lahore. Pakistan lies in a subtropical region primarily at 30° North and 70° East with an esoteric mixture of geographical features and landscape. Pakistan is a long corridor between Central Asia in the east, India in the west, the Himalayas in the north, and the Arabian Sea in the south, overlapping with the Eurasian and Indian tectonic plates. It consists of the Indus River plains, the Thar desert, the Balochistan plateau, and the northern highlands. Climatically, it is situated in the temperate zone, just above the tropic of Cancer.

Lahore lies 217 m above sea level between 31°N and 74°E and remains hot and dry for most of the year. The winds are hot and full of dust during the daytime and cooler at night. Its solar irradiance ranges from 5.5 kWh/m^2 at an annual average. The city receives an average of 8.5 sunshine hours per day annually, which is one reason for its semi-arid climate.

Lahore experiences five seasons of semi-arid climate: dry autumn, foggy winter, pleasant spring, summer with dust storms, rainstorms, heatwaves, and rainy monsoon season (Köppen climate classification BSh). Lahore experiences hot and dry weather during March and April, peaking with hot, dry winds from May through June. The region receives monsoon rains from July to September, increasing relative humidity from 25% to more than 60%. After the end of the monsoon season, the weather remains warm and dry during October and November, which causes a lot of dust and haziness. The region experiences pleasant, cool, and sunny weather from December to February, with a peak high of 20°C and a low of 6°C.

Changing Patterns in Climate

According to the ADB report on the climate change profile of Pakistan, the temperature has increased by 0.5°C in the past 50 years, with a fivefold rise in heatwave days and a slight increase in precipitation. The report also shows a 0.97°C rise in winter temperature and a 0.22°C rise in summers, ranking it as one of the top ten countries on the climate risk index. It is also predicted that by the end of this century, the temperature will rise from 3°C to 5°C, including inter-annual variability in the precipitation pattern. Following are the key findings of past observed trends of climate change indicators and future projections for Pakistan.

During the last century, Pakistan's average annual temperature increased by 0.57°C compared to 0.75°C for South Asia, and average annual precipitation increased by 25%. The warming is mainly due to an increase in winter temperature. Heat wave days per year increased by 31 days from 1980 to 2007. Cold waves decreased in the northeastern and southern parts and increased in the western and northwestern parts of the country.

During 1960–2007, the following changes were noted:

- An increase of 0.6–1.0°C in the mean temperature over the hyper-arid plains, arid coastal areas, and mountains regions of Pakistan;
- A decrease of 10–15% in winter and summer rainfall in the arid plains and coastal areas;
- A rise of 18–32% in the summer rainfall over the core monsoon region of Pakistan;
- A decrease of 3–5% in cloud cover over central parts of Pakistan and a consequent increase of 0.9°C in temperature;
- A decrease of 17–64% in rainfall was observed during the seven strong El Niño events in the last 100 years.

Future Projections

Climate change is believed to bring to Lahore an increase in average temperature, primarily affecting winter temperatures and increasing summer heat waves. While there is no systematic increase in precipitation, variations in precipitation are expected, with a decrease in precipitation and increased aridity in some areas. Lahore, located in the northeast monsoon area, is expected to experience an increase in winter temperatures and summer rain and storms. Apart from the temperature, precipitation, changing cycles of various kinds of weather, and increased heatwave days, environmental pollution is also a concern. Lahore is not excluded from environmental pollution with the additional phenomenon of smog.

Construction Techniques and Materials

Lahore Fort

Major building materials used in the construction of the Lahore Fort include burnt brick masonry, lime plaster, red sandstone, white marble, glazed tiles, mirror inlay, precious stone inlay, wooden rafters, beams, purlins, etc. The red sandstone was obtained from Fatehpur Sikri, marble from Rajasthan. Most of the wood types, including Shisham (Dalbergia sissoo) and Bakayin (Melia indica) were obtained locally, while Diyar (Cedrus deodara) and Kail (Pinus wallichiana) were transported from Kashmir. Many of the wooden beams were coated with lime plaster on bamboo lath brought from the nearby districts. The bamboo is still widely grown in nearby districts of Kasur, Sheikhupura, Narowal, Gujranwala, Sargodha, and Mandi Bahauddin. Three types of bamboo are native to Pakistan, of which Bambusa bamboo is cultivated in the districts of Punjab and Sindh.

The Mughal imperial buildings, including Kala Burj and Jahangir's bedchamber, were constructed in burnt brick masonry in lime mortar coated with lime plaster. The openings were provided with the help of arches instead of old Indian traditional flat stone or corbelled masonry lintel. The roof was usually constructed in brick domes and lime mortar brick masonry vaults. Persia and central Asia mostly inspired the shapes and types of domes. However, Akbar widely used flat lintels, ceilings, and coved ceilings, which transitioned and developed into a distinctive style. The surface decoration of Mughal buildings is composed of a wide array of materials, techniques, and finishing.

Figure 4.3.3 Detail of brick masonry and plaster coating at Kala Burj. (Photo by Ibrahim and Usman, 2023.)

Jahangir's bedchamber is the focal point of a majestic quadrangle started by Akbar and completed during Jahangir's reign in 1617–1618 among Sehdara pavilion, arcade, Harem, and a Mughal Garden with fountains and Mahatabi in the center. The current chamber was reconstructed during British rule in 1903 and is now used as a Mughal antiquities' museum. The building was used as a private residence by Ranjit Singh, which was used as an armory by the British Army, and they converted the Mughal Garden into badminton courts.

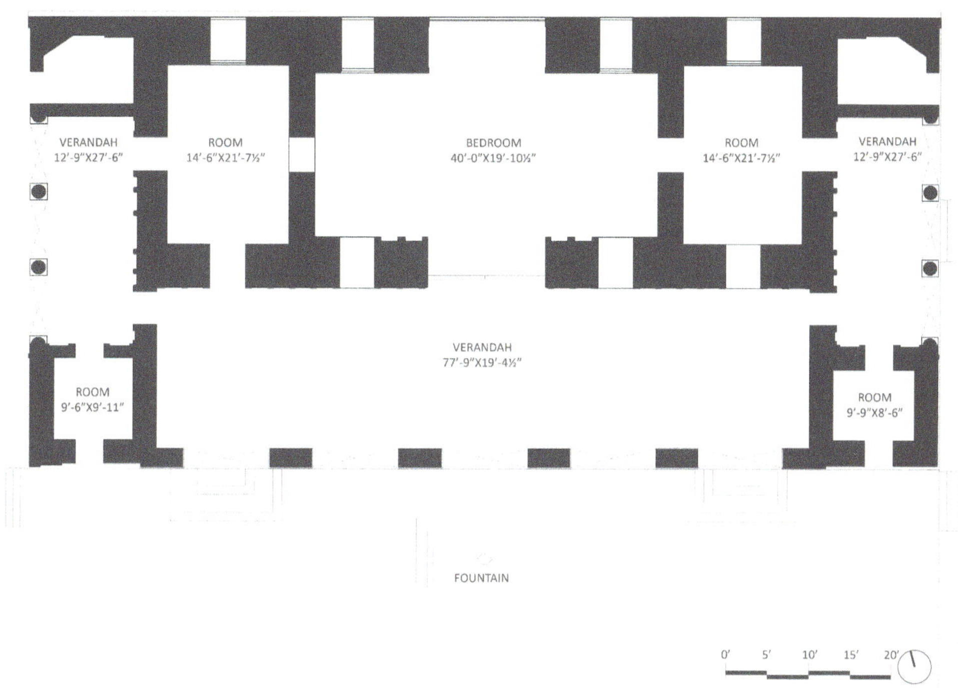

Figure 4.3.4 Floor Plan of Jahangir's Bedchamber. (Drawing by Ibrahim and Usman, 2023.)

Kala Burj (Black Pavilion) is a summer pavilion built during Jahangir era in the northwest corner of the private garden designed by Khawaja Jahan Dost Mohammad. It is built on top of multistoried summer chambers with vaults coated in lime plaster and finished with intricate figurative frescos. An octagonal pavilion was added to the burj during the Sikh period.

Figure 4.3.5 Floor Plan of Kala Burj Pavilion Floor. (Drawing by Ibrahim and Usman, 2023.)

Burnt Brickwork and Lime Plaster Coating at Lahore Fort

Historical and archaeological evidence establishes that Lahore Fort was originally constructed in mud bricks and was rebuilt by Mughal emperor Akbar in the 1560s using burnt bricks. It is also believed that the Fort was substantially enlarged by Akbar in 1566 and Jehangir in 1626. During archaeological excavations in 1959, twenty stratified layers of cultural periods were revealed. Abul Fazl mentions three types of bricks used in India during the 16th century. The types were namely burnt brick, semi-burnt, and unburnt. The two later types were mostly used by the middle and lower class, wherein two brick sizes were common: 8″ × 7½″ × 1¾″ and 7″ × 4½″ × 1″.

There is no scientific evidence of the foundations' depth, size, materials, and technique. However, the foundations of the building would have been the same as other buildings of similar nature and scale. Usually, the foundation was laid with stone and lime mortar with the addition of iron bar grating wherever needed.

The bricks used in Jahangir's Quadrangle and Kala Burj are 8″ × 4½″ × 1″ burnt bricks. The recent scientific investigation of these bricks reveals that it contains SiO_2, Al_2O_3, and Fe_2O_3 with low amounts of Na_2O, K_2O, MgO, and calcium compounds fired at 850–900°C temperature.

Lime or gypsum was widely used as a binding agent. During the Mughal era, there were three sources of lime, i.e., limestone, gravel (Kankar), and marine shells. The mortar used by the Mughals contains white lime, natural hydraulic lime, and coarse gravel (Kankar) lime. The layer of the mortar used in Lahore Fort brick masonry is mostly 1″ thick though traces of much thicker layers can be seen. The gravel (Kankar) lime in Pakistan contains impurities, making it suitable for use as fine and coarse aggregate in mortar mixes. The most common composition of lime found locally has 30% clay, 70% calcium carbonate and sand, and other elements.

A thick layer of Kahgil (a mortar mix of earth, straw, and water) was applied before the plastering of walls and ceiling. Lime alone is not sufficient for making an appropriate mortar or plaster mix. It requires numerous glutinous, gummy, non-resinous, and gelatinous cementing agents, including suitable water. Surkhi (a coarse aggregate of ground-fired bricks) was mixed with lime mortar. Wood ash called Saruj was also added to waterproof the lime mortar. Scientific investigation of the plaster indicates several other admixtures to increase its efficacy and efficiency. Possible admixtures include straw, glue, raw sugar, gum, jute fiber, pulse, and yogurt.

Deterioration and Physical Changes

The present location of the Fort was originally erected by Mahmud of Ghazni in 1021 AD. Lahore Fort stands on a mound 50 feet from the base. River Ravi flowing at the northern boundary of the mound was a defense against the invaders, but at the same time, it is the oldest documented reason for the structural deterioration. The Fort went through a series of destructions and rebuilding till 1923. In the 1560s, Emperor Akbar started the structural consolidation of the Fort. The first step was to rebuild the external fortification by digging the foundation down to the riverbed level and filling it with stones. The superstructure used burnt bricks slanted towards the river wider at the bottom.

Akbar spent 12 years of his reign in Lahore. During this era of extension, structural stabilization of the Fort occurred, particularly in these underground chambers and the external wall. A 12-foot-thick burnt brick buttress wall measuring 1,450 feet long and 50 feet high with a strong foundation of 25 feet wide and 15 feet deep using lime mortar was added. The Fort was rebuilt during the rule of Emperor Jahangir in 1618. Numerous buildings were added, and the Fort was also enlarged. His successor Emperor Shahjahan continued the extension through additions of gardens, chambers, and structural consolidation primarily in three phases, i.e., 1631, 1633, and 1640.

Sikh Era Additions

A Sikh general Lehna Singh took control of the walled city of Lahore and the Fort in 1764 from Afghan occupation, ultimately conquered by Ranjit Singh in 1799. The city remained under the control of the successors to Ranjit Singh. In 1840, a strong earthquake struck Lahore, damaging most parts of the Badshahi Mosque. Although the damages to the buildings inside the Fort are not known, it can be assumed that it might have affected the structures, and during the battle with Chand Kaur, Sher Singh, another son of Ranjit Singh, destroyed numerous parts of the Fort through bombardment from cannons fixed on the minarets of Badshahi Mosque in 1841. The Fort was also damaged and plundered badly during the siege by Heera Singh in 1843.

British Era Additions

The Fort remained under the control of British troops from 1849 till 1923, when possession was handed over to the Archaeological Survey of India. In 1903, major work was carried out to restore the quadrangle. Most of the buildings in the Fort were critically damaged by the 1905 earthquake. This led to repairing of the Shish Mahal, Naulakha Pavilion, and other buildings. Most of the work comprised stabilizing the roofs with iron rails, repairing burnt brick masonry walls, and resurfacing with lime plaster coatings. After taking possession of the Fort in 1927, the Archaeological Survey of India carried out major restoration works and demolished numerous British additions to the Fort. Conservation of the Shish Mahal and Naulakha Pavilion was carried out between 1935 and 1936.

Post-1947 Additions

Undocumented repair was performed incrementally before the Shah of Iran visited in March 1950 and the 1960s. The southwest part of Naulakha Pavilion was restored, including lime plaster coating of the Shah Burj Quadrangle. Consequent to declaring the Fort a world heritage site by UNESCO in 1981, a conservation project was prepared in 1986 by the Department of Archaeology emphasizing Shah Burj Quadrangle, Athdara, and Naulakha Pavilion. The work included structural stabilization and restoration of floors, ceilings, and roofs.

Post-1990, archaeological investigations revealed that the deterioration process started during the reign of Jahangir and was caused by water saturation and the dead load of the structures on top. The shift in the river's course also caused damage to the structures in the north. During documentation of the Fort in the 1970s and recently in 2007, numerous brick columns were added haphazardly in the underground chambers as a support, but no record is available. These columns, sized 2′ × 2′, 3′ × 2′, and even larger sizes, are erected using Mughal-era bricks.

The possession of the Fort was transferred to the Walled City of Lahore Authority (WCLA) in 2014. WCLA, with technical support from Aga Khan Trust for Culture (AKTC), initiated digital documentation and conservation works in an organized manner on a much larger scale. The conservation of Lal Burj was carried out from 2019–2021. Conservation of the northern part of the Picture Wall, Dewan-e-Aaam, Daulat Khana Khas-o-Aam, Kharak Singh Haveli, and Akbari Gate was in progress in April 2022.

Previous Preservation and Alterations to Burnt Brick and Lime Plaster Coatings

The details of preservation efforts before the British period are not known. However, the Archaeological Survey of India started restoration work in 1904 and 1922. Conservation efforts were performed in 1963 and 1973. In 1981, Lahore Fort was declared a World Heritage Site, and in 2000, it appeared on the Heritage Site Danger list. After that, conservation works were realized in 2003, 2008, and 2009, and major works have been underway since 2012 under the newly formed WCLA umbrella. The WCLA, with technical support from AKTC, developed numerous projects from 2017–2020. The recent works undertaken by WCLA include reversing the British era alterations, structural stabilization, and brick masonry, and lime plaster repair.

Shah Burj Complex comprises a portion of the picture wall, Naulakha Pavilion, Shah Burj Gate, and Sheesh Mahal (Mirror Palace). The conservation works include restoring picture walls, roof repair, waterproofing, surface water drainage, structural consolidation, brick masonry, and lime plaster coatings.

Response of the Building Materials to Weather Conditions

The primary threat to the materials used in the Lahore Fort is physical. It is established through scientific studies that apart from photochemical or thermo-oxidation, mechano-degradation also contributes significantly to changes in materials. Researchers also noted that milder conditions, like indoor environments may affect the morphology of the materials.

Experimental studies have shown that the correlation of two or more weather conditions significantly impacts the materials. The temperature has always been influential in the degradation of exterior materials and surfaces. Another facet of temperature affecting material is the fluctuation and cycles of temperature. Studies show that substantial changes occur in the material composition in several seasons when the fluctuation is extremely high in Lahore though the tensile strength is unaffected.

It isn't easy to correlate precipitation with geographic conditions successfully. Precipitation is a perplexing phenomenon influencing materials in numerous ways and cycles. The relative humidity constantly changes in Lahore in daily and weather cycles, which is its prominent feature. The increase in volume and rate of volume change caused by moisture absorption is one of the major factors in the reduction of tensile strength of lime plaster coatings. The combination of high relative humidity and temperature reduction causes condensation on the surface of bricks and lime plaster coatings. It is a major force for deterioration in burnt bricks and lime plaster coatings.

Mass Reduction

The erosion of the surface due to any weathering phenomena results in a reduction of the mass. Usually, it is difficult to calculate the exact mass reduction during the degradative process in contrast to thermal degradation. According to scientific studies on other materials, it is found that mass reduction happens fast at the beginning of the degradation, while it will expedite if the temperature rises. It is mostly caused by the evaporation of the moisture and vapors absorbed by the material.

Mechanical Properties

Frequently, the material is either eroded or suffers a change in properties caused by phased transfer from the solid to the liquid state. If such changes are only surface related and there is no crack formation on the surface, the material's tensile strength decreases, and elongation stays almost constant. If there is crack formation, the notches rapidly reduce tensile strength and elongation. Physical-mechanical changes depend not only on the photochemical changes but also on their combination with the geometry of the material.

Optical Properties

Most optical changes are due to the formation of colored substances from photoinitiated radicals. Following the loss of mechanical properties, changes in color and other optical properties, such as loss of gloss and transparency, are the next most important weathering issues to be resolved. The yellowness of the material depends on radiation wavelength, and the degree of yellowness also depends on the chemical composition of the material.

The discoloration mechanism is always related to the formation of products capable of absorbing blue light. Discoloration is not the only change in the optical properties of polymeric materials. The absorption spectrum of exposed material changes dynamically, and this causes a new range of photochemical changes in the exposed material. It frequently leads to increased absorption of UV radiation and faster degradation.

Surface Change

The drying process causes gloss deterioration in burnt brick and lime plaster coatings. This process is causing more impact on glazed tile work, pietra dura, marble inlay, and other surface decoration elements at the Fort. The surface of the materials becomes uneven by solvent evaporation due to shrinkage. This phenomenon will occur in materials that have solvent-less formulations, particularly in baked materials like burnt bricks and glazed tiles.

Surface Composition

Two types of changes are important to the surface of brick and lime plaster coatings: the change in the chemical composition of the surface and the change in the depth of photochemical reactions. Describing surface changes in general terms is easy because photodegradation is dominated by products of oxidation, which are qualitatively similar in most exposed materials. Large differences exist in their concentrations, the properties of different photodegradation products, and their impact on the material's physical properties.

Brick and Brick Masonry

The brick used in the construction of major buildings in the Fort is a low-density, high porosity, duly fired 8″ × 4½″ × 1″ modular unit. The primary threats to the brick masonry at Lahore Fort are changes in freeze-thaw cycles, heat waves, and solar radiation. These are causing shrinkage in bricks resulting in cracks, spalling, and shattering.

Dampness from groundwater, rainfall, and moisture has seriously damaged the brick masonry, primarily the walls that are less exposed to sunlight due to its direction or vegetation. Deterioration of plaster and mortar in joints is also a major cause of the decay of the brick masonry. The mortar is mainly weakened by moisture, which gives passage to dampness to the whole structure. Growth of plants in the masonry joints and salt crystallization are evident in most structures. The plant growth is due to the bird fecal matter and roots deep inside the masonry leading to cracks and increased water penetration. Fungus flourishes in damp, high humidity and with less ventilation.

Figure 4.3.6 Kala Burj, View from Southeast, Lahore Fort. (Photo by Ibrahim and Usman, 2023.)

Lime Plaster Coating

The increased heat results in the Lahore Fort's shrinkage and flaking of lime plaster coatings. Internal cracks and surface spalling are due to freeze-thaw. It is also observed that solar radiation causes shrinkage due to excessive evaporation, thus resulting in cracking. Vibrations produced by earthquakes and vehicular traffic are another cause of cracking and settlement of the brickwork and lime plaster coatings. Biological factors like plant growth, insects, and animal presence cause a small amount of damage.

Figure 4.3.7 Detail of lime plaster coating at Jahangir's Quadrangle. (Photo by Ibrahim and Usman, 2023.)

Projected Impacts of Climate on Materials

As the climate warms, it is expected that heating degree day values will decrease and that the boundary between the severe-cold and cold zones will shift northward, decreasing some risks for interior moisture risks. The projected threats to the buildings can be:

Phenomena of causes	Projected effects on building and infrastructure
Rise in temperature and increased chemical reactions	Intensified deterioration
Acute precipitation	Accelerated decay, particularly in wood
Rise in the freeze-thaw cycle	Early decay of lime concrete, damage due to moisture, and ice damming on roofs
Alteration in seasonal precipitation, length of season, and rainfall cycles	Higher moisture penetration, weakening of materials, efflorescence in brick and lime mortar, and flooding
UV radiation	Photodegradation such as chalking, yellowing, and cracking of wood
	Biodegradation by increased surface hydrophilicity in wooden components

Figure 4.3.8 Projected threats. (Table by Ibrahim and Usman, 2023.)

The frequency and intensity of freeze-thaw cycles alone do not deteriorate the bricks, brick masonry, and lime plaster coatings. However, blending the freeze-thaw cycle with moisture is a critical factor. Wind during precipitation also becomes threatening because it determines the wetting time of the walls. Smog, acid rain, air-borne pollutants, and radiation also need to be considered due to complications of the decaying process due to the blending of two or more phenomena acting simultaneously. High atmospheric relative humidity would develop irradiation of materials after wetting through precipitation.

Ultraviolet B is proven to be the primary agent in damaging building materials, including lime coatings and wood. Increased ultraviolet radiation due to stratospheric ozone layer depletion will reduce the life of building materials. Climate change will potentially increase the weathering of materials through a rise in absolute humidity and increased temperature. Both phenomena will expedite the photodegeneration of building materials and decrease their life.

Recommendations for Future Maintenance and Protection

The foremost vital concern in the maintenance and protection of the buildings at Lahore Fort is to identify the weathering processes and develop appropriate conservation actions. In some cases, excessive precipitation may be the prime threat, while rain-driven intensity may be the principal cause of the deterioration of external surfaces.

Moisture infiltration can be declared as one of the major impacts imposed on the buildings at Lahore Fort. It is also credited with the generation of molds and its related problems. Avoiding moisture infiltration and ensuring quick drying of the buildings is a primary strategy to reduce degradation. Articulating water management procedures into the principles mentioned below can help in controlling moisture penetration.

Strategies	**Design and adaptation**
Diversion	Systems, assemblies, and materials to stop rainwater penetration
Discharge	Discharging rainwater once it has penetrated the structure
	Treatment and ridging of the roof at horizontal parts of the structure
	Develop gravity flow of the water and moisture accumulated inside the structure or in its elements
Dehydration	Aeration and diffusion of penetrated water
	Envelope treatment for drainage and dehydration of water
Robust materials	The drying potential of materials
Combination with radiation	Damages due to UV-B are high
	Using performance-oriented materials and surface protection measures
	Thermal or chemical treatment of wooden members

Figure 4.3.9 Water and moisture management strategies for historic buildings. (Table by Ibrahim and Usman, 2023.)

Measures needed for adaptation to climate change would mainly consist of developing different formulations of bricks, brick masonry, and lime plaster coating, including allied engineering practices for ensuring improved durability in preventive maintenance.

Site-Specific Research

Collecting more accurate site-specific contributors to weathering materials in Lahore Fort is important. Temperature-humidity criteria shall be developed to understand the corrosion of materials, including burnt brick and lime plaster, in Lahore Fort. Scientific studies have shown that vapor is condensed when the temperature is above or equal to 0°C and relative humidity is more than 80%. Consequently, it is recommended that a chart may be developed for Lahore that would not only help conserve the Fort but other monuments in Lahore and its adjacent areas.

The mechanical deterioration index of building elements at the Fort must be developed by taking daily temperature differences. It will help create better strategies for protecting the exterior surface. An hourly solar UV energy chart falling horizontally outdoors will be beneficial. Conducting a scientific investigation on the correlation of the atmospheric weathering process and solar radiation with the building materials at Lahore Fort is critical. It will help develop maintenance measures, the degree of intervention, and their schedule.

Conclusions

The city and Fort of Lahore have a long history of adaptation and change. In the past, changes have been driven by war and political landscape. Today, adaptation must also occur in response to the increasing pressure of the climate. Increased maintenance and management of moisture is the most immediate goal. However, detailed research is also required to understand how changes in UV radiation, temperature, and moisture affect historical materials and how the next-generation materials can respond to these new stresses.

Biographical Note

Khalid Ibrahim AIAP, MPCATP

Khalid Ibrahim is an Associate Professor at the National College of Arts, Lahore, Pakistan. He has taught in the Department of Architecture for the last 28 years. He is a registered architect and a member of the International Council on Monuments and Sites (ICOMOS), the Pakistan Council of Architects and Town Planners (PCATP), Institute of Architects Pakistan (IAP) and Anjuman-e-Mimaran, Pakistan. He worked as the project director for the development of National College of Arts Rawalpindi campus.

He holds a bachelor's degree in architecture from the National College of Arts, Lahore, and a master's degree from the University of Engineering & Technology, Lahore, Pakistan. He completed his second master's degree in Historic Preservation from Boston Architectural College, Boston, Massachusetts in 2018. Khalid has been engaged in restoration projects with the Nayyar Ali Dada and Aga Khan Programs for Islamic Architecture and Urban Planning, such as the Lahore Railway Station, selected areas of the walled city of Lahore, and the National College of Arts. He is currently working on illuminating a world heritage Buddhist site, "Takht-e-Bahi".

Zahid Usman AIAP, MPCATP

Zahid Usman is a Ph.D. scholar at International Islamic University Kuala Lumpur and an associate professor of Architecture at National College of Arts Lahore,

Pakistan. He holds M.Sc. in Built Environment from KTH Stockholm. Usman is also the principal of an architecture and design firm in Pakistan. He has worked in conservation, restoration, and adaptive re-use of historical buildings in Pakistan, Afghanistan, and Sweden. His research interests focus on the historic buildings, spatial planning, built environment institutional policies, and architecture design.

Bibliography

Abul Fazl, Ain-i-Akbari, and Tran. H. Blochmann. 1873. Asiatic Society of Bengal, Calcutta.

Annual Progress Report of the Superintendent, Muhammadan and British Monuments, Northern Circle. 1916. Government of India Press, Allahabad.

Archaeological Survey of India: Annual Report 1907–8. 1911. Superintendent Government Printing, Calcutta.

Archaeological Survey of India: Annual Report 1935–36. 1938. Superintendent Government Printing, Calcutta.

Audouin, L. et al. 1998. Agew. Makromol. Chem., 25–34, 261–262.

Aziz, Mian Abdul. 1963. "Engineer – The Nation Builder. *Quarterly Journal of the West Pakistan Engineering Congress*," VIII (4).

Bergin, M., S. N. Tripathi, J. Jai Devi, T. Gupta, M. Mckenzie, K. S. Rana, M. M. Shafer, A. M. Villalobos, and J. J Schauer. 2015. "The Discoloration of the Taj Mahal Due to Particulate Carbon and Dust Deposition." *Environmental Science & Technology* 49 (2): 808–12.

Blockmann, H. (tr.). (Vol. II, 1873 reprint 1989). Ain-i-Akbari by Abu 'l-Fazl ibn Mubarak.

Charola, A. E., J. Pühringer, and M. Steiger. 2007. "Gypsum: A Review of Its Role in the Deterioration of Building Materials." *Environmental Geology* 52: 339–352.

Chaudhary, N.A. 2004. *A Guide to Lahore Fort*. Lahore: Sang-e-Meel Publications.

Chaudhary, N. A. 1999. *Lahore Fort: A Witness to History*. Lahore: Sang-e-Meel Publications.

Chaudhry, Q. Z. 2017. Climate Change Profile of Pakistan. Asian Development Bank Report, Manila.

Farooqi, A. B., A. H. Khan, and H. Mir. 2005. "Climate Change Perspective in Pakistan." *Pakistan Journal of Meteorology* 2 (3): 1121.

Gulzar, S. et al. 2013. "Characterization of Mughal Bricks from Jahangir Tomb, Lahore-Pakistan," *Asian Journal of Chemistry* 25, no. 6.

Hussain, Talib. 2017. History of Interventions on Prototype Areas of Picture Wall, Lahore Fort. Agha Khan Cultural Services Pakistan.

Kabir, Najma, and Khizer Hayat. 2016. Analysis of Mortar Used in Shish Mahal in Lahore Fort, Pakistan. Elsevier Limited.

Kassim, P. S., M. K. fauzi, Y. A. Abakr, and N. S Latip. 2016. The architectural integration of passive systems through thermal flywheel effect with evaporative cooling in Shish Mahal, Lahore Fort, Pakistan.

Khan, A. N. 1997. *Studies in Islamic Archaeology of Pakistan*. Lahore: Sang-e-Meel Publications.

Khan, M. Ishtiaq 1958. Lahore and its Important Monuments, Superintendent of Archaeology West Pakistan, Lahore.

Khan, M. Ishtiaq. 1974. Lahore Fort. Department of Archaeology & Museums, Government of Pakistan.

Khan, R. 2011. *Archaeology below Lahore Fort, UNESCO World Heritage Site*. Pakistan: The Mughal Underground Chambers, For Global Heritage Fund Preservation Fellowship.

Khan, W. M. 1959. Lahore and its Important Monuments, Department of Archaeology Government of Pakistan, Lahore.

Koch, Ebba. 1983. "Jahangir and the Angels: Recently Discovered Wall Paintings Under European Influence in the Fort of Lahore." In *India and the West: Proceedings of a Seminar Dedicated to the Memory of Hermann Goetz*, edited by Joachim Deppert, 173–195. New Delhi: Manohar Publications.

Lal, K. 1990. *Tarikh-e-Lahore (pp.345–347)*. Lahore: Sang-e-Meel Publications.

Maurya, Rahul. 2018. "A Brief History of Materials and Construction Techniques of Mughal Architecture." *IJAR* 4-12: 75–78.

Nath, R. 1972. Taj Mahal: The Evolution of the Tomb in Mughal Architecture, Bombay: Taraporevala & Sons.

Pakistan Archaeology No. 5. 1968.

Qaiser, Ahsan Jan. 1988. Building Construction in Mughal India. Centre of Advanced Study in History. Delhi.

Raheem, A. A., and M. A. Tahir. 2008. "Mughal Architecture under the Cloud of Demolition: Environmental Effects on Monuments." *Journal of Building Appraisal* 4: 103–114.

Rogers, P., and Y. Lari. 2006. *Lahore Fort Master Plan 2006–2011*. Islamabad: UNESCO.

Sönke, K. et al. 2015. Global Climate Risk Index. Germanwatch eV.

Suleman, K. M. 2005. Pakistan: Country Report on Bamboo Resources, Global Forest Assessment. Food and Agriculture Organization of the United Nations, Peshawar, Pakistan.

Tomiita, T. 1992. "Solar U.V., Wetness and Thermal Degradation Maps of Japan." *Construction and Building Materials* 6: 195–9.

Vogel, Jean Philippe. 1920. *Tile-Mosaics of the Lahore Fort*. Calcutta: Superintendent Government Printing.

Wescoat, James, and Joachim Wolschke-Bulmahn. 1996. *Mughal Gardens*. Washington: Dumbarton Oaks Research Library and Collection.

Wypych, George. 2013. *Handbook of Material Weathering*. Toronto: ChemTec Publishing.

Zerlaut, G. A. 1994. Accelerated and Outdoor Durability Testing of Organic Materials, ASTM STP 1202. Grossman D. Philadelphia.

CHAPTER 4.4

Wood and Coatings in Japan

Introduction

Japanese society has long reconciled the ephemeral nature of their natural and built environment with their desire to protect and preserve their resources and cultural assets. Their gardens, shrines, and temples have been preserved in their original form over centuries, even though the gradual effects of weather and insects, and the sudden destruction of fire and earthquakes have required their complete replacement. Preservation philosophy in Japan includes a strong tradition of reconstruction, the deliberate rebuilding of historic structures using identical materials and craftsmanship, as well as a high regard for the traditions of craftsmanship, and the craftsmen who embody these skills.

In the 21st century, Japan is facing increasing challenges from changes in the climate, vanishing resources, and a general loss of traditional knowledge and skills. Increased UV exposure, increased insect activity, and fluctuations in humidity and temperature are all affecting traditional materials in new ways. Japan is responding to these challenges with increased education, training, and research, as illustrated by their work at shrines and temples in Nikkō and the historic Machiya homes in Kyoto.

History, Geography and Climate

History

Sacred places were built to honor gods and political leaders, who often became gods, and were maintained by a combination of military rulers and religious organizations. By the 17th century, construction and renovation work was

DOI: 10.4324/9781003160489-26

being carefully documented by monks, and governmental statutes were established to protect important structures and monuments. The Association for Protection of Cultural Properties provides oversight and establishes restrictions for National Treasures or Important Cultural Properties while local governing bodies determine the type of materials, the conservation methods, and the extent of replacement that is appropriate for each shrine or temple location.

Geography and Climate

Japan is an archipelago of 6,852 islands, most of them uninhabited, over a distance of 2,361 miles from north to south. Climate varies from subarctic in the north to subtropical in the south, with distinct differences between the west coast on the Sea of Japan and the east coast facing the Pacific. Most of Japan, including Honshu Island, is classified Cfa, humid subtropical, or humid continental climate. Nikkō, in the mountains of eastern Japan, is hot and humid in the summer and cold in the winter, with heavy snow in the mountains. Heavy rainfall from the Pacific falls during the late summer monsoon season. Kyoto, in a bowl surrounded by mountains on three sides in western Japan, is even hotter and more humid in the summer, with average summer temperatures of 95°F. Typhoon season can bring cold winds from the west from August to September.

In 2015, the fourth warmest year since 1898, the mean surface temperature was 1.3°F higher in Japan than the period between 1981 and 2010. Japan experienced heat waves in 2018 and again in 2022. In July 2022, temperatures exceeded 95°F for nine consecutive days and peaked at 104°F.

Temperatures in Japan are expected to continue to rise at a higher rate than the global average. The average mean temperature in the EJP/Cfa zone is expected to be 4.2°F higher in the period from 2076 to 2095 than it was in the period from 1980 to 1999, with five to ten more dry days. Summer precipitation is expected to increase by 19% while humidity will decrease by 4% in the summer and increase in the winter.

Traditional Materials and Techniques

Wood

Although there is a range of construction styles and materials, the most predominant features are wood, with flexible connections, deep overhangs, and decorative coatings. The pole and platform construction of pagodas and the interlocking brackets at temples and shrines provide a stable but flexible structure that can withstand the shaking of most earthquakes. Deep overhangs protect the wood and finishes below from rain and UV light.

Traditional construction used domestic wood, including Japanese pine, hinoki (cypress), tsuga (hemlock), turbinata (chestnut), and sugi (red cedar). The tremendous amounts of lumber required to build or reconstruct large temples and shrines quickly depleted local supplies and had to be transported from great distances across Japan. Plantations were established during the 1950s reconstruction after World War II but were abandoned for imported wood that was less expensive. At the same time, harvesting of natural stands of cypress and cedar was largely prohibited. Current policies are now promoting the use of farmed Japanese hinoki and cedar to increase employment and opportunities in the forest industry, but the relative durability and pest resistance of the farmed wood is unknown.

Roofs

Roofs were originally constructed with multiple layers of thatch, cypress, cedar, or bark. Many were replaced around 1800 when industrial copper became available and fashionable as a sign of wealth, although cypress remained traditional for most Shinto shrines. Nails were rare in traditional wood construction, but bamboo or corrosion-resistant "Wakugi" nails were used to attach both copper and wood roofing. "Wakugi" nails were made from low-carbon "Hocho-tetsu" or "Wari-tetsu" steel. that was fabricated prior to the Meiji period from decarburized pig iron.. Over-oxygenation during fabrication creates a reaction of iron and oxygen that forms a thin surface film of protective magnetite on the surface without corroding (oxidizing) the steel.

Figure 4.4.1 Red lead paint and cedar bark roof fascia at Wakamiya Shrine, Nara, Japan. (Photo by Daniel Thomas, 2022.)

Urushi Lacquer

Traditional Urushi lacquer has been used in Japan for over 8,000 years to protect and enhance wood and as an adhesive. The resin from native Japanese Sumac trees is painstakingly harvested and processed to produce Urushi. Mineral pigments, usually cinnabar, copper oxide, and iron oxide, or a mixture of cinnabar and mercuric sulfide are used to produce black, Japanese red, and red lacquers. Gold leaf is applied over lacquer.

Figure 4.4.2 Sanding to reveal existing layers of black, red, red, and gold-leaf lacquer, Workshop, Nikkō, Japan. (Photo by Daniel Thomas, 2022.)

Figure 4.4.3 Layers of black, Japan red, red, and gold-leaf lacquer, Nikkō Workshop, Japan. (Photo by Daniel Thomas, 2022.)

As many as 30 coats of Urushi lacquer might be applied to wood over a period of months. Each layer is cured for one to two days under controlled temperatures from 77 to –95°F and from 65% to 80% humidity before sanding. Oxygen and humidity, both plentiful in Japan's forests, oxidize and chemically harden the catechol in the Urushiol. Low humidity can allow the coating to dry out and become brittle, less strong, and vulnerable to cracks that permit water, insects, or fungi. A properly cured coating is permeable and waterproof, resistant to acid, alkali, and alcohol, and can last for centuries, especially in Japan's warm, humid climate. However, use of poor-quality materials, imported resins, improper curing, low humidity, or excess UV exposure can cause cracking and delamination. Variations in the extraction process and curing affect the final properties and drying time, increasing the challenges for contemporary craftsmen.

Both UV and visible light cause color fading by breaking the molecular bonds of the Urushi polymer. The extent of fading varies with the type of pigment, the wavelength of the UV light, and solar exposure. Black lacquer shows less deterioration than red. Northern exposures fade noticeably less than southern exposures. UV light can also cause "post curing", shrinkage, and cracking caused by an increase in density of the chemical cross links.

Urushiol has some termiticidal property that decreases as the resin dries. Cured urushi, especially multiple layers, can deter termites by creating a smooth, hard surface, but provides no protection once the termites have penetrated the coating, and can hide damage. Extensive interior deterioration discovered under intact exterior columns confirmed this hypothesis.

The Agency for Cultural Affairs is actively promoting the use of domestic Japanese lacquer at cultural properties and has invested in human resource training to ensure a stable supply. Joboji is still the main production area, but other prefectures are beginning to plant new Sumac trees and increase the number of Japanese Urushi-kaki (sap collectors).

Research continues to better understand traditional methods of production, installation, and repair. New repair methods include the use of Urushi diluted with a hydrocarbon solvent to fill in shrinkage cracks from UV damage and the use of polymerizable UV stabilizers.

Color Paints

Color painting uses different materials and techniques but is often used in combination with lacquer. Exterior paints use a mix of paulownia wood oil and sesame oil, which are boiled together in different percentages based on humidity and other requirements.

Figure 4.4.4 Protected Lacquer and Paint, Tōshōgū Shrine, Nikko, Japan. (Photo by Daniel Thomas, 2022.)

Deterioration and Climate Risks

Japan has a long history of natural disasters including earthquakes, fires, floods, falling trees, and lightning strikes, followed by faithful restorations of historic buildings to their original design, using traditional materials and techniques. There have also been long periods of neglect when climate conditions have led to deterioration and a weak structure that is more vulnerable to natural disasters and biological damage.

Earthquakes and Flood Damage

Traditional Japanese buildings are designed to absorb the shaking from earthquakes through flexible wood connections or mass masonry. Earthquakes can cause minor damage to masonry walls, stairs, and sculptures but the greatest damage is from fires, flooding, or landslides that are triggered by earthquakes. There is no evidence of an increase in earthquake activity; however, timbers deteriorated by insect or fungal decay are more likely to be damaged by earthquakes.

Fire Damage

There has always been a risk of fire triggered by human activity, earthquakes, forest fires, or lightning strikes. Fires are most frequent in the early spring and late summer, correlating to low precipitation, low relative humidity, and high human activity. Lightning during a thunderstorm may damage individual properties but rarely causes widespread fire damage. The frequency and severity of wildfires are expected to increase with the projected increases in temperature and variation in rainfall, although improved technology has allowed many important buildings to be protected with discrete alarms, protection systems, fire hydrants, and lightning arresters.

UV Deterioration of the Urushi Coatings

In spite of its sensitivity to UV and visible light, Urushi varnish has been used extensively on the exterior of shrines and temples, protected by broad overhangs, shade from the forest environment, protection from the mineral pigments, and regular periods of restoration. However, UV intensity and related photodegradation of wood and coatings are expected to increase. There is already evidence of more rapid wear and fading of lacquer and coatings at Nikkō on south facing exposures.

Figure 4.4.5 UV damage to red lacquer at Toshugu Shrine, Nikkō, Japan. (Photo by Susan Pranger, 2022.)

Biological Damage

Termite damage to wooden buildings has been documented since the 17th-century Edo period but has increased over the last 100 years. Camphor oil was traditionally used to deter termites. Beetle damage is a more recent discovery, although earlier beetle damage may have been attributed to termites, or simply undetected.

The impact of increasing temperatures on fungal and insect growth in Nikkō will more than compensate for the impact of decreased humidity. The Scheffer index measures the risk of fungal and insect damage based on the joint impact of temperature and humidity, has increased in the last 20 years from an index of 82 for the period from 1961 to 1990) to and idex of 84.5 for the period from 1981 to 2010). The index is expected to increase dramatically to 100 by the end of the century (2076–2095).

Temperature increases decrease the time required for insects to hatch and lay new eggs. Shorter life cycles could result in multiple cycles in a given year. Hatching could begin one or two weeks earlier by the end of the century. Increases in temperature also increase fungal growth and decay, which makes wood more vulnerable to the attack of insects. Increased photodegradation of protective coatings by increased UV light also increases the risk of biological damage to wood.

Figure 4.4.6 Insect damage at Rinnoji Jigendo Haiden, Nikkō, Japan. (Photo by Margaret Rosewitz, 2022.)

Machiya Townhouses of Kyoto

The old town of Kyoto has streets lined with Machiya, traditional wooden townhouses where merchants and artisans lived and worked. Hundreds have been torn down to be replaced with more comfortable, modern replacements, but the establishment of historic districts now provides some protection.

Shishokai, a non-profit organization, was established by traditional building masters, including Tomohiro Naito, a fifth-generation Kyoto master carpenter to expand the knowledge of Machiya construction methods. Young working carpenters receive free weekend classes in the traditional woodworking and plastering skills needed to repair or replicate the townhouses.

Machiya townhouses are built with a post and beam construction infilled with earth walls and a wood lattice storefront. The roofs were constructed with "kumki" joinery and no nails, which made the structures very earthquake resistant. Many of these townhouses are now being restored using traditional woodworking and plaster techniques but reinforced to meet new structural and comfort demands.

Figure 4.4.7 Roof reconstruction at Machiya townhouse, Kyoto, Japan. (Photo by Daniel Thomas, 2022.)

The Temples and Shrines at Nikkō

By the end of the century, cherry trees in Nikkō are expected to bloom a full month before the traditional cherry festival date as temperatures continue to increase. Recent studies have documented the escalation of weather-related damage over the last half century, primarily from flooding and insects. In response, maintenance and restoration by the Association for the Preservation of the Nikkō World Heritage Site Shrines and Temples has included training and a 15-month residential seminar program allows educated artists and craftsmen to develop traditional skills under the supervision of master craftsmen. A multi-year research and pilot test program is pursuing alternatives to chemical fumigation.

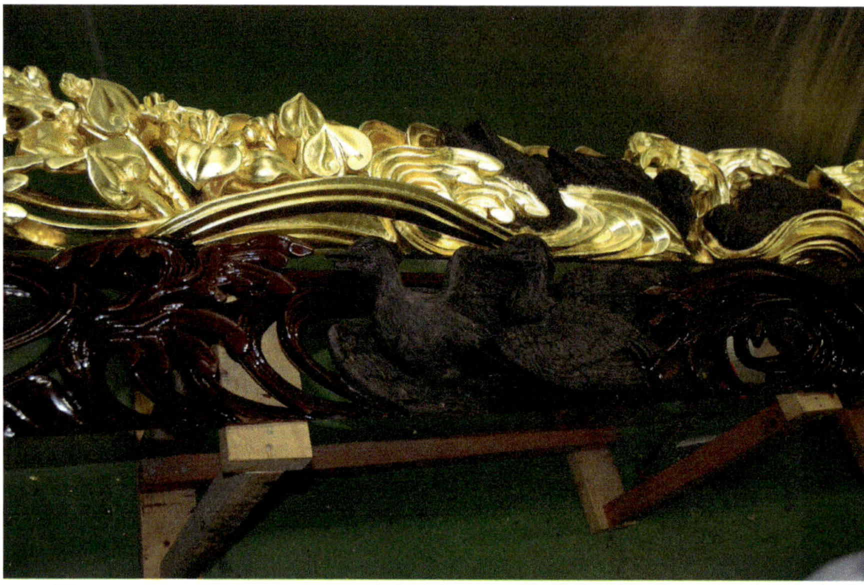

Figure 4.4.8 Stages of lacquer and gold leaf restoration, Nikkō Workshop, Japan. (Photo by Daniel Thomas, 2022.)

History

The mountain forest of Nikkō was an important religious and cultural destination before the construction of temples or shrines. The first buildings at Rinnō-ji temple were built by Buddhist monk Shodo: the Betsugū Hongu-jinja worship hall in the 8th century, and the Betsugū Taki-no-o-jinja Honden shrine in the 9th century. Many temples and shrines were added in the following centuries but the area was largely abandoned during the upheavals of the 16th-century Muromachi period. During the 17th century, the Shogun Tokugawa Ieyasa rebuilt or restored almost all of the existing shrines, planted thousands of Japanese cedars, and constructed many structures in the Edo style including the Shinkyo bridge and the Tōshōgū shrine.

In keeping with the Decree of 1871, the entire Nikkō complex was divided into three separate complexes, each controlled by a different religious entity: Futarasan-jinja and Tōshōgū by the Shinto sect, and Rinnō-ji by the Buddhist. Additional oversight was provided by the Association for Conservation of the Shrines and Temples of Nikkō, the Hokokai, established in 1879 and the Law on the Preservation of Shrines and Temples, established in 1897. After decades of neglect in the early 20th century, there were a series of major renovations from 1950 through 1986 during the Showa period. The 1957 Natural Parks Law restricted construction of new buildings and removal of trees and created a 922-acre buffer zone that extends to the ridges of most mountains surrounding the properties. In 1999, the 125-acre Nikkō Shrines and Temples complex was designated a World Heritage Site by UNESCO.

In 2007, the Association for the Preservation of the Nikkō World Heritage Site Shrines and Temples began a 20-year Heisei Restoration project. In 2018, the project focused on the repair of Rinnoji Jigendo Haiden (Jigen-do Hall of Worship) and the Tōshōgū Shrine. Renovations are expected to continue until 2024. A pilot survey in 2009 and a more extensive survey in 2016 revealed extensive concealed damage from Deathwatch beetles, which led to an ongoing pilot program to investigate a heat treatment as an environmentally friendly alternative to chemical fumigation.

The Religious Organizations of Futarasan-jinga, Tōshōgū, and Rinnō-ji continue to own and manage the temples and shrines of Nikkō. They are responsible for daily operations and receive financial and technical assistance from the Agency for Cultural Affairs, Tochigi Prefecture, and Nikkō City. Regular maintenance is the responsibility of the Agency of Cultural Affairs, while repairs and restorations are completed by the qualified conservation architects and engineers of the Foundation for Preserving Nikkō Shrines and Temples.

Restoration of the Toshugu Shrine and Fence

The Toshugu shrine is usually restored on a 60-year rotation. However, the temple lacquer work restored seven years ago has already faded on the west exposure from increased UV intensity, while the north exposure is still shiny. The lacquer and colored paint on the fence were recently restored, and covered with transparent plastic to protect the exterior paint from visitor contact. It appears that the plastic is also providing UV protection, as the paint has not faded as much as in unprotected areas.

Figure 4.4.9 Lacquer at Tōshōgū Shrine, Nikkō, Japan. (Photo by Margaret Rosewitz, 2022.)

Figure 4.4.10 Tōshōgū Fence, Nikkō, Japan. (Photo by Margaret Rosewitz, 2022.)

Restoration of the Rinnoji Jigendo Haiden (Jigen-do Hall of Worship)

The restoration of Rinnoji Jigendo Haiden combines new and old technology. Traditional materials and methods were used throughout, under a modern metal scaffolding enclosure. Extensive restoration was required to wood that had been decayed by insects and fungi, but every effort was made to retain as much original wood as possible. Dutchmen or patches were used where possible. Columns were split vertically so deteriorated wood could be removed from the interior, and the solid exterior reused to enclose new structural columns. New wood members are shaped using traditional methods by carpenters who are specially trained and licensed, using traditional axe (Yoki), adze (Chona), and a double-sided Japanese spear plane (Yari ganna) and are date stamped, as was done in an 1819 renovation. The copper roof pans are carefully removed and salvaged, repaired by hand and reinstalled.

In 2018, craftsmen at Nikkō carefully examined sanded layers of the original lacquer finishes under microscope to identify the methods used over the past 400 years at Rinnoji Jigendo Haiden. They developed a 40-step process to apply, dry, and sand 17 layers of lacquer to replicate the original Edo Period lacquer.

Figure 4.4.11 Roof work at Rinnoji Jigendo Haiden, Nikkō, Japan. (Photo by Margaret Rosewitz, 2022.)

Figure 4.4.12 Column repair at Rinnoji Jigendo Haiden, Nikkō, Japan. (Photo by Margaret Rosewitz, 2022.)

Heat Fumigation Test Program

Insect damage by termites and beetles is not always visible. The extensive damage to interior heartwood of the columns at Rinnoji Jigendo Haiden was concealed by what appeared to be an intact wood and lacquer exterior. Following the confirmation that beetles existed in the Nikkō temple complex, in 2010, all 72 buildings were monitored for the emergence of adult insects, using 27,000 adhesive traps to determine the location and extent of infestations. Tests in 2011 indicated that CO_2 treatment and sulfuryl fluoride (Vikane®) fumigation were effective in eradicating beetles inside untreated wood but did not achieve 100% mortality of insects in dense wood treated with thick lacquer.

The pilot heat fumigation testing program at Nikkō is the culmination of a 10-year research project conducted by the Association for Preservation of the Nikkō World Heritage Site Shrines and Temples in cooperation with Kyoto University and the National Institute for Cultural Properties, Center for Conservation Science and Restoration Techniques. The program is testing the effectiveness of prolonged high heat as an alternative to the more environmentally toxic chemical insecticides. The Association is working with Kansai Electric Power to develop ways to reduce energy requirements and provide clean electricity more efficiently by using line voltage instead of fossil fuel generators.

The test program is studying the impact of heat treatment on insects at all stages of their life cycle (larvae through adults), and the effect on historic wood and coatings. Phase 3, treatment of a replica of a small, lacquered wood temple, began in June 2022 and was completed in September 2022. If the test is determined to be successful, and subsequently approved, the next step will be to use the process on an existing building.

A temporary insulated enclosure was constructed around the replica temple, adjacent to the Association's administration building at Nikkō. To reduce stress on the wood, the enclosure is heated slowly, less than 1°F per hour, to approximately 140°F. The temperature is maintained for three days to allow the wood to stabilize before being gradually cooled. Humidity in the air and in the wood is carefully monitored and controlled to maintain a constant relative humidity to reduce swelling and other adverse effects to the wood and lacquer coatings. A monitoring system uses probes to measure the temperature and humidity in the air and within the wood of the replica temple and a series of 30 cm wooden cubes with a variety of finishes. Sensors also measure the shape and expansion of the wood. The cubes are used to test the impact on insects and larvae that have been introduced to the wood.

Within the enclosure, four fan units blow air counterclockwise within the enclosure to match the normal flow direction in the Northern Hemisphere. Each fan unit includes electric heating coils, steam coils to add humidity, and cooling coils to reduce temperature as required. Dehumidifiers reduce humidity when the treatment area is cooled after treatment. A chiller, water treatment, equipment and electrical transformer are located outside the enclosure. The process currently uses 6,600 KW of electricity from a generator. Improved insulation and more efficient equipment are required so that line voltage can be used for future installations.

Conclusions

Japan has an exceptional history of taking a long view, documenting and restoring its built heritage, and honoring its cultural traditions and building technology. In the past, buildings were faithfully restored using both original plans and techniques. However, Japan is facing new challenges in their efforts.

Changes in the weather and climate are increasing damage and deterioration, especially from insects and UV light. The lack of traditional materials and qualified craftsmen makes complete replacement and reconstruction of the temples technically impossible, as well as economically infeasible. Japan is responding to these challenges with immediate and long-term action by investing in the education and training of craftsmen, supporting the production of traditional materials, and researching sustainable alternatives to insect control.

Bibliography

UNESCO. 2021. *Shrines and Temples of* Nikkō. https://whc.unesco.org/en/list/913/.

2019. "Application of Humidified Warm-Air Treatment." *Integrated Pest Management (IPM), Proceedings from the 4th International Conference.* Stockholm: Swedish National Heritage Board. 242–250.

Bartus, Jan, William J. Jr Simonsick, and Otto Vogl. 1995. "Oriental Lacquers IX. Polymerizable Ultraviolet Stabilizers for Oriental Lacquers." *Polymer Journal* 27 (7): 703–771.

Brimblecombe, Peter, and Mikiko Hayashi. 2018. "Pressures from Long Term Environmental Change at the Shrines and Temples of Nikkō." *Herigage Science.* doi:10.1186/s40494-018-0186-1.

CG, Carll. 2009. "Decay hazard (Scheffer) i." ndex values calculated from 1971–2000 climate normal dataDepartment of Agriculture FPL-GTR-179LESS, Madison.

Climate-Data.org. 2021. Nikkō *Climate Summary.* https://en.climate-data.org/asia/japan/tochigi/nikko-47714/#:~:text=Nikk%C5%8D%20climate%20summary,classified%20as%20warm%20and%20temperate.&text=The%20climate%20here%20is%20classified,mm%20%7C%2065.7%20inch%20per%20year.

Global Facility for Disaster Reduction and Recovery. 2023. *Japan Wildfire.* https://thinkhazard.org/en/report/126-japan/WF.

Harada, Masahiko, interview by Susan Pranger. 2022. *Restoration of* Nikkō *World Heritage Site* Translated by Yoshido Ichiro. Nikkō, (August 5).

Harada, Masahiko, Marito Nomura, Rika Kigawa, Yukio Komine, Mikiko Hayashi, Wataru Kawanobe, and Takeshi Ishizaki. 2011. "Large-scale Survey of Wood-boring Anobiids by Sticky Insect Ribbons in Historic Buildings in Nikkō World Heritage Site." https://www.tobunken.go.jp/ccr/pdf/50/5014.pdf.

Hirota, Kouichi. 2022. "Chief Lacquer Technician, Nikkō." *Lacquer at* Nikkō *World Heritage Site*, translated by Yoshido Ichiro. Nikkō, (August 5).

Houlton, Sarah. 2004. *Illuminating Lacquer Layers.* August. https://www.chemistryworld.com/features/illuminating-lacquer-layers/3004637.article.

ICOMOS. 1999. " Nikkō (Japan) No 913." Nomination by UNESCO Advisory Board (UNESCO).

IPPC. 2000. "Emissions Scenarios, Summary for Policy Makers." A Special Report of IPCC Working Group III. https://www.ipcc.ch/site/assets/uploads/2018/03/sres-en.pdf.

Japan City Tour. n.d. Nikkō *Futarasan Shrine*. https://www.tctour.co.jp/japan-area-guide/nikko-futarasan-shrine.html.

Japan Meteorological Society. 2023. *General Information on Climate of Japan.* https://www.data.jma.go.jp/gmd/cpd/longfcst/en/tourist.html.

Murata, Yasuko, and Judy Evans. 2018. "Interview: Noritake Satō, expert urushi laquer craftsman leading restoration work at UNESCO World Heritage Site, Nikkō Shrines and Temples." *Manabi Japan*, July 15. Nikkō.

Naito, Akihiro. 2022. Interview with Naito Akihiro, Machiya carpenter for the non profit organization Shishokai on site at a machiya townhouse restoration in the Higashiyama District of Kyoto, Japan by Susan Pranger

Nakagoshi, Kosuke, and Kunio Yoshizumi. 2011. "Degradation of Japanese Lacquer under Wavelength Sensitivity of Light Radiation." *Materials Science and Applications* (Science Research) 2: 1507–1515. Variations in the extraction process and curing will impact the final properties and drying time of the finished lacquer, increasing the challenge for contemporary craftsmen. (Murata and Evans 2018).

New World Encyclopedia contributors. 2019. *Shinto*. https://www.newworldencyclopedia.org/p/index.php?title=Shinto&oldid=1026695.

Okahisa, Yoko, Chieko Narita, and Tsuyoshi Yoshimura. 2019. "Resistance of Wood Coated with Oriental Lacquer (Urushi) against Damage Caused by Subterranean Termite." *Journal of Wood Science* (Springer Nature) 65 (41). doi:10.1186/s10086-019-1820-6.

Tanihata. 2012. *Japanese Cedar/Japanese Cypress*. https://www.tanihata.co.jp/english/takumi/kumiko/beihiba.htm.

Tezuka, Shigeyuki. 2022. *Color Painting at* Nikkō *Temples*.

Tsukamoto, Showzi. n.d. http://urushiart.com/eng_about_urushi.html.

UNESCO map. 1998. "Map indicating the Nominated Property and the Surrounding Natural and Built Environment."

Urushi Nation Joboji. n.d. *Beyond the Time, Japanese Elegance*. http://en.urushi-joboji.com/urushi.

weatherbase. 2023. *Japan*. https://www.weatherbase.com/weather/weather-summary.php.

Worlddata. 2023. *Geography*. https://www.worlddata.info/asia/japan/index.php.

Yasuko, Furunushi, and Nagata Kazuhiro. 2014. *Over-saturated Oxygen in Japanese Iron Nails of "Wakugi" for Wooden Structures*. Iron and Steel Institute of Japan (journal) 10.2355/isijinternational.54.1074

Zorn, Tobias, Kouji Nakayama, and Osamu Hashiramoto. 2000. *3.4.3 Fire Situation in Japan*. Global Forest Fire Assessment 1990-2000, Global Fire Monitoring Center & Forestry Agency. http://www.fao.org/3/ad653e/ad653e56.htm.

PART 5

Moving Forward

CHAPTER 5.1

A Comprehensive Preservation Plan

Introduction

Making decisions about the stabilization, retention, or restoration of historic materials is an exercise that must take place within a larger context of the goals, priorities, and available resources for the materials as well as the building, or in some cases the community, as a whole. It must consider the aspirations for preservation, sustainability and resiliency, the uncertainties of climate and weather, and the reality of limited technical and financial resources.

The projected changes in climate from global warming will affect all existing structures but presents particular risks to traditional materials and historic buildings. The nature of the materials, the age of the buildings, and the limitations that preservation goals can place on modifications, all affect the ability to implement strategies to retain, protect, or modify historic fabric. The risks to historic fabric are also compounded by the historic and cultural value of the materials, embodied technology, and craft, as well as of the overall value of the buildings. Appropriate changes may be necessary to historic fabric if historic structures are to survive more severe weather conditions.

If our historic assets are going to continue to survive, it is important that caretakers and advocates identify goals, assess the threats and opportunities, and develop a strategic plan – *A Comprehensive Preservation Plan*. The plan should incorporate traditional preservation planning as well as resiliency planning and emergency planning. It must be sustainable and resilient: *sustainable* in the sense that it can be sustained in the future without compromising other goals, and *resilient* in the sense that both the plan, and the people implementing the plan, must be able to respond quickly and effectively to changing conditions and unanticipated situations.

Many organizations have capital improvement plans for new and existing buildings that prepare for upgrades in response to wear and tear, outdated systems, or

DOI: 10.4324/9781003160489-28

operational changes. A few organizations also have an *Emergency Preparedness Plan* that focuses on disaster response for occupants and collections. Very few have a proactive plan to protect the building or building materials from flooding, extraordinary heat or rain, and other climate related disasters. Very few plans include a risk assessment, which is necessary to determine whether preservation priorities dictate a need to protect a material, feature, or building, or whether the risk of future replacement and repair is acceptable.

Resiliency Planning

Resiliency planning recognizes the incredible level of uncertainty. In some ways, preparing for climate change is like planning for a wilderness canoe trip. Even though we may study the river maps and scout the river before we depart, we never have complete control over the river. The rainstorm upstream, the hidden rocks, the wind on the surface, the shift in weight from your partner all affect the movement of the canoe in ways that are complex and difficult to predict. But, if we have practiced and prepared, we can react quickly and instinctively in the sudden moment of chaos.

1. Take an inventory of the potential climate threats, the condition of the building, and of the stakeholders' priorities, and tolerance for risk (both risk of damage and financial risk).
2. Identify options that are technically feasible.
3. Evaluate options through the filter of preservation goals, available resources, and risk tolerance to find the most appropriate choices under the current conditions and information.
4. Develop a strategy for long-term implementation, which might initially involve no physical response.
5. Establish a process for re-evaluation that monitors changes in climate conditions, building condition, changes in available resources or technology.
6. Most important, periodically reassess and adjust the plan.

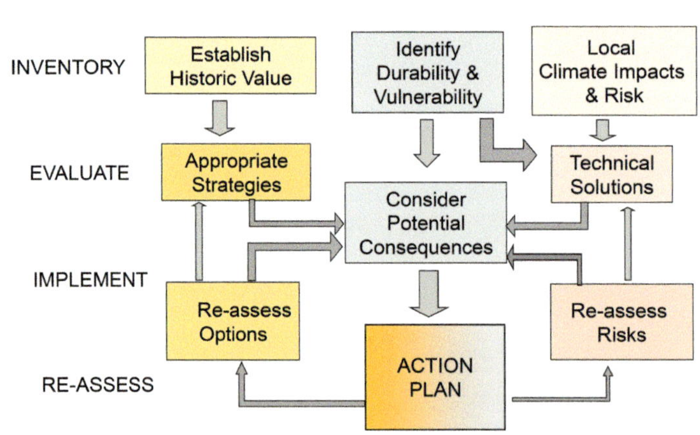

Figure 5.1.1 Developing a resilient plan. (Diagram by Susan Pranger, 2015.)

Risk Assessment

Clear preservation goals provide owners, caretakers, preservationists, architects, and engineers with a benchmark that can be used to evaluate the risks and benefits of various actions. Climate risk is based both on the impact of climate change on a building or material and the probability that a climate change will occur in a given timeframe. Part of the risk assessment is understanding the tolerance for risk. Tolerance is based on the value of the asset, the likelihood of irreparable damage, the cost and benefits of proactive solutions, and the potential results of deferred action.

The risk to a historic asset is defined by the known and unknown conditions:

- The historic or cultural value of the asset.
- The durability or vulnerability of an asset.
- The probability and timing of a climate event.
- The availability of resources and alternative regional approaches.

Tolerance for risk is based on preservation priorities:

- Is the material, assembly, or building irreplaceable?
- Is replacement of damaged or destroyed materials an acceptable alternative?
- Will proactive preventive action compromise the historic value or integrity?
- Will damage put other materials or assemblies at risk? (For example, a leaking roof that allows interior damage or storm surge that undermines foundations.)

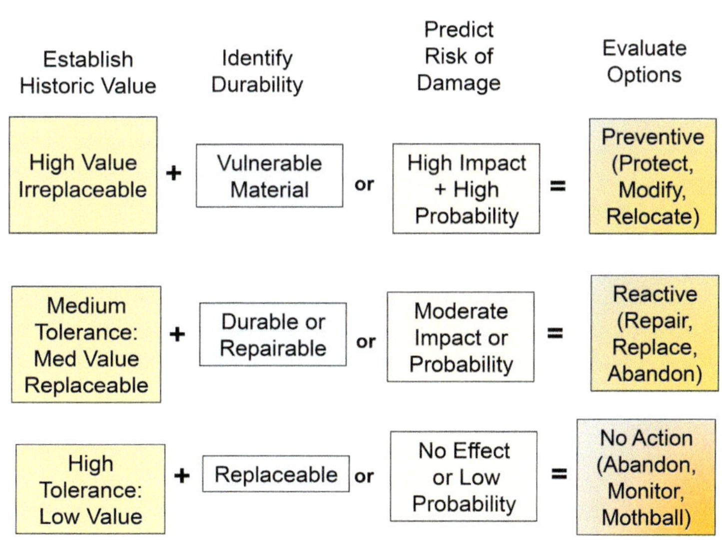

Figure 5.1.2 Risk assessment. (Diagram by Susan Pranger, 2015.)

Low tolerance (or no tolerance) for any risk of damage may reflect the difficulty of replacement or repair. It may also reflect the historic value of the asset. In these situations, preventive or protective action is preferred. When there is low risk of changes to the climate, or those changes are in the distant future, then it may be appropriate to take no action or to monitor the situation. The risk may be low because the materials are especially durable or easily repaired. A lack of action may be appropriate if the original fabric is not very valuable, or if replacement is both easy to achieve and appropriate from a preservation perspective.

Scenario Planning

Scenario planning considers the risk and consequences of different approaches including no action, preventive action, emergency action, and repair. A more advanced analysis would consider the relative costs of preventive action versus repair. Preventive action is desired where there is a high risk of damage from climate change to a valuable asset that is irreplaceable or difficult to repair. The risk of damage may be high due either to a high risk of climate impact, such as a high likelihood of heavy rains, or to the high vulnerability of a material, such as adobe that is particularly susceptible to heavy rain. If resources are not immediately available, what are the risks of monitoring the situation until resources are available or repairs are required?

More complicated scenarios combine low tolerance of risk with low or unknown risk of climate impacts. There may be a low probability of a severe storm, but it would only take one such event to destroy an irreplaceable asset.

Implementing a Comprehensive Plan

Each historic building will face its own challenges based on its location, condition, and the resources available to it. Recognizing that the future is uncertain, the comprehensive plan, and the people charged with implementing the plan, must also be resilient. The uncertainty around climate change, the continuing growth of knowledge about materials, and the availability of new technologies, all require that plans be re-evaluated and adjusted as new information about climate risks and new technology becomes available. Implementation of a comprehensive plan is also dependent on developing a body of knowledge and of experienced craftspersons capable of maintaining, stabilizing, or restoring historic material.

Bibliography

Pranger, Susan. 2015. "Resiliency: Preparing Historic Buildings for the Impact of Climate Changes" Presented to the Faculty of the School of Design Studies in Partial Fulfillment of the Requirements For the Degree of Master of Design Studies. Boston Architectural College. Thesis.

CHAPTER 5.2

Conclusions

Deterioration of materials is often attributed to the wear and tear of weather: the abrasion of rain and wind, aggravated by UV, heat, frost, and insects. While direct damage from weather is significant, materials are also changed by chemical reactions with two compounds: water (H_2O) and carbon dioxide (CO_2). Water and carbon dioxide combine with components in wood, stone, concrete, and metal to cause deterioration including corrosion of metals and carbonation of limes and concrete. They combine in the atmosphere to create a mild acid and with other elements and compounds to cause more potent acids. Moisture also absorbs and transports salts that can cause crypto-florescence or delamination.

Chemical reactions, both natural and man-made, are sensitive to heat and moisture. Changes in temperature and humidity accelerate chemical reactions and create pressure changes that move water, carbon dioxide, and other elements through the pores in materials. The global climate has already experienced small but significant changes in temperature. These changes are not consistent across the planet and are already more impactful in several areas. Hot areas are getting hotter. Cold areas are experiencing freeze-thaw cycles and biological changes. Changes are increasing at a gradual but exponential rate and are expected to trigger sudden shifts as impacts hit various tipping points.

These small changes can lead to big changes, not only in the ecology of plants and animals but also in the pace of chemical deterioration. Damage from acid rain has decreased as the atmosphere has become "cleaner", but increased temperature can accelerate damage from the natural acidity of rain. The earliest examples of "modern" reinforced concrete have reached a critical stage of carbonation that is also accelerated by higher temperatures and increased moisture penetration.

DOI: 10.4324/9781003160489-29

Bacteria, fungi, plants, and animals create both physical stress and chemical change in the right circumstances. Wood is particularly sensitive to increased water saturation, but stone, masonry, and concrete are all vulnerable to the mechanical stress of biological growth, and metals react to acids excreted by biological growth.

Precipitation is a complex phenomenon that causes mechanical damage from abrasion and flooding as well as chemical deterioration. In many areas, the amount of rainfall will decrease or remain the same, but the interaction between humidity and heat in the climate is already triggering more intense wind and rain events. These events can increase abrasion of weaker materials, trigger river and surface flooding, and drive moisture deeper into porous materials. Rapid changes in humidity, precipitation, and heat will increase the number of wet/dry and freeze/thaw cycles in many climate zones, increasing mechanical and chemical damage.

Vigilance has always been required to keep man-made materials in their manipulated state. We have learned to protect building materials with coatings and overhangs and a variety of details that divert water and block UV rays. Understanding the natural changes that occur as materials interact with the natural and man-made environment is the first step in developing strategies to stabilize and protect historic fabric.

Maintenance and repair are labor intensive. Effective maintenance benefits from a higher level of expertise and knowledge, which requires education, training, and recognition of the value of skilled labor. Limited economic and physical resources and a lack of experienced craftsperson are factors that will influence the ability to implement philosophical and technical decisions. Understanding how climate affects materials and understanding how man's attempt to divert weather affects materials are both critical regardless of the chosen approach to maintenance, stabilization, or restoration.

Index

Note: - Page references in *italics* denotes figures.